Mathematical Methods Applied in Artificial Intelligence and Multi-Agent Systems

Mathematical Methods Applied in Artificial Intelligence and Multi-Agent Systems

Editors

Jiangping Hu
Zhinan Peng

Basel • Beijing • Wuhan • Barcelona • Belgrade • Novi Sad • Cluj • Manchester

Editors

Jiangping Hu
School of Automation
Engineering
University of Electronic
Science and Technology
of China
Chengdu
China

Zhinan Peng
School of Automation
Engineering
University of Electronic
Science and Technology
of China
Chengdu
China

Editorial Office
MDPI AG
Grosspeteranlage 5
4052 Basel, Switzerland

This is a reprint of articles from the Special Issue published online in the open access journal *Mathematics* (ISSN 2227-7390) (available at: https://www.mdpi.com/journal/mathematics/special_issues/Mathematical_Methods_Applied_Artificial_Intelligence_Multi_Agent_Systems).

For citation purposes, cite each article independently as indicated on the article page online and as indicated below:

Lastname, A.A.; Lastname, B.B. Article Title. *Journal Name* **Year**, *Volume Number*, Page Range.

ISBN 978-3-7258-1895-2 (Hbk)
ISBN 978-3-7258-1896-9 (PDF)
doi.org/10.3390/books978-3-7258-1896-9

© 2024 by the authors. Articles in this book are Open Access and distributed under the Creative Commons Attribution (CC BY) license. The book as a whole is distributed by MDPI under the terms and conditions of the Creative Commons Attribution-NonCommercial-NoDerivs (CC BY-NC-ND) license.

Contents

About the Editors vii

Preface ix

Yuri Tavares dos Passos, Xavier Duquesne and Leandro Soriano Marcolino
On the Throughput of the Common Target Area for Robotic Swarm Strategies
Reprinted from: *Mathematics* **2022**, *10*, 2482, doi:10.3390/math10142482 1

Zeyang Lin, Jun Lai, Xiliang Chen, Lei Cao and Jun Wang
Learning to Utilize Curiosity: A New Approach of Automatic Curriculum Learning for
Deep RL
Reprinted from: *Mathematics* **2022**, *10*, 2523, doi:10.3390/math10142523 39

Shuoting Wang and Kaibo Shi
Mixed-Delay-Dependent Augmented Functional for Synchronization of Uncertain Neutral-
Type Neural Networks with Sampled-Data Control
Reprinted from: *Mathematics* **2023**, *11*, 872, doi:10.3390/math11040872 59

Rui Luo, Zhinan Peng and Jiangping Hu
On Model Identification Based Optimal Control and It's Applications to Multi-Agent Learning
and Control
Reprinted from: *Mathematics* **2023**, *11*, 906, doi:10.3390/math11040906 80

Jiangbo Zhang and Yiyi Zhao
Dynamics Analysis for the Random Homogeneous Biased Assimilation Model
Reprinted from: *Mathematics* **2023**, *11*, 1661, doi:10.3390/math11071661 99

Long Jian, Yongfeng Lv, Rong Li, Liwei Kou and Gengwu Zhang
Distributed Disturbance Observer-Based ContainmentControl of Multi-Agent Systems via an
Event-Triggered Approach
Reprinted from: *Mathematics* **2023**, *11*, 2363, doi:10.3390/math11102363 116

Shoubo Jin and Guanghui Zhang
Adaptive Consensus of the Stochastic Leader-Following Multi-Agent System with Time Delay
Reprinted from: *Mathematics* **2023**, *11*, 3517, doi:10.3390/math11163517 128

Rong Li, Hengli Wang, Gaowei Yan, Guoqiang Li and Long Jian
Robust Model Predictive Control for Two-DOF Flexible-Joint Manipulator System
Reprinted from: *Mathematics* **2023**, *11*, 3593, doi:10.3390/math11163593 146

Qiuzhen Wang and Jiangping Hu
Modeling and Control of Wide-Area Networks
Reprinted from: *Mathematics* **2023**, *11*, 3984, doi:10.3390/math11183984 171

Yongwen Liu, Dongqing Liu and Shaolin Zhu
Bilingual–Visual Consistency for Multimodal Neural Machine Translation
Reprinted from: *Mathematics* **2024**, *12*, 2361, doi:10.3390/math12152361 195

About the Editors

Jiangping Hu

Dr. Jiangping Hu received a B.S. in Applied Mathematics and an M.S. in Computational Mathematics from Lanzhou University, Lanzhou, China, in 2000 and 2004, respectively. He obtained a Ph.D. in modelling and control of complex systems from the Academy of Mathematics and Systems Science, Chinese Academy of Sciences, Beijing, China, in 2007. He has held various positions with the Royal Institute of Technology, Stockholm, Sweden; The City University of Hong Kong, Hong Kong; Sophia University, Tokyo, Japan; and Western Sydney University, Sydney, NSW, Australia. He is currently a Professor at the School of Automation Engineering, University of Electronic Science and Technology of China, Chengdu, China.

His current research interests include the modeling and control of complex systems and decision and optimization of unmanned systems. Dr. Hu has served as an Associate Editor of the *Journal of Systems Science and Complexity*, *Kybernetika*, *Mathematics*, and *PLOS Complex Systems*.

Zhinan Peng

Dr. Zhinan Peng received a B.S. in Information and Computing Science from Fuyang Normal University, Fuyang, China, in 2014, and an M.S. in Computational Mathematics and a Ph.D. in Control Science and Engineering from the University of Electronic Science and Technology of China (UESTC), Chengdu, China, in 2016 and 2020, respectively. He is currently a lecturer with the School of Automation Engineering, UESTC, Chengdu, China.

His research interests include multi-agent systems, adaptive dynamic programming, and reinforcement learning.

Preface

This Special Issue on "Mathematical Methods Applied in Artificial Intelligence and Multi-Agent Systems" addresses the integration of AI and multi-agent systems in improving the efficiency of optimization and control over complex systems in different domains. The collection, featuring contributions from leading researchers, investigates AI-driven optimization, swarm intelligence, agent-based modeling, and distributed control. Aimed at researchers and practitioners, it offers valuable insights into AI applications in distributed optimization and control. We acknowledge the contributions of the authors and the efforts of the reviewers and editorial team in bringing this work to completion.

Jiangping Hu and Zhinan Peng
Editors

mathematics

Article

On the Throughput of the Common Target Area for Robotic Swarm Strategies

Yuri Tavares dos Passos [1,2,*], Xavier Duquesne [2] and Leandro Soriano Marcolino [2]

1. Centro de Ciências Exatas e Tecnológicas, Universidade Federal do Recôncavo da Bahia, Rua Rui Barbosa, 710. Centro., Cruz das Almas 44380-000, Brazil
2. School of Computing and Communications, Lancaster University, Bailrigg, Lancaster LA1 4WA, UK; duquesne.xavier.13@gmail.com (X.D.); l.marcolino@lancaster.ac.uk (L.S.M.)
* Correspondence: yuri.passos@ufrb.edu.br

Abstract: A robotic swarm may encounter traffic congestion when many robots simultaneously attempt to reach the same area. This work proposes two measures for evaluating the access efficiency of a common target area as the number of robots in the swarm rises: the maximum target area throughput and its maximum asymptotic throughput. Both are always finite as the number of robots grows, in contrast to the arrival time at the target per number of robots that tends to infinity. Using them, one can analytically compare the effectiveness of different algorithms. In particular, three different theoretical strategies proposed and formally evaluated for reaching a circular target area: (i) forming parallel queues towards the target area, (ii) forming a hexagonal packing through a corridor going to the target, and (iii) making multiple curved trajectories towards the boundary of the target area. The maximum throughput and the maximum asymptotic throughput (or bounds for it) for these strategies are calculated, and these results are corroborated by simulations. The key contribution is not the proposal of new algorithms to alleviate congestion but a fundamental theoretical study of the congestion problem in swarm robotics when the target area is shared.

Keywords: robotic swarm; common target; throughput; congestion; traffic control

MSC: 68T40; 70-10

Citation: dos Passos, Y.T.; Duquesne, X.; Marcolino, L.S. On the Throughput of the Common Target Area for Robotic Swarm Strategies. *Mathematics* **2022**, *10*, 2482. https://doi.org/10.3390/math10142482

Academic Editors: Jiangping Hu and Zhinan Peng

Received: 12 June 2022
Accepted: 14 July 2022
Published: 16 July 2022

Copyright: © 2022 by the authors. Licensee MDPI, Basel, Switzerland. This article is an open access article distributed under the terms and conditions of the Creative Commons Attribution (CC BY) license (https://creativecommons.org/licenses/by/4.0/).

1. Introduction

Swarms of robots are systems composed of a large number of robots that can only interact with direct neighbours and follow simple algorithms. Interestingly, complex behaviours may emerge from such straightforward rules [1,2]. An advantage of such systems is the usage of low-priced robots instead of a few expensive ones to solve problems. Robotic swarms accurately projected for simple robots may solve complex tasks with greater efficiency and fault-tolerance, while being cheaper than a small group of complex robots oriented for a specific problem domain. They can also be seen as a multi-agent system with spatial computers, which is a group of devices displaced in the space such that its objective is defined in terms of spatial structure and its interaction depends on the distance between them [3]. Swarms have recently been receiving attention in the multi-agents systems literature in problems such as logistics [4], flocking formation [5], pattern formation [6] and the coordination of unmanned aerial vehicle swarms [7]. In such problems relating to spatial distribution, conflicts may be created by the trajectories of the robots, which may slow down the system, especially when a group is intended to go to a common region of the space. Some examples where this happens are waypoint navigation [8] and foraging [9].

The topic of robotic traffic control has been studied for a long time [10–12], but with the premise that autonomous cars navigate on delimited lanes and that coordination is needed only at junctions. Even recent related works on multi-agent systems [13–15] also deal with this problem in a similar way. In [16,17], they also deal with multi-agents and pathfinding, but not in a situation where the target of every agent is the same area. Furthermore, distributed solutions are considered here where agents only have local information, while [16,17] propose centralised solutions. Xia et al. [18] investigate the topology of the neighbourhood relations between multiple unmanned surface vehicles in a swarm. They deal with maintaining formation in swarms, but they have to keep virtual leaders, and their goal is not to minimise congestion.

Moreover, there has not been much research on the problem of reducing congestion when a swarm of robots is aimed at the same target. Surveys about robotic swarms [19–24] do not provide information regarding these situations. Even a recent survey on collision avoidance [25] does not address this issue, though it provides insights into multi-vehicle navigation. Congestion in robotic swarms is mostly managed by collision avoidance in a decentralised fashion, allowing for improved algorithm scalability.

However, solely avoiding collisions does not necessarily lead to a good performance in problems with a common target. For example, Marcolino et al. [26] showed that the ORCA algorithm [27] reaches an equilibrium where robots could not arrive at the target despite avoiding collisions. That paper also presented three algorithms using artificial potential fields for the common target congestion problem, but no formal analysis of the cluttered environment was conducted. Hence, congestion is still not well understood, and more theoretical work is needed to measure the optimality of the algorithms. A better understanding of this topic should lead to a variety of new algorithms adapted to specific environments. Thus, this paper aims to introduce the first theoretical study on this problem, which should lead to future enhancements in handling congestion in robotic swarms.

Therefore, this work fits in the literature on mathematical models of swarm robotics, such as the works by Lima and Oliveira [28], which models a cellular automata ant memory to control a robot swarm for foraging tasks; Varghese and McKee [29], for pattern transformation modelling; Li and Chen [30], for box-pushing; Taylor-King et al. [31], which studies the effect of turning delays on the behaviour of groups of robots; Galstyan et al. [32], for microscopic robots that reside in a fluid and can detect chemicals; Khaluf and Dorigo [33], which models swarm performance measures using the integral of linear birth–death processes; and Mannone et al. [34], which uses category theory and quantum computing to model the development of robotic swarm systems. However, as mentioned, these theories do not yet allow one to better understand swarm congestion.

Furthermore, any elaborated analysis on that subject must investigate the effect of the increase in the number of individuals on the swarm congestion, as it is desirable for the system to perform well as it grows in size. If one has a finite measure that abstracts the optimality of any algorithm as the number of robots goes to infinity, this can be used as a metric to compare different approaches to the same problem. Thus, this work presents as a metric the common target area throughput. That is, a measure of the rate of arrival in this area is proposed as the time tends to infinity as an alternative approach to analyse the congestion in swarms with a common target area. In network and parallel computing studies [35,36], asymptotic throughput is used to measure the throughput when the message size is assumed to have infinite length. The same idea is used here, but instead of message size, it is applied with infinite time, as if the algorithms run forever. As it will be presented in the next section, this implies dealing with an infinite number of robots. Thus, time is being used here instead of message size or bytes, as in computer network studies.

Therefore, the contributions in this paper are the following.

1. A method for evaluating algorithms for the common target problem in a robotic swarm by using the throughput in theoretical or experimental scenarios is proposed.
2. An extensive theoretical study of the common target problem is presented, allowing one to better understand how to measure the access to a common target using a metric not yet used in other works on the same problem.
3. Assuming a circular target area and that the robots are constantly moving at the maximum linear speed and have a fixed minimum distance from each other, theo-

retical strategies for entering the area are developed, and their maximum theoretical throughput for a fixed time and their maximum asymptotic throughput when time goes to infinity are calculated (or bounds for it). Additionally, the correctness of these calculations is verified by simulations.

The presented theoretical strategies are based on forming a corridor towards the target area or making multiple curved trajectories towards the boundary of the target area. For the corridor strategy, the throughput when the robots are moving towards the target in square and hexagonal packing formations is also discussed. The theoretical strategies are evaluated by realistic Stage [37] simulations with holonomic and non-holonomic robots. These experiments corroborate that whenever an algorithm makes a swarm take less time to reach the target region than another algorithm, the throughput of the former is higher than the latter.

Note that the key contribution of this work is not the proposal of new algorithms to alleviate congestion but a fundamental theoretical study of the congestion problem in swarms having the same target. The presented strategies are the theoretical grounding for new distributed algorithms for robotic swarms in our concurrent work [38]. When we assume that the robots are constantly moving at maximum linear speed and maintaining a fixed minimum distance, we can provide analytical calculations of the maximum possible throughput for a given time and bounds or exact value of the maximum asymptotic throughput for the different theoretical strategies. Based solely on these calculations, we can compare which strategy is better. However, for robots using artificial potential fields, it is not straightforward to obtain explicit throughput equations due to the changeability of those quantities previously assumed constant. Then, in the lack of closed asymptotic equations, simulations were performed in [38] for the algorithms inspired by our strategies in order to obtain experimental throughput and compare algorithms for varying linear speeds and inter-robot distances. As shown by these experimental data, their variation and the effect of the other robots in the trajectory does affect the throughput. However, the analytically calculated maximum throughput in this work serves as an upper bound to the ones obtained from the simulations in more realistic conditions when considering the mean speed and mean distance between the robots in place of the constant values on the obtained equations.

This paper is organised as follows. The next section briefly explains the mathematical notation being used. Section 3 formally defines the common target area throughput and proves statements about this measure for theoretical strategies that allow robots to enter the common target area. Section 4 describes the experiments and presents their results to verify the correctness of the theoretical strategies results. Finally, Section 5 summarises the results and gives final remarks.

2. Notation

Geometric notation is used as follows. $\overleftrightarrow{AB}, \overrightarrow{AB}$ and $\overline{AB}$ represent a line passing through points A and B, a ray starting at A and passing through B and a segment from A to B, respectively. $|\overline{AB}|$ is the size of $\overline{AB}$. $\overleftrightarrow{AB} \parallel \overleftrightarrow{CD}$ means $\overleftrightarrow{AB}$ is parallel to $\overleftrightarrow{CD}$. If a two-dimensional point is represented by a vector P_1, its x- and y-coordinates are denoted by $P_{1,x}$ and $P_{1,y}$, respectively.

$\triangle ABC$ expresses the triangle formed by the points A, B and C. $\triangle ABC \cong \triangle DEF$ and $\triangle ABC \sim \triangle DEF$ mean the triangles ABC and DEF are congruent (same angles and same size) and similar (same angles), respectively. Depending on the context, the notation is omitted for brevity.

$\widehat{AOB}$ means an angle with vertex O, one ray passing through point A and another through B. Depending on the context, if only one $\triangle EFG$ is being dealt with, its angles will be named only by $\widehat{E}, \widehat{F}$ and $\widehat{G}$. All angles are measured in radians in this paper.

3. Theoretical Analysis

This paper considers the scenario where a large number of robots must reach a common target. After reaching the target, each robot moves towards another destination which may or may not be common among the robots. It is assumed that the target is defined by a circular area of radius s. A robot reaches the target if its centre of mass is at a distance below or equal to the radius s from the centre of the target. In addition, it is supposed that there is no minimum amount of time to stay at the target. Additionally, the angle and the speed of arrival have no impact on whether the robot reached the target or not. In this section, theoretical strategies are constructed to solve that task and show limits for the efficiency of real-life implementations, which we developed in a concurrent work [38]. To measure performance, the following definition is presented.

Definiton 1. *The* throughput *is the inverse of the average time between arrivals at the target.*

Informally speaking, the throughput is measured by someone located on the common target (i.e., on its perspective). It is considered that an optimal algorithm minimises the average time between two arrivals or, equivalently, maximises throughput. The unit for throughput can be in s^{-1}. It will be noted f (as in frequency). The rest of the paper focuses on maximising throughput.

Assume an experiment was run with $N \geq 2$ robots for T units of time, such that the time between the arrival of the i-th robot and the $i+1$-th robot is t_i, for i from 1 to $N-1$. Then, by Definition 1, $f = \frac{1}{\frac{1}{N-1}\sum_{i=1}^{N-1} t_i} = \frac{N-1}{\sum_{i=1}^{N-1} t_i} = \frac{N-1}{T}$, because $\sum_{i=1}^{N-1} t_i = T$. Thus, an equivalent definition of throughput is given:

Definiton 2. *The* throughput *is the ratio of the number of robots that arrive at a target region, not counting the first robot to reach it, to the arrival time of the last robot.*

The target area is a limited resource that must be shared between the robots. Since the linear speeds of the robots have an upper bound, a robot needs a minimum amount of time to reach and leave the target before letting another robot in. Let the *asymptotic throughput* of the target area be its throughput as the time tends to infinity. Because any physical phenomenon is limited by the speed of light, this measure is bounded. Then, the asymptotic throughput is well suited to measure the access of a common target area as the number of robots grows.

One should expect that the asymptotic throughput depends mainly on the target size and shape, the speed of the robots, and the distance between robots. As any bounded target region can be included in a circle of radius s, only circular target regions will be dealt with hereafter. If the robots are moving at maximum speed and keeping the distance between each at a minimum value all the time, then it is also expected that the throughput and asymptotic throughput reach their maximum value. Thus, it is assumed hereafter that the robots move at a constant maximum linear speed, v, and the distance between each other is either constant when possible or no lower than a fixed value, d.

To efficiently access the target area, two main cases are identified: $s \geq d/2$ and $s < d/2$. There are targets that several robots can simultaneously reach without collisions. That is the case if the radius $s \geq d/2$. Thus, one approach is making lanes arrive in the target region so that as many robots as possible can simultaneously arrive. After the robots arrive at the target, they must leave the target region by making curves. However, we discovered [38] that this approach does not obtain good results in realistic simulations due to the influence of other robots, although it is theoretically the best approach if the robots could run at a constant speed and maintain a fixed minimum distance between each other.

The case where $s < d/2$, when only one robot can occupy the target area simultaneously, is of interest. Making two queues and avoiding the inter-robot distance being less than d is good guidance to work efficiently. Particularly, the case $s = 0$ offers interesting insights, so this is discussed next.

Some lemmas and propositions need a long technical treatment to be proven. In order to avoid the reader missing the main idea of this paper, only their statements are provided. All proofs are available in the Supplementary Materials.

3.1. Common Target Point: s = 0

Consider the case where robots are moving in straight lines at constant linear speed v, maintaining a distance of at least d between each other. A robot has reached the target when its centre of mass is over the target. When $s = 0$, the target is a point. The first result is the optimal throughput when robots are moving in a straight line to a target point. It is illustrated in Figure 1. This section constructs a solution to attain the optimal throughput.

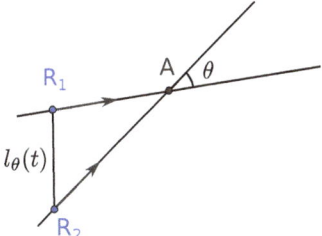

Figure 1. Two robots, R_1 and R_2, are moving in straight lines toward a target at A. The angle between their trajectory is θ. The distance between the two robots over time is denoted by $l_\theta(t)$.

First, consider two robots, Robot 1 and Robot 2. Their trajectories are straight lines towards the target. Assume the straight-line trajectory of Robot 1 has an angle θ_1 with the x-axis and the one of Robot 2 has θ_2. Define $\theta_2 - \theta_1 = \theta$ as the angle between the two lines. The positions of the robots are described by the kinematic Equation (1) below, where $(x_1(t), y_1(t))$ and $(x_2(t), y_2(t))$ are the positions of Robot 1 and Robot 2, respectively, and $t \in \mathbb{R}$ is an instant of time. Without loss of generality, the origin of time is set when Robot 1 reaches the target, and the target is located at $(0,0)$. Thus, $(x_1(0), y_1(0)) = (0,0)$. τ is the delay between the two arrivals at the target. Then, $(x_2(\tau), y_2(\tau)) = (0,0)$, and

$$\begin{bmatrix} x_1(t) \\ y_1(t) \end{bmatrix} = \begin{bmatrix} vt\cos(\theta_1) \\ vt\sin(\theta_1) \end{bmatrix} \text{ and } \begin{bmatrix} x_2(t) \\ y_2(t) \end{bmatrix} = \begin{bmatrix} v(t-\tau)\cos(\theta_2) \\ v(t-\tau)\sin(\theta_2) \end{bmatrix} \quad (1)$$

In order to find the optimal throughput, this paper starts with its first lemma:

Lemma 1. *To respect a distance of at least d between the two robots, the minimum delay between their arrival is $\frac{d}{v}\sqrt{\frac{2}{1+\cos(\theta)}}$.*

This result leads to Proposition 1.

Proposition 1. *The optimal throughput f for a point-like target ($s = 0$) is $f = \frac{v}{d}$. It is achieved when robots form a single line, i.e., the angle between the trajectories of the robots must be 0.*

The insight derived from Proposition 1 implies that one should increase the speed of the robots or decrease the minimum distance between them to increase the throughput. It is also noted that the optimal trajectory for all the robots is to form a queue behind the target and Robot 1. As a result, the optimal path is to create one lane to reach the target. When the angle θ between the path of a robot and the next one is increased, a delay from the optimal throughput is introduced. For instance, Figure 2 shows the normalised delay for different angles θ (normalised by dividing τ by $\tau_{min} = d/v$) between two robots, according to Lemma 1. This figure shows that for an angle of $\pi/3$, the minimum delay is 15% higher than for an angle of 0, and the minimum delay is 41% higher for an angle of $\pi/2$.

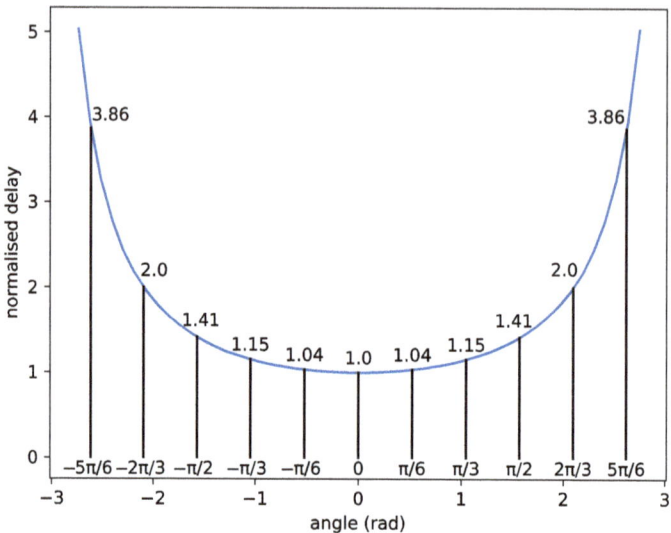

Figure 2. Normalised delay versus the angle between the trajectories of the robots.

3.2. Small Target Area: $0 < s < d/2$

This section supposes a small target area where $0 < s < d/2$; hence, two lanes with a distance d cannot fit towards the target yet. The next results are based on a strategy using two *parallel lanes* as close as possible to guarantee the minimum distance d between robots. Figure 3 describes these two parallel lanes. This strategy is called *compact lanes* hereafter. Proposition 2 considers a target area with radius $0 < s \leq \frac{\sqrt{3}}{4}d$, and Proposition 3 assumes $\frac{\sqrt{3}}{4}d < s < \frac{d}{2}$.

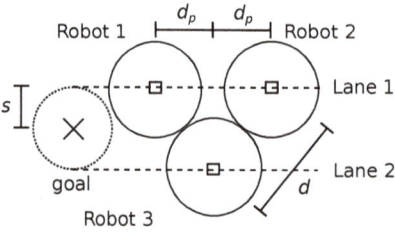

Figure 3. Two parallel robot lanes for a small target, illustrating the compact lanes strategy.

Proposition 2. *Assume two parallel lanes with robots at constant speed v and maintaining a constant distance d between them. The throughput of a common target area with radius $0 < s \leq \frac{\sqrt{3}}{4}d$ at a given time T after the first robot has reached the target area is*

$$f(T) = \frac{1}{T}\left(\left\lfloor \frac{vT}{2\sqrt{d^2-(2s)^2}} \right\rfloor + \left\lfloor \frac{vT}{2\sqrt{d^2-(2s)^2}} + \frac{1}{2} \right\rfloor\right) \quad (2)$$

and is limited by

$$f = \lim_{T \to \infty} f(T) = \frac{v}{d\sqrt{1-(\frac{2s}{d})^2}}. \quad (3)$$

Proposition 3. *Assume two parallel lanes with robots at constant speed v and maintaining a constant distance d between them. The throughput of a common target area with radius $\frac{\sqrt{3}}{4}d < s < \frac{d}{2}$ at a given time T after the first robot has reached the target area is*

$$f(T) = \frac{1}{T}\left(\left\lfloor \frac{vT}{d} \right\rfloor + \left\lfloor \frac{vT}{d} + \frac{1}{2} \right\rfloor\right) \tag{4}$$

and is limited by

$$f = \lim_{T \to \infty} f(T) = \frac{2v}{d}. \tag{5}$$

Observe that if $T = k\frac{d}{v}$ for any $0 < k \in \mathbb{Z}$ is used in (4), the compact lanes strategy can achieve the throughput of two parallel lanes of robots going in the direction of the target region when $T = k\frac{d}{v}$ for any $k \in \mathbb{Z}$ or when $T \to \infty$, even though two robots cannot reach the target region at the same time.

3.3. Large Target Area: $s \geq d/2$

This section focuses on situations where more than two robots can simultaneously touch the target. Three feasible strategies are presented.

The simplest strategy is to consider several parallel lanes being at a distance d from each other. However, it is possible to obtain higher throughput. In particular, two other strategies are identified: (a) using parallel straight line lanes that may be distanced lower than d and (b) robots moving towards the target following curved trajectories. Strategy (a) uses more than two compact lanes, extending the strategy presented in the previous section. By doing this, the robots fit in a hexagonal packing arrangement moving toward the target region. Strategy (b) uses a touch and run approach. In it, robots do not cross the target area, they only reach it and return in the opposite direction using curved trajectories which respect the minimum distance d.

The next section starts with the parallel lanes strategy, which has the lowest asymptotic throughput over the strategies presented in this section, for comparison with the other strategies. In particular, it will be used later as a justification for the lowest number of lanes used in the strategy (b) in (14) in Proposition 7. Following their description and properties, a discussion comparing them is provided.

3.3.1. Parallel Lanes

It is considered here that the robots are moving inside lanes. The lanes are straight lines, and the linear speed v of the robots is constant. The lanes are separated by a distance d, and each robot maintains a distance d from each other. Figure 4 illustrates an example of this strategy. The first lane, Lane 1, is at the top. The first robot of each lane is located at $(s, s - (i-1)d)$ for the Lane i. The next proposition states the throughput for a given time and the asymptotic throughput for this strategy.

Proposition 4. *Assume a circular target region with its centre at $(0,0)$ and radius $s \geq \frac{d}{2}$ and parallel lanes starting at $(s, s - (i-1)d)$ for $i \in \{1, \ldots, \left\lfloor \frac{2s}{d} \right\rfloor + 1\}$. At each Lane i, the first robot is located at the point $(s, s - (i-1)d)$ in the starting configuration. Then, the first robot to reach the target is located at $(s, s - (J-1)d)$, for $J = \left\lfloor \frac{s}{d} \right\rfloor + 1$, if $|s - \left\lfloor \frac{s}{d} \right\rfloor d| \leq |s - \left\lceil \frac{s}{d} \right\rceil d|$, otherwise $J = \left\lceil \frac{s}{d} \right\rceil + 1$. The throughput for a given time T after the first robot reaches the target region is:*

$$f_p(T) = \frac{1}{T}\left(\sum_{i=1}^{\left\lfloor \frac{2s}{d} \right\rfloor + 1} N_i(T)\right) - \frac{1}{T}, \tag{6}$$

for $N_i(T) = \left\lfloor \frac{vT - d_i + d_J}{d} + 1 \right\rfloor$, if $T \geq \frac{d_i - d_J}{v}$, otherwise, $N_i(T) = 0$, $d_j = s - \sqrt{s^2 - (s - (j-1)d)^2}$, and

$$f_p = \lim_{T \to \infty} f_p(T) = \left\lfloor \frac{2s}{d} + 1 \right\rfloor \frac{v}{d}. \quad (7)$$

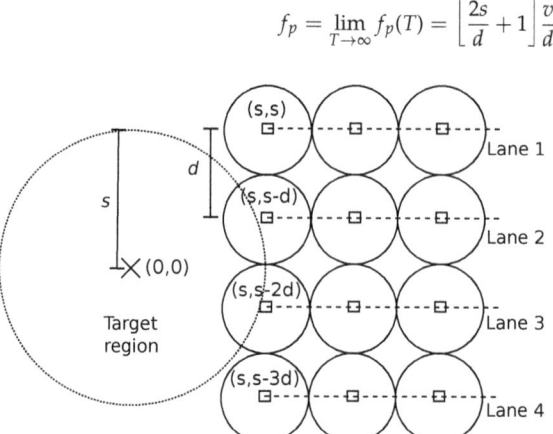

Figure 4. Example of the parallel lanes strategy.

3.3.2. Hexagonal Packing

By extending the compact lanes to more than two lanes, the robots will be packed in a hexagonal formation. An illustration of this strategy is shown in Figure 5. As it can be seen, robots from different lanes are still able to move towards the target keeping a distance d from each other, even though the lanes have a distance lower than d.

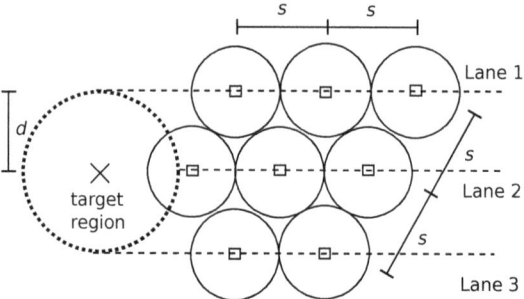

Figure 5. Robot lanes for hexagonal packing.

An upper bound of the asymptotic throughput for the *hexagonal packing* strategy is first computed, then the throughput for a given time using this strategy is calculated.

Proposition 5. *Assume robots moving at speed v, going to a circular target of radius s. The upper bound of the asymptotic throughput for the hexagonal packing strategy is*

$$f_h^{max} = \frac{2}{\sqrt{3}} \left(\frac{2s}{d} + 1 \right) \frac{v}{d}. \quad (8)$$

Proposition 5 presents an upper bound of the asymptotic throughput using hexagonal packing, but it does not tell us which is the best placement of the robots inside a corridor since the hexagonal formation can be rotated by different angles. Hence, the results about the throughput considering the placement of the hexagonal packing inside a corridor of robots going to the target region will be presented. First, however, the following definition will be needed.

Definiton 3. *The hexagonal packing angle θ is the angle formed by the x-axis and the line formed by any robot at position (x,y) and its neighbour at $(x + d\cos(\theta), y + d\sin(\theta))$ under the target region reference frame.*

Observe that any robot at (x,y) under the hexagonal packing has at most six neighbours located at $\left(x + d\cos(\theta), y + d\sin(\theta)\right)$, $\left(x + d\cos\left(\theta + \frac{\pi}{3}\right), y + d\sin\left(\theta + \frac{\pi}{3}\right)\right)$, ..., $\left(x + d\cos\left(\theta + \frac{5\pi}{3}\right), y + d\sin\left(\theta + \frac{5\pi}{3}\right)\right)$ (Figure 6). If $\theta = \frac{\pi}{3}$, putting this value in the previous series results in the first neighbour robot being at $(x + d\cos(\pi/3), y + d\sin(\pi/3))$ and the last neighbour robot at $(x + d\cos(0), y + d\sin(0))$. This is the same result if $\theta = 0$ was used. Consequently, due to this periodicity, hexagonal packing angles in $[0, \frac{\pi}{3})$ are assumed.

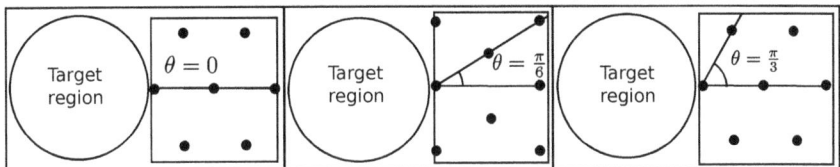

Figure 6. Example of hexagonal packing with different angles. The robots are the black dots.

The next proposition states the bounds of the throughput in the limit towards the infinity for hexagonal packing using an arbitrary, but fixed, hexagonal packing angle θ. A fixed θ is assumed because normally in a robotic swarm the robots rely on local sensing. In order to obtain the maximum number of robots inside the corridor, all robots should know the size of the corridor and communicate by local-ranged message sending. It would take time to send information, and for all robots to adjust their orientation each time a new robot joins the swarm when using this local sensing approach.

In other words, if the corridor where the robots are going in the direction of the target is increasing over time, then θ should change over time for the optimal throughput. However, in practice, changing the hexagonal packing angle implies all robots must turn to a hexagonal packing angle θ^* depending on the size of the new rectangle based on the added robots to it to maximise the number of robots inside the corridor. In addition to the time to send messages with this parameter, more time would be needed for every robot to adapt to the updated computed θ^* because the turning speed of the robots is finite. Therefore, this paper does not handle this adjustable scenario.

Proposition 6. *Assume the robots using hexagonal formation coming to a circular target area with radius s such that the first robot to reach it was at time 0 at $(x_0, y_0) = (w, 0)$, for any $w \geq s$. For a given time T, the robots are going to the target at linear speed v, keeping a distance d between neighbours ($0 < d \leq 2s$), using fixed hexagonal packing angle $\theta \in [0, \pi/3)$. The throughput for a given time is given by*

$$f_h(T, \theta) = \frac{1}{T} \sum_{x_h = -n_l^-}^{n_l^+ - 1} \left(\lfloor Y_2^R(x_h) \rfloor - \lceil Y_1^R(x_h) \rceil + 1 \right) + \\ \frac{1}{T} \sum_{x_h = B}^{U} \left(\lfloor Y_2^S(x_h) \rfloor - \lceil Y_1^S(x_h) \rceil + 1 \right) - \frac{1}{T}, \quad (9)$$

for $\lfloor Y_2^R(x_h) \rfloor \geq \lceil Y_1^R(x_h) \rceil$ and $\lfloor Y_2^S(x_h) \rfloor \geq \lceil Y_1^S(x_h) \rceil$ (if for some x_h, either of these conditions are false, it is assumed that the respective summand for this x_h is zero). Additionally, $n_l^- = \left\lfloor \frac{2s \sin(|\pi/6 - \theta|)}{\sqrt{3}d} \right\rfloor$, $n_l^+ = \left\lfloor \frac{2(vT-s)\cos(\pi/6-\theta) + 2s\sin(|\pi/6-\theta|)}{\sqrt{3}d} + 1 \right\rfloor$,

$$Y_1^R(x_h) = \begin{cases} \max\left(\dfrac{\sin(\frac{\pi}{3}-\theta)x_h - \frac{s}{d}}{\cos(\theta - \frac{\pi}{6})}, \dfrac{-\cos(\frac{\pi}{3}-\theta)x_h}{\sin(\frac{\pi}{6}-\theta)}\right), & \text{if } \theta < \pi/6, \\ \max\left(\dfrac{\sin(\frac{\pi}{3}-\theta)x_h - \frac{s}{d}}{\cos(\theta - \frac{\pi}{6})}, \dfrac{\frac{vT-s}{d} - \cos(\frac{\pi}{3}-\theta)x_h}{\sin(\frac{\pi}{6}-\theta)}\right), & \text{if } \theta > \pi/6, \\ \dfrac{x_h}{2} - \dfrac{s}{d}, & \text{if } \theta = \pi/6, \end{cases}$$

$$Y_2^R(x_h) = \begin{cases} \min\left(\dfrac{\sin(\frac{\pi}{3}-\theta)x_h + \frac{s}{d}}{\cos(\theta - \frac{\pi}{6})}, \dfrac{\frac{vT-s}{d} - \cos(\frac{\pi}{3}-\theta)x_h}{\sin(\frac{\pi}{6}-\theta)}\right), & \text{if } \theta < \pi/6, \\ \min\left(\dfrac{\sin(\frac{\pi}{3}-\theta)x_h + \frac{s}{d}}{\cos(\theta - \frac{\pi}{6})}, \dfrac{-\cos(\frac{\pi}{3}-\theta)x_h}{\sin(\frac{\pi}{6}-\theta)}\right), & \text{if } \theta > \pi/6, \\ \dfrac{x_h}{2} + \dfrac{s}{d}, & \text{if } \theta = \pi/6, \end{cases}$$

$$B = \begin{cases} \left\lceil \dfrac{2(\sin(\pi/3 - \theta)(c_x - l_x) + \cos(\pi/3 - \theta)(y_0 - l_y - s))}{\sqrt{3}d} \right\rceil, & \text{if } T > \dfrac{s}{v}, \\ \left\lceil -\dfrac{2\sqrt{2svT - (vT)^2}}{\sqrt{3}d} \sin\left(\theta + \dfrac{\pi}{6}\right) \right\rceil, & \text{otherwise}, \end{cases}$$

for $c_x = x_0 + vT - s$ and $(l_x, l_y) = \operatorname{argmin}_{(x,y) \in Z} |vT - s + x_0 - x| + |y_0 - y|$, if $T > \frac{s}{v}$, otherwise, $(l_x, l_y) = (x_0, y_0)$, where Z is the set of robot positions inside the rectangle measuring $vT - s \times 2s$ for $vT - s > 0$. If $T > \frac{s}{v}$ or $\arctan\left(\dfrac{\frac{s}{2} - \sin(\theta)(vT - s)}{\frac{\sqrt{3}s}{2} + \cos(\theta)(vT - s)}\right) < \frac{\pi}{2} - \theta$,
$U = \left\lfloor \dfrac{2(\sin(\pi/3 - \theta)(c_x - l_x) + \cos(\pi/3 - \theta)(y_0 - l_y) + s)}{\sqrt{3}d} \right\rfloor$, otherwise, $U = \left\lfloor \dfrac{2\sqrt{2svT - (vT)^2}}{\sqrt{3}d} \cos\left(\theta - \dfrac{\pi}{3}\right) \right\rfloor$. In addition, $Y_1^S(x_h) = \dfrac{dx_h - C_{-\theta,x} + \sqrt{3}C_{-\theta,y} - \sqrt{\Delta(x_h)}}{2d}$ and

$$Y_2^S(x_h) = \begin{cases} \min(L(x_h), C_2(x_h)) - 1, & \text{if } \min(L(x_h), C_2(x_h)) \\ & = \lfloor L(x_h) \rfloor \text{ and } T > \dfrac{s}{v}, \\ \min(L(x_h), C_2(x_h)), & \text{otherwise}, \end{cases} \quad (10)$$

$C_{-\theta} = \begin{bmatrix} \cos(-\theta) & -\sin(-\theta) \\ \sin(-\theta) & \cos(-\theta) \end{bmatrix} \begin{bmatrix} c_x - l_x \\ y_0 - l_y \end{bmatrix}$, $\Delta(x_h) = 4s^2 - \left(\sqrt{3}(dx_h - C_{-\theta,x}) - C_{-\theta,y}\right)^2$,
$C_2(x_h) = \dfrac{dx_h - C_{-\theta,x} + \sqrt{3}C_{-\theta,y} + \sqrt{\Delta(x_h)}}{2d}$, $L(x_h) = \dfrac{\sin(\frac{\pi}{2} - \theta)(dx_h - C_{-\theta,x}) + \cos(\frac{\pi}{2} - \theta)C_{-\theta,y}}{d\sin(\frac{5\pi}{6} - \theta)}$, if $T > \frac{s}{v}$, otherwise $L(x_h) = \dfrac{\sin(\frac{\pi}{2} - \theta)x_h}{\sin(\frac{5\pi}{6} - \theta)}$, and

$$\lim_{T \to \infty} f_h(T, \theta) \in \left(\dfrac{4vs}{\sqrt{3}d^2} - \dfrac{2v\cos(\theta - \pi/6)}{\sqrt{3}d}, \dfrac{4vs}{\sqrt{3}d^2} + \dfrac{2v\cos(\theta - \pi/6)}{\sqrt{3}d} \right]. \quad (11)$$

The upper and lower bounds presented on (11) are below or equal the maximum asymptotic throughput presented by the Proposition 5, Equation (8). The result of the Proposition 5 only concerns the maximum asymptotic throughput and does not consider the hexagonal packing angle θ, while Proposition 6 gives a lower bound and tightens the bounds for a given θ. Figure 7 presents an example comparison of these equations for two different values of s. As expected, the maximum asymptotic throughput under the optimal density assumption (in (8)) is a possible value of the throughput using hexagonal packing and is above or equal to the interval in (11) for any given θ. However, for practical robotic swarms applications, a certain hexagonal packing angle must be fixed depending on the expected height of the corridor, target size and the minimum distance between the robots, resulting in a throughput below or equal to the upper value presented in Proposition 5.

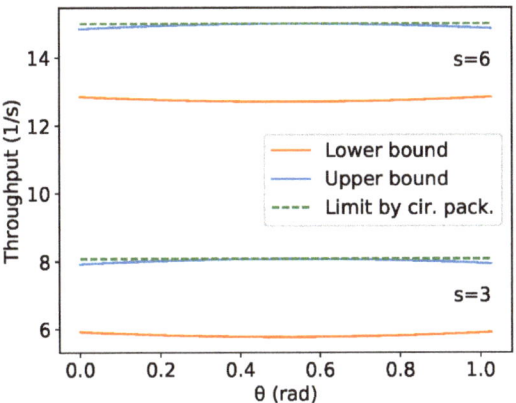

Figure 7. Limit given by (8) using the circle packing results and the lower and upper bounds of the hexagonal packing limit by (11) for $\theta \in [0, \pi/3)$, $d = 1$ m, $v = 1$ m/s and $s \in \{3, 6\}$ m.

On the other hand, due to the discontinuities of (9), it is difficult to obtain an exact θ that maximises the throughput given the other parameters. In addition, there is no specific value of θ that achieves the maximum throughput for all possible values of the other parameters. Interestingly, given a fixed sub-interval of θ, depending on the number of sample values, new local maxima and minima can arise from these discontinuities. Additionally, a different parity of the number of samples can produce a global maximum in even or odd interval points. To illustrate this, Figures 8–11 present the result of this equation for some randomly generated parameters and a different number of samples of θ equally spaced and taken from the domain interval, that is, from 0 to $\pi/3$, including these values. Two different orders of magnitude are chosen for the number of equally spaced points in each plot (a small one, about two orders, and a large one of seven orders), and different parities are also given (99 and 100 for the small order, and 10^7 and $10^7 + 1$ for the large one).

In Figures 8–11, θ is over the x-axis, and the number of robots inside the given rectangle is over the y-axis. These plots use $v = 1$ m/s. The maximum value in each image is represented by an orange circle, and a rectangle represents the maximum between the left and the right image. No square means the maximum values in both sides are equal. Each one of the Figures 8–11 presents two different sets of parameters. In Figures 8 and 9, 99 equally spaced values are shown for $\theta \in [0, \pi/3)$ on the left-hand side images and 100 on the right-hand side; then, the maximum on each side is compared, and the best one is chosen. The same is performed in Figures 10 and 11, but using 10^7 and $10^7 + 1$. Figures 8a, 9a, 10b and 11b show an example that $\theta \approx \pi/6$ reaches the maximum throughput, and in Figures 8c,d and 10c,d, the maximum is at $\theta = 0$. Moreover, Figure 9c,d have their maximum for θ different from the other examples. Figure 8c,d have the same maximum, despite the plots being different. This also occurs in Figures 10c,d and 11c,d. If the parameters are known, one can find an approximate best candidate for θ by searching several values, as presented. However, as far as the authors know, obtaining the true value which maximises that equation by a closed-form is an open problem.

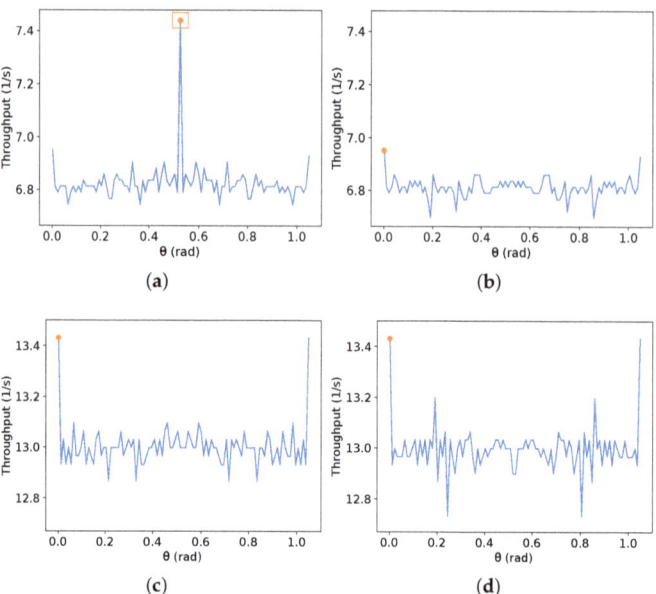

Figure 8. Examples of (9) varying θ from 0 to $\frac{\pi}{3}$ for different and randomly generated values of T, s, and d. It continues in Figure 9. (**a**) For 99 samples, $T = 43$ s, $s = 3$ m, $d = 1$ m. (**b**) For 100 samples, $T = 43$ s, $s = 3$ m, $d = 1$ m. (**c**) For 99 samples, $T = 30$ s, $s = 2.5$ m and $d = 0.66$ m. (**d**) For 100 samples, $T = 30$ s, $s = 2.5$ m and $d = 0.66$ m.

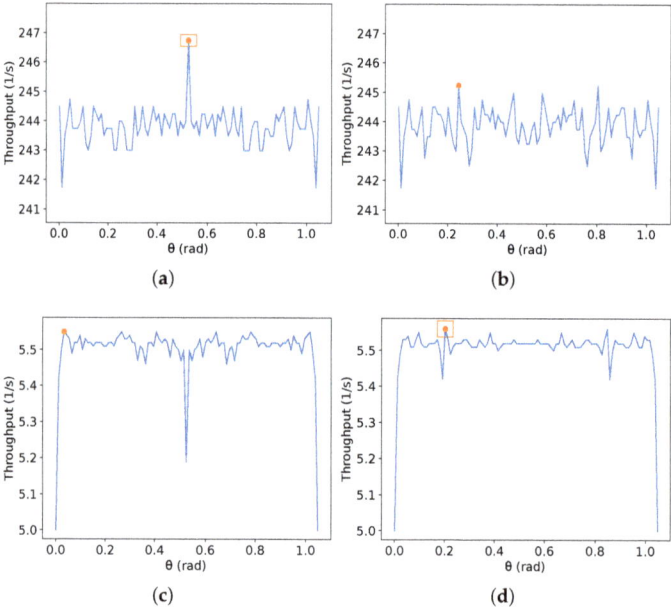

Figure 9. Continuation of Figure 8: examples of (9) varying θ from 0 to $\frac{\pi}{3}$ for different and randomly generated values of T, s and d. (**a**) For 99 samples, $T = 4$ s, $s = 2$ m and $d = 0.13$ m. (**b**) For 100 samples, $T = 4$ s, $s = 2$ m and $d = 0.13$ m. (**c**) For 99 samples, $T = 100$ s, $s = 2.40513$ m and $d = 1$ m. (**d**) For 100 samples, $T = 100$ s, $s = 2.40513$ m and $d = 1$ m.

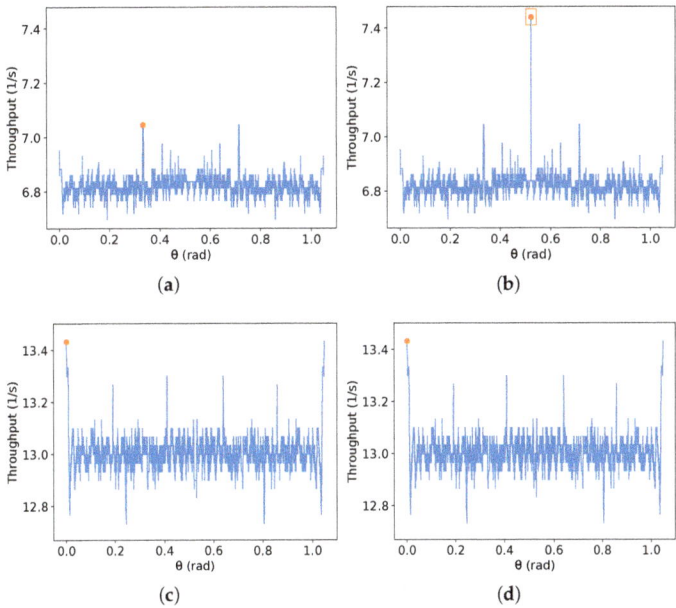

Figure 10. Similar to Figures 8 and 9 but using 10^7 and $10^7 + 1$ equally spaced points for $\theta \in [0, \pi/3]$. It continues in Figure 11. (**a**) For 10^7 samples, $T = 43$ s, $s = 3$ m, $d = 1$ m. (**b**) For $10^7 + 1$ samples, $T = 43$ s, $s = 3$ m, $d = 1$ m. (**c**) For 10^7 samples, $T = 30$ s, $s = 2.5$ m and $d = 0.66$ m. (**d**) For $10^7 + 1$ samples, $T = 30$ s, $s = 2.5$ m and $d = 0.66$ m.

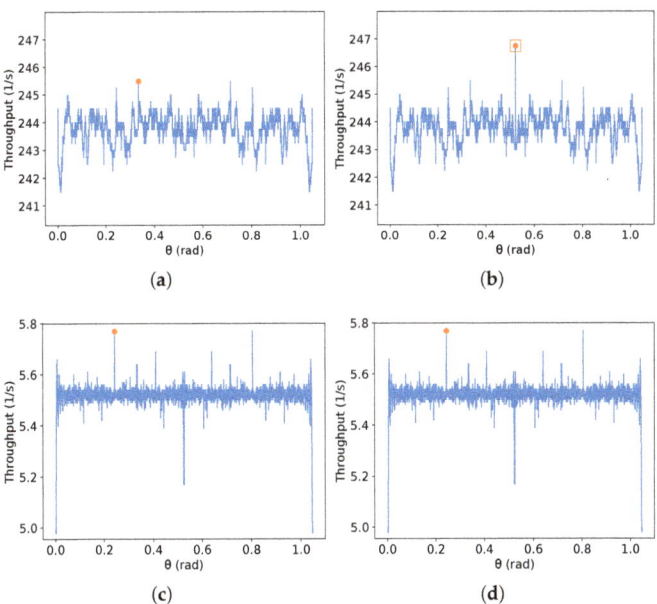

Figure 11. Continuation of Figure 10: examples similar to Figures 8 and 9 but using 10^7 and $10^7 + 1$ equally spaced points for $\theta \in [0, \pi/3]$. (**a**) For 10^7 samples, $T = 4$ s, $s = 2$ m and $d = 0.13$ m. (**b**) For $10^7 + 1$ samples, $T = 4$ s, $s = 2$ m and $d = 0.13$ m. (**c**) For 10^7 samples, $T = 100$ s, $s = 2.40513$ m and $d = 1$ m. (**d**) For $10^7 + 1$ samples, $T = 100$ s, $s = 2.40513$ m and $d = 1$ m.

Additionally, notice that whenever the number of samples is odd, the value $\theta = \pi/6$ is sampled. Observe in these figures that when the maximum is at $\theta = \pi/6$, it tends to be higher than the maximum found without considering it. For instance, compare the maximum found on the pairs (a) and (b) in Figures 8–11. On the other hand, $\theta = \pi/6$ is not always the optimal value. Thus, the authors suggest to compute first the value for $\theta = \pi/6$, then compare it with the result for a search for the maximum for any chosen number of samples in the interval from $\theta \in [0, \pi/3)$.

3.3.3. Touch and Run Strategy

Now, the *touch and run* strategy is discussed. Since a robot should spend as little time as possible near the target, a simple scenario is imagined where robots travel in predefined curved lanes and tangent to the target area where they spend minimum time on the target. To avoid collisions with other robots, the trajectory of a robot nearby the target is circular, and the distance between each robot must be at least d at any part of the trajectory. Hence, no lane crosses another, and each lane occupies a region defined by an angle in the target area, denoted by α and shown in Figure 12a.

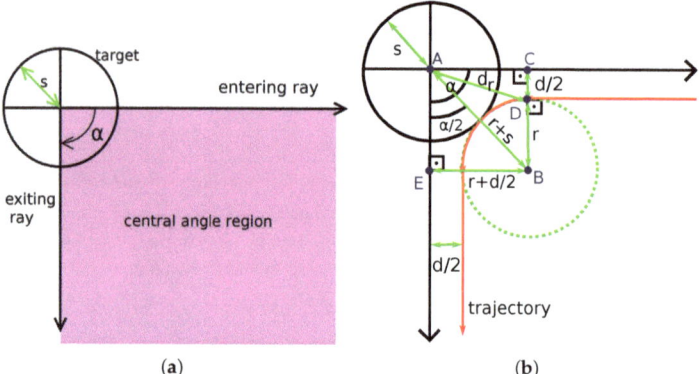

Figure 12. Illustration of the touch and run strategy. (**a**) Central angle region and its exiting and entering rays defined by the angle α. (**b**) Trajectory of a robot next to the target in red.

Figure 12b shows the trajectory of a robot towards the target region following that strategy. This figure also shows the relationship between the target area radius (s), the minimum safety distance between the robots (d), the turning radius (r), the central region angle (α) and the distance from the target centre for a robot to begin turning (d_r)—used as justification for (12) and (13). The green dashed circle represents the whole turning circle. The robot first follows the boundary of the central angle region—that is, the entering ray—at a distance of $d/2$. Then, it arrives at a distance of s of the target centre using a circular trajectory with a turning radius r. Due to the trajectory being tangent to the target shape, it is close enough to consider that the robot reached the target region.

Finally, the robot leaves the target by following the second boundary of the central angle region—that is, the exiting ray—at a distance of $d/2$. Depending on the value of α, it is possible to fit several of these lanes around the target. For example, in Figure 13, when $\alpha = \pi/2$, it is possible to fit four lanes. In this figure, robots are black dots, and d_o is the desired distance between the robots in the same lane—which is calculated depending on the values of d, s, r and the number of lanes K as shown later. When robots of all lanes simultaneously occupy the target region, their positions are the vertices of a regular polygon—it is represented in the figure by a grey square inside the target region.

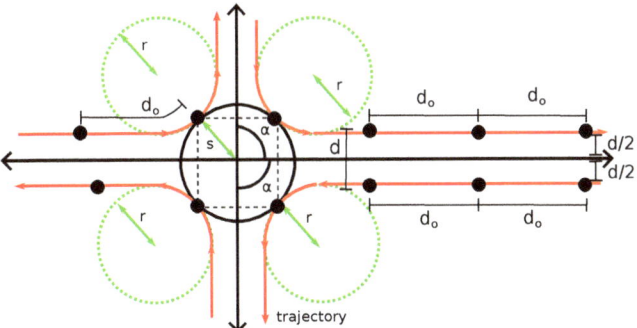

Figure 13. Theoretical trajectory in red, for $\alpha = \pi/2$ and $K = 4$.

The lemma below concerns the distance to the target centre where the robots start turning on the curved path. It will also be useful in the discussion about experiments using this strategy in Section 4.4.

Lemma 2. *The distance d_r to the target centre for the robot to start turning is*

$$d_r = \sqrt{s(2r+s) - rd}. \tag{12}$$

Now, a lemma about the turning radius is presented, and then the domain of K and α are defined in order to calculate the throughput for the touch and run strategy.

Lemma 3. *The central region angle α, the minimum distance between the robots d and the turning radius r are related by*

$$r = \frac{s \sin(\alpha/2) - d/2}{1 - \sin(\alpha/2)}. \tag{13}$$

Proposition 7. *Let K be the number of curved trajectories around the target area, α be the angle of each central area region, and r the turning radius of the robot for the curved trajectory of this central area region. For a given $d > 0$ and $s \geq d/2$, the domain of K is*

$$3 \leq K \leq \frac{\pi}{\arcsin\left(\frac{d}{2s}\right)}, \text{ and} \tag{14}$$

$$\alpha = \frac{2\pi}{K}. \tag{15}$$

Now that the correct parametrisation has been determined for the touch and run strategy, its throughput is obtained in the next proposition.

Proposition 8. *Assuming the touch and run strategy and that the first robot of every lane begins at the same distance from the target, given a target radius s, the constant linear robot speed v, a minimum distance between robots d, and the number of lanes K, the throughput for a given instant T is calculated by*

$$f_t(K, T) = \frac{1}{T}\left(K\left\lfloor \frac{vT}{d_o} + 1 \right\rfloor - 1\right), \text{ for} \tag{16}$$

$$d_o = \max(d, d'), \text{ and} \tag{17}$$

$$d' = \begin{cases} r(\pi - \alpha) + \frac{d - 2r\cos(\alpha/2)}{\sin(\alpha/2)}, & \text{if } 2r\cos(\alpha/2) < d, \\ 2r\arcsin\left(\frac{d}{2r}\right), & \text{otherwise,} \end{cases} \quad (18)$$

with r obtained from (13). In addition,

$$f_t(K) = \lim_{T \to \infty} f_t(K, T) = \frac{Kv}{d_o}. \quad (19)$$

Figure 14 presents examples of (19) for some parameters. Observe that the maximum throughput for different values of s, d and v can be found by a linear search in the interval obtained by (14).

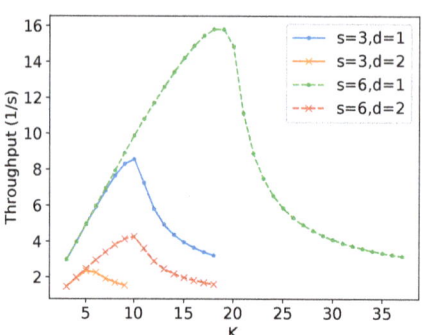

Figure 14. Plot of the asymptotic throughput of the touch and run strategy (given by (19)) for some values of s and d, in metres, and $v = 1$ m/s, for the interval of values for K obtained by (14).

3.3.4. Comparison of the Strategies

The parallel lanes strategy has the lowest of the limits concerning $u = \frac{s}{d}$, the ratio between the radius of the target region and the minimum distance between the robots. However, its asymptotic value is still higher than the minimum possible asymptotic throughput for hexagonal packing just for some values of u. This section will make explicit the dependence on the argument u in every throughput function defined previously to compare them to this ratio. Let $f_p(u) = \lim_{T \to \infty} f_p(T, u)$ and $f_h^{min}(u)$ be the asymptotic throughput for the parallel lanes strategy and the lower asymptotic throughput for the hexagonal packing strategy for a ratio u, respectively. Hence, by Proposition 4, $f_p(u) = \lfloor 2u + 1 \rfloor \frac{v}{d}$, and by (11) using $\theta = \pi/6$ as it minimises the lower bound of $\lim_{T \to \infty} f(T, \theta)$ in Proposition 6, $f_h^{min}(u) = \frac{2}{\sqrt{3}}(2u - 1)\frac{v}{d}$.

Proposition 9. *There are some* $u < \frac{\sqrt{3}+2}{4-2\sqrt{3}}$ *such that* $f_p(u) > f_h^{min}(u)$, *and for every* $u \geq \frac{\sqrt{3}+2}{4-2\sqrt{3}}$, $f_p(u) \leq f_h^{min}(u)$.

Figure 15 shows an example of $f_h^{min}(u)$, $f_p(u)$ and the maximum possible asymptotic throughput of the hexagonal packing $f_h^{max}(u) = \frac{2}{\sqrt{3}}(2u + 1)\frac{v}{d}$ for $u \in [0, 10]$. Observe that, from the left side of $u = 7$, $f_p(u)$ has some values above $f_h^{min}(u)$ even though they are below $f_h^{max}(u)$ for every u.

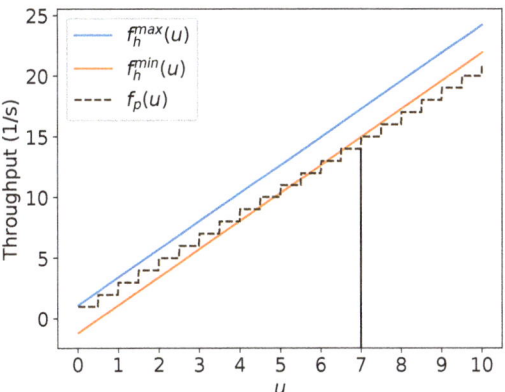

Figure 15. Example of u values such that $f_h^{max}(u) > f_p(u)$ for $v = 1$ m/s and $d = 1$ m.

Because of this proposition, for values of $u \geq \frac{\sqrt{3}+2}{4-2\sqrt{3}} \approx 7$, the hexagonal packing strategy at the limit will have higher throughput than parallel lanes. However, for values $u < \frac{\sqrt{3}+2}{4-2\sqrt{3}}$, there is the possibility of the parallel lanes strategy being better than hexagonal packing. As there is not an exact asymptotic throughput for the hexagonal packing strategy for a given angle θ, one can numerically find the best θ using large values of T on (9); then, after choosing θ, the numerical approximation of the asymptotic throughput using this fixed θ and those T values is calculated. This result can be compared with the throughput for the same large values of T for the parallel lanes strategy using (6). Furthermore, in a scenario with the target region only being accessed by a corridor with a finite height, the maximum time T can be inferred by its size, and then the exact throughput for this specific value can be calculated by (9) and (6) as stated before, but using only this specific value T, instead of a set of large values, to decide which strategy is more suitable.

Let $f_h(T, \theta, u)$ and $f_p(T, u)$ be (9) and (6) making explicit the parameter u. Let θ^* be the outcome from the search of the θ, which maximises $f_h(T, \theta, u)$ by numeric approximation. Thus, define $f_h(T, u) = f_h(T, \theta^*, u)$. Figure 16 illustrates the result of the procedure mentioned above for $T = 10,000$ for 100 equally spaced values of $u \in [0, 7]$ and seeking the maximum throughput using 1000 evenly spaced points between $[0, \pi/3)$ to find the best θ for the hexagonal packing strategy. Then, it is compared with the result for $\theta = \pi/6$ as explained previously when Figures 8–11 were discussed. Observe that for $u \in [0.5, 0.9]$ there is some values for which $f_h(10,000, u) < f_p(10,000, u)$. Figure 17 shows this by 100 equally spaced values of $u \in [0.4, 1]$ for different values of v. This occurs because, for such values of u, using square packing fits more robots inside the circle over the time than hexagonal packing, as shown in Section 4.5.

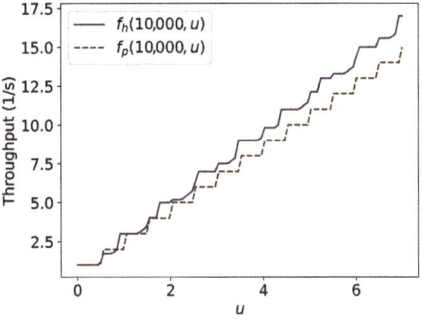

Figure 16. Comparison of $f_p(T, u)$ and $f_h(T, u)$ for $u \in [0, 7]$, $T = 10,000$ s, $v = 1$ m/s and $d = 1$ m.

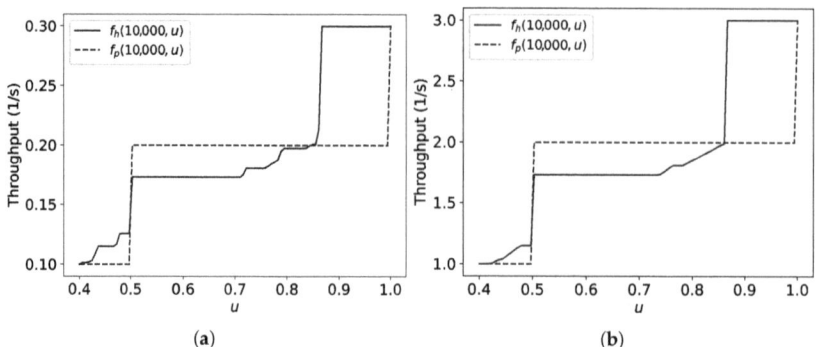

Figure 17. Comparison of f_p and f_h for $u \in [0.4, 1]$, $T = 10{,}000$ s, $v \in \{0.1, 1\}$ m/s and $d = 1$ m. The difference in the lines of f_h is due to θ^* being different for each value of v. (**a**) $v = 0.1$ m/s; (**b**) $v = 1$ m/s.

Additionally, the asymptotic throughput of the touch and run strategy, $f_t(u) = \lim_{T \to \infty} f_t(T, u)$, for higher values of u is greater than the maximum possible asymptotic value of the hexagonal packing $f_h^{max}(u) = \frac{2}{\sqrt{3}}(2u+1)\frac{v}{d}$, as shown later by numeric experimentation. Before presenting this result, it is necessary to verify which values of u are allowed by $f_t(u)$ and to express the asymptotic throughput of the touch and run strategy from Proposition 8 in terms of the ratio u.

From Proposition 7, the possible number of lanes K is in $\{3, \ldots, K(u)\}$ with $K(u) = \left\lfloor \frac{\pi}{\arcsin\left(\frac{1}{2u}\right)} \right\rfloor$. Consequently, $f_t(u)$ is only allowed for any $u \geq \frac{1}{\sqrt{3}}$. In fact, by Proposition 7, $K \geq 3$, then $\frac{\pi}{\arcsin\left(\frac{1}{2u}\right)} \geq \left\lfloor \frac{\pi}{\arcsin\left(\frac{1}{2u}\right)} \right\rfloor \geq 3 \Rightarrow \frac{\pi}{3} \geq \arcsin\left(\frac{1}{2u}\right) \Leftrightarrow \sin\left(\frac{\pi}{3}\right) \geq \frac{1}{2u} \Leftrightarrow \frac{\sqrt{3}}{2} \geq \frac{1}{2u} \Leftrightarrow u \geq \frac{1}{\sqrt{3}}$.

The algebraic manipulations for expressing the asymptotic throughput of the touch and run strategy from Proposition 8 is shown below in terms of the ratio u. The asymptotic throughput expressed in (19) is

$$\frac{Kv}{d_o} = \frac{K}{\frac{d_o}{d}} \frac{v}{d} = \frac{K}{\frac{\max(d,d')}{d}} \frac{v}{d} = \frac{K}{\max(1, \frac{d'}{d})} \frac{v}{d'}, \qquad (20)$$

for an integer $K \in \{3, \ldots, K(u)\}$. From (15), $\alpha = \frac{2\pi}{K}$, and, from (13), $\frac{r}{d} = \frac{\frac{s}{d}\sin(\alpha/2) - \frac{d}{2d}}{1-\sin(\alpha/2)} = \frac{u\sin\left(\frac{\pi}{K}\right) - \frac{1}{2}}{1-\sin\left(\frac{\pi}{K}\right)} \stackrel{\text{def}}{=} r(u, K)$, resulting in

$$\frac{d'}{d} = \begin{cases} \frac{r}{d}(\pi - \alpha) + \frac{d - 2r\cos(\alpha/2)}{d\sin(\alpha/2)}, & \text{if } 2r\cos(\alpha/2) < d, \\ 2\frac{r}{d}\arcsin\left(\frac{d}{2r}\right), & \text{otherwise,} \end{cases} \quad \text{[by (18)]}$$

$$= \begin{cases} \frac{r}{d}\left(\pi - \frac{2\pi}{K}\right) + \frac{1 - 2\frac{r}{d}\cos\left(\frac{\pi}{K}\right)}{\sin\left(\frac{\pi}{K}\right)}, & \text{if } 2\frac{r}{d}\cos\left(\frac{\pi}{K}\right) < 1, \\ 2\frac{r}{d}\arcsin\left(\left(2\frac{r}{d}\right)^{-1}\right), & \text{otherwise,} \end{cases}$$

$$= \begin{cases} r(u,K)\left(\pi - \frac{2\pi}{K}\right) + \frac{1 - 2r(u,K)\cos\left(\frac{\pi}{K}\right)}{\sin\left(\frac{\pi}{K}\right)}, \\ \qquad \text{if } 2r(u,K)\cos\left(\frac{\pi}{K}\right) < 1, \\ 2r(u,K)\arcsin\left(\frac{1}{2r(u,K)}\right), \text{ otherwise,} \end{cases} \qquad (21)$$

$$\stackrel{\text{def}}{=} d'(u, K).$$

Thus, from (20) and (21), $f_t(u, K) = \frac{K}{\max(1, d'(u,K))} \frac{v}{d'}$, and the upper throughput for the touch and run strategy in terms of u is given by $f_t(u) = \max_{K \in \{3, \ldots, K(u)\}} f_t(u, K) =$

$\max_{K\in\{3,\ldots,K(u)\}} \frac{K}{\max(1,d'(u,K))} \frac{v}{d} = \frac{K^*(u)}{\max(1,d'(u,K^*(u)))} \frac{v}{d}$, for some function $K^*(u)$ that finds this maximum in $\{3,\ldots,K(u)\}$. Similarly, for a fixed maximum time T, by (16), $f_t(T,u) = \max_{K\in\{3,\ldots,K(u)\}} f_t(K,T,u)$.

Figure 18 presents a comparison of the asymptotic throughput $f_t(u)$ and the lower and upper values of the asymptotic throughput of the hexagonal packing $f_h^{min}(u)$ and $f_h^{max}(u)$ for values of u ranging from $1/\sqrt{3}$ to 1000. Observe that the asymptotic throughput of the touch and run strategy is greater than the maximum possible asymptotic throughput of the hexagonal packing strategy for almost all values of u, except for some in $(1.12, 1.25)$ (Figure 18b).

Additionally, numerical experiments for $f_t(T,u)$ and $f_h(T,u)$ are performed using fixed time $T = 10,000$ in (16), (9) and $u \in [1/\sqrt{3}, 7]$. For finding θ^*, the same procedure is applied, which was described before to compare $f_h(T,u)$ and $f_p(T,u)$. Figure 19 shows the result. It suggests the touch and run strategy has higher throughput than hexagonal packing for large values of T. Although hexagonal packing has lower asymptotic throughput than the touch and run strategy for almost all u values, it is suitable for $u < \frac{1}{\sqrt{3}}$ whenever it surpasses the parallel lanes strategy.

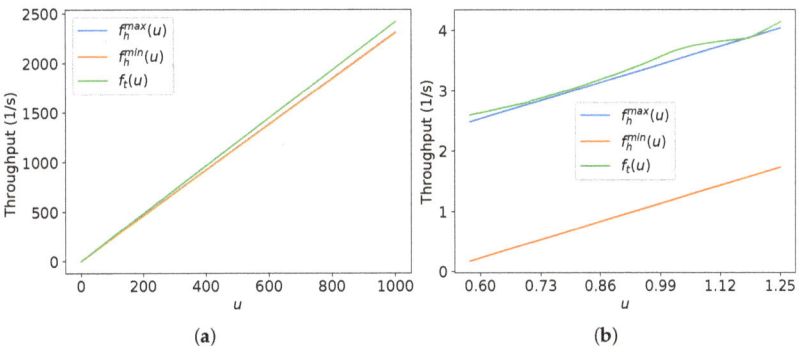

Figure 18. Graph varying u for $f_h^{min}(u)$, $f_h^{max}(u)$ and $f_t(u)$ with $v = 1$ m/s and $d = 1$ m for different intervals of u. In (**a**), $f_h^{min}(u)$ and $f_h^{max}(u)$ are almost overlapped. In (**b**), $f_t(u) > f_h^{max}(u)$ for all u, except in an interval within $(1.12, 1.25)$. (**a**) $u \in [1/\sqrt{3}, 1000]$; (**b**) $u \in [1/\sqrt{3}, 1.25]$.

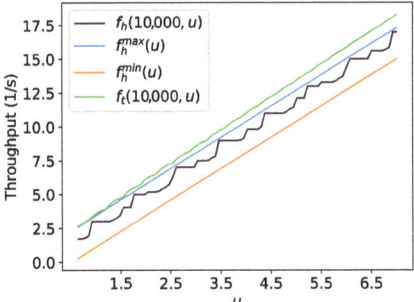

Figure 19. Example for $T = 10,000$ s, $v = 1$ m/s, $d = 1$ m and 100 equally spaced points of $u \in [1/\sqrt{3}, 7]$. $f_h(T,u) < f_t(T,u)$, albeit $f_h^{max}(u) \geq f_t(T,u)$ for a few values of $u < 1.5$.

For real-world applications and assuming the robots are constantly at maximum linear speed and at fixed distance between other robots, the hexagonal packing strategy is adequate for a situation where the target is placed in a constrained region, for example, walls in north and south positions. In this example, the number of lanes used in the touch and run strategy would be reduced because of the surrounding walls. In an unconstrained scenario, if the ratio u and the maximum time T are known, the throughput value of the hexagonal packing strategy from (9) (for the θ which maximises it) can be compared with the throughput of the touch and run strategy from (16) (for $K^*(u)$) to choose which strategy should be applied. However, assuming a constant speed and a fixed minimum distance between robots in a swarm is not practical because other robots influence the movement in the environment. Hence, these strategies are the inspiration to propose novel algorithms based on potential fields for robotic swarms in [38].

4. Experiments and Results

To evaluate this approach, several simulations were executed using the Stage robot simulator [37] for testing the equations presented in the theoretical section (Section 3). Hyperlinks to the video of executions are available in the captions of each corresponding figure. They are in real-time so that the reader can compare the time and screenshots presented in the figures in this section with those in the supplied videos (The source codes of each experimented strategy are in https://github.com/yuri-tavares/swarm-strategies, accessed on 12 June 2022).

Experiments were executed for all strategies considering $s > 0$. We could not make experiments for point-like targets because a point with a fixed value is nearly impossible to be reached by a moving robot in Stage computer simulations due to the necessity of exact synchronization of the sampling frequency of positions made by the simulator and the speed of the robot. Hence, a circular area with a radius $s > 0$ around the target must be used to identify that a robot reached it. After presenting the experiments and results for all strategies for circular target region with radius $s > 0$, they are compared experimentally considering the analysis previously discussed in Section 3.3.4.

It is saved for each robot its arrival time in milliseconds since the start of the experiment. The arrival time of every robot is subtracted by the arrival time of the first robot. By doing so, the experiment is assumed to begin in time $T = 0$ without worrying about the initial inertia. After this, the number of robots (N) is registered for each time value (T).

To alleviate some of the numerical errors caused by the floating-point representation, rounding on the 13th decimal place was used before using floor and ceiling functions on the equations presented. For example, in contemporary computers, by using double variables in C or float in Python, if you divide 9.6 by 1.6, the result is 5.999999999999999 for 15 decimal places formatting, but it should be 6. If the floor function was applied to the previous result, the outcome would be 5 instead of the expected 6.

For all experiments in this section, the robots are distant from each other by $d = 1$ m. In the figures of this section, black robots indicate they reached the target, and red did not. In addition, the experiments shown on this section were not repeated because the linear speed and initial positions are constant, so there is no random aspect, and the same results are obtained for different runs.

4.1. Compact Lanes

For compact lanes simulations, $v = 1$ m/s, and the first robot to reach the target is at the bottom lane and starts at the target. For a target area radius s, such that $0 < s < \sqrt{3}d/4$, $s = 0.3$ m, and for $\sqrt{3}d/4 \leq s < d/2$, $s = 0.45$ m. Figure 20 shows screenshots of the simulation using $s = 0.3$ m during $T = 7.1$ s and Figure 21 for $s = 0.45$ m and $T = 10.1$ s.

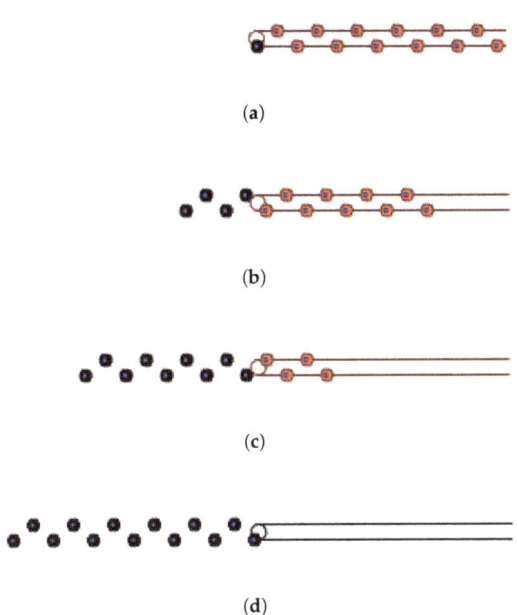

Figure 20. Simulation on Stage for compact lanes strategy using $s = 0.3$ m, $d = 1$ m during $T = 7.1$ s. Available on https://youtu.be/e1cWJzWhQmQ, accessed on 12 June 2022. (**a**) 0 s: beginning of the simulation; (**b**) After 2.7 s; (**c**) After 6.7 s; (**d**) 5 s: ending of the simulation.

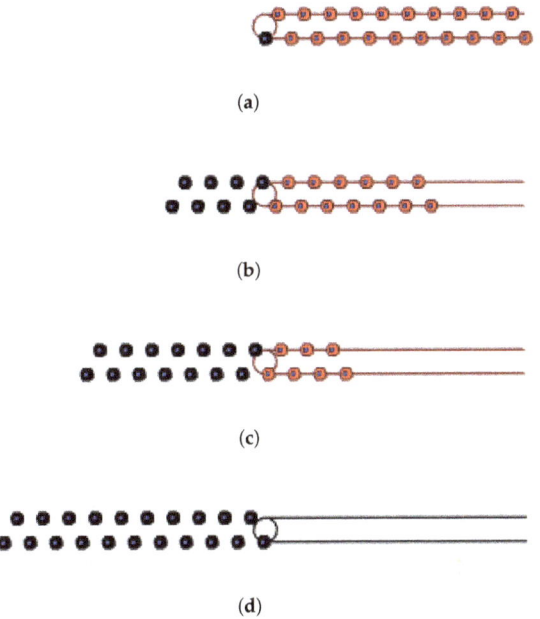

Figure 21. Simulation on Stage for compact lanes strategy using $s = 0.45$ m, $d = 1$ m during $T = 10.1$ s. Available on https://youtu.be/9OXGC1w83j0, accessed on 12 June 2022. (**a**) 0 s: beginning of the simulation; (**b**) After 3.5 s; (**c**) After 7 s; (**d**) 10.1 s: ending of the simulation.

Experiments were run in order to verify the throughput for a given time and the asymptotic throughput calculated by (2) to (5). Figure 22 shows the throughput for different values of time obtained by the experiments in Stage, i.e., $(N-1)/T$, in comparison with the calculated value by (2) and (3) for $s = 0.3$ m and by (4) and (5) for $s = 0.45$ m. "Simulation" stands for the data obtained from Stage, "Instantaneous" for the equations of the throughput for a given time calculated in (2) and (4) and "Asymptotic" for the asymptotic throughput obtained from (3) and (5). The mentioned results of the equations match the data obtained from simulations. These figures confirm that the equations presented in the theoretical section agree with the throughput obtained by simulations.

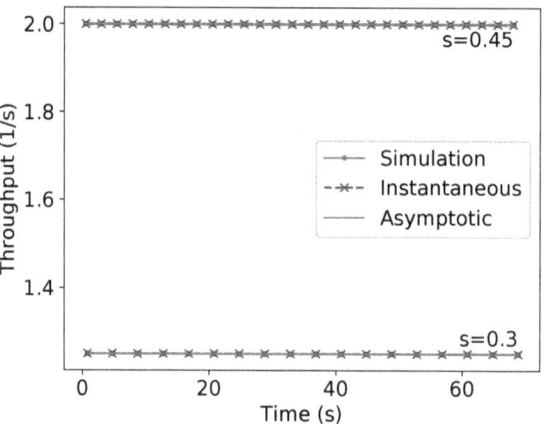

Figure 22. Throughput versus time plot for compact lanes strategy for different values of s.

4.2. Parallel Lanes

The parallel lanes strategy was experimented for $v = 1$ m/s and $s \in \{3, 6\}$ m. Figures 23 and 24 present screenshots from executions using these parameters.

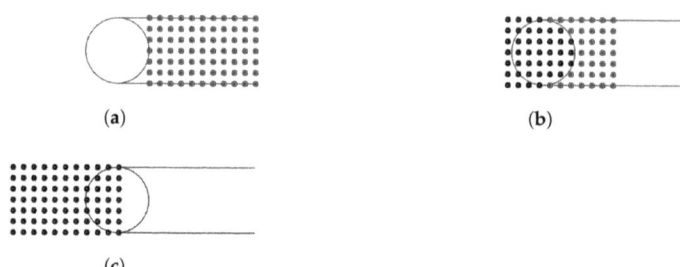

Figure 23. Simulation on Stage for parallel lanes strategy using $s = 3$ m, $d = 1$ m during $T = 13$ s. Available on https://youtu.be/2Y1RHc9YVaw, accessed on 12 June 2022. (**a**) 0 s: beginning of the simulation; (**b**) After 6.5 s; (**c**) 13 s: ending of the simulation.

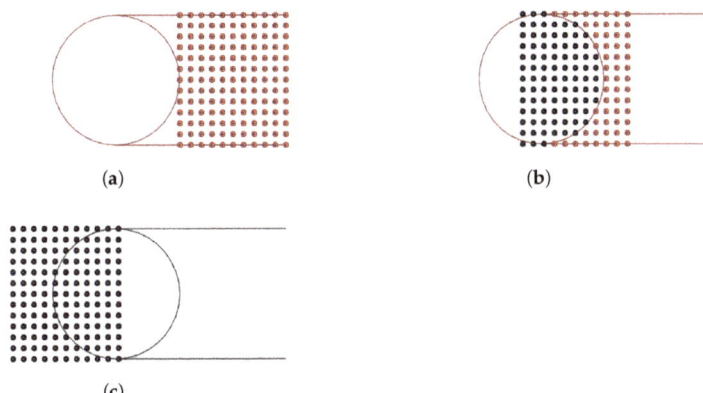

Figure 24. Simulation on Stage for parallel lanes strategy using $s = 6$ m, $d = 1$ m during $T = 16$ s. Available on https://youtu.be/TVdka65fi1g, accessed on 12 June 2022. (**a**) 0 s: beginning of the simulation; (**b**) After 8 s; (**c**) 16 s: ending of the simulation.

To verify the throughput for a given time calculated by (6) and its asymptotic value as in (7), they are compared with the throughput obtained from Stage simulations. Figure 25a presents these comparisons. "Simulation" stands for the data obtained from Stage, "Instantaneous" for the equations of the throughput for a given time calculated in (6), and "Asymptotic" for the asymptotic throughput obtained from (7). As expected, the values of (6) approximate to (7) as time passes. Additionally, observe that the values from (6) are almost aligned with the values from the simulation, except for some points. The difference in those points is due to the floating-point error discussed at the beginning of Section 4 that happens in the division before the use of floor or ceiling functions used on (6). Figure 25b shows the number of robots versus the time of arrival of the last robot for the same data used in Figure 25a. As the running time is proportional to the number of robots in the experiments, observe that the higher throughput per time is reflected as a lower arrival time of the last robot per the number of robots. In addition, note that the values tend to infinity as the horizontal axis values grow.

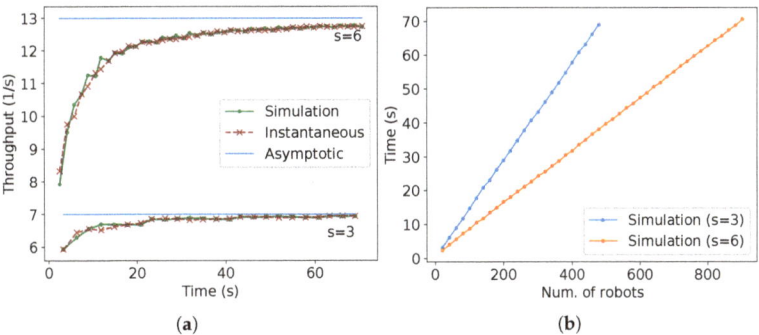

Figure 25. Plots for the experiments of parallel lanes strategy for $s \in \{3,6\}$ m. (**a**) Number of robots versus throughput. (**b**) Number of robots versus the time of arrival of the last robot.

4.3. Hexagonal Packing

The hexagonal packing was experimented for $v = 1$ m/s and the combination of the following variables and values: $s \in \{3,6\}$ m and $\theta \in \{0, \pi/12, \pi/6, 5\pi/18\}$. Figures 26–33 present screenshots from executions using these parameters.

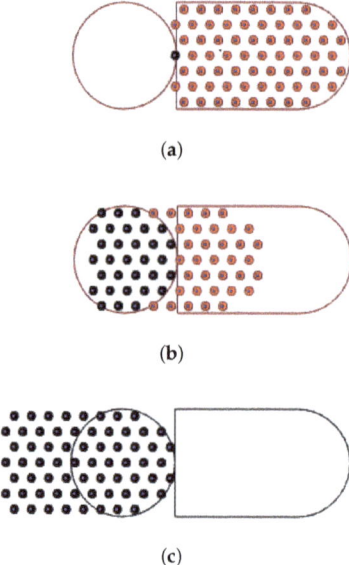

Figure 26. Simulation on Stage for hexagonal packing strategy using $s = 3$ m, $\theta = 0$ during $T = 9.8$ s. Available on https://youtu.be/6_LgZWFOWd0, accessed on 12 June 2022. (**a**) 0 s: beginning of the simulation; (**b**) After 4.9 s; (**c**) 9.8 s: ending of the simulation.

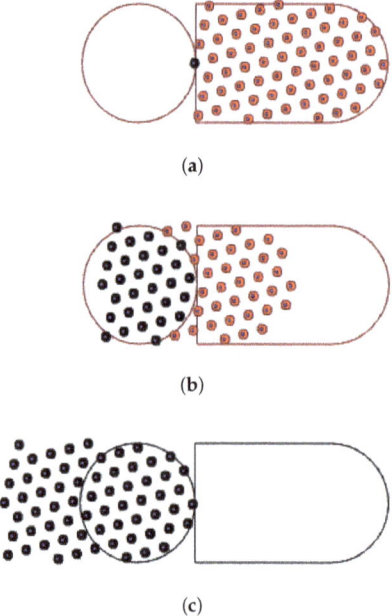

Figure 27. Simulation on Stage for hexagonal packing strategy using $s = 3$ m, $\theta = \pi/12$ during $T = 10$ s. Available on https://youtu.be/Wji8XlSQJBQ, accessed on 12 June 2022. (**a**) 0 s: beginning of the simulation; (**b**) After 5 s; (**c**) 10 s: ending of the simulation.

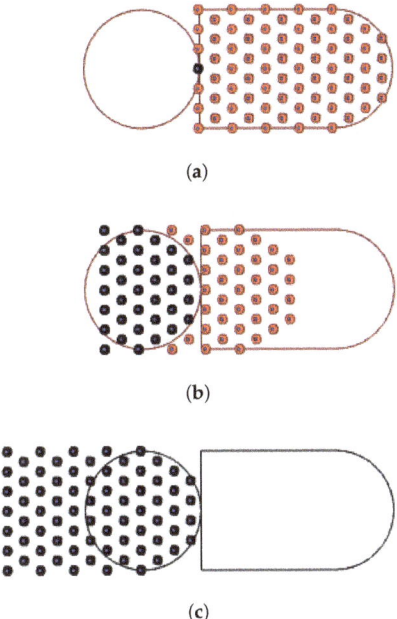

Figure 28. Simulation on Stage for hexagonal packing strategy using $s = 3$ m, $\theta = \pi/6$ during $T = 10$ s. Available on https://youtu.be/szOBU8no_sU, accessed on 12 June 2022. (**a**) 0 s: beginning of the simulation; (**b**) After 4.9 s; (**c**) 10 s: ending of the simulation.

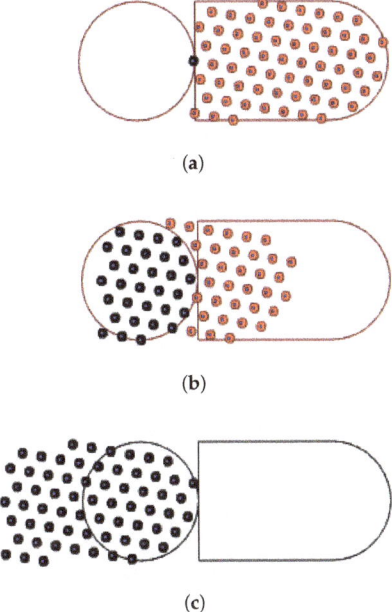

Figure 29. Simulation on Stage for hexagonal packing strategy using $s = 3$ m, $\theta = 5\pi/18$ during $T = 10$ s. Available on https://youtu.be/jRLgaF7Te1Q, accessed on 12 June 2022. (**a**) 0 s: beginning of the simulation; (**b**) After 4.9 s; (**c**) 10 s: ending of the simulation.

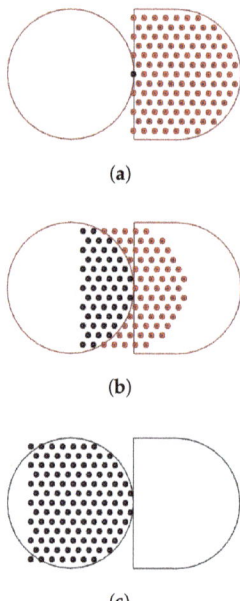

Figure 30. Simulation on Stage for hexagonal packing strategy using $s = 6$ m, $\theta = 0$ during $T = 9.8$ s Available on https://youtu.be/v0FK8YpGrL8, accessed on 12 June 2022. (**a**) 0 s: beginning of the simulation; (**b**) After 4.9 s; (**c**) 9.8 s: ending of the simulation.

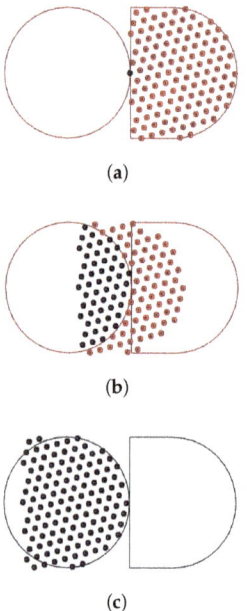

Figure 31. Simulation on Stage for hexagonal packing strategy using $s = 6$ m, $\theta = \pi/12$ during $T = 10.1$ s. Available on https://youtu.be/OBS_HADH5OE, accessed on 12 June 2022. (**a**) 0 s: beginning of the simulation.; (**b**) After 5 s; (**c**) 10.1 s: ending of the simulation..

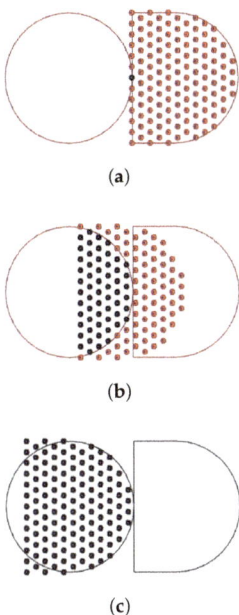

Figure 32. Simulation on Stage for hexagonal packing strategy using $s = 6$ m, $\theta = \pi/6$ during $T = 10$ s. Available on https://youtu.be/-KX7ziOp8b0, accessed on 12 June 2022. (**a**) 0 s: beginning of the simulation.; (**b**) After 4.9 s; (**c**) 10 s: ending of the simulation.

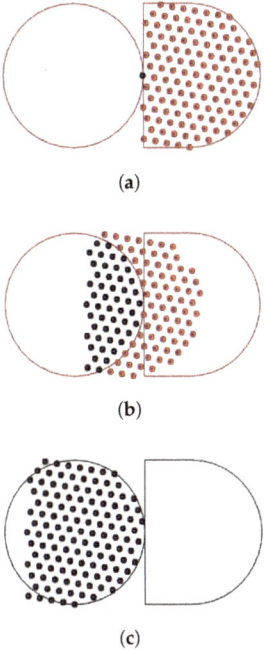

Figure 33. Simulation on Stage for hexagonal packing strategy using $s = 6$ m, $\theta = 5\pi/18$ during $T = 10$ s. Available on https://youtu.be/GRYRnH5CrhU, accessed on 12 June 2022. (**a**) 0 s: beginning of the simulation; (**b**) After 4.9 s; (**c**) 10 s: ending of the simulation.

To evaluate the throughput for a given time and angle calculated in (9) and the bounds on the asymptotic throughput as in (11), they are compared with the throughput obtained from Stage simulations. Figure 34 presents these comparisons. Observe that the values from (9) are almost aligned with the values from the simulation, except for some points. The difference in those points is also due to the floating-point error—discussed in the introduction of Section 4—over the divisions and trigonometric functions performed before the use of floor or ceiling functions used on (9). In addition, due to the floating-point error, in the computation of (10), instead of using $\min(L(x_h), C_2(x_h)) = \lfloor L(x_h) \rfloor$, $|\min(L(x_h), C_2(x_h)) - \lfloor L(x_h) \rfloor| < 0.001$ was checked.

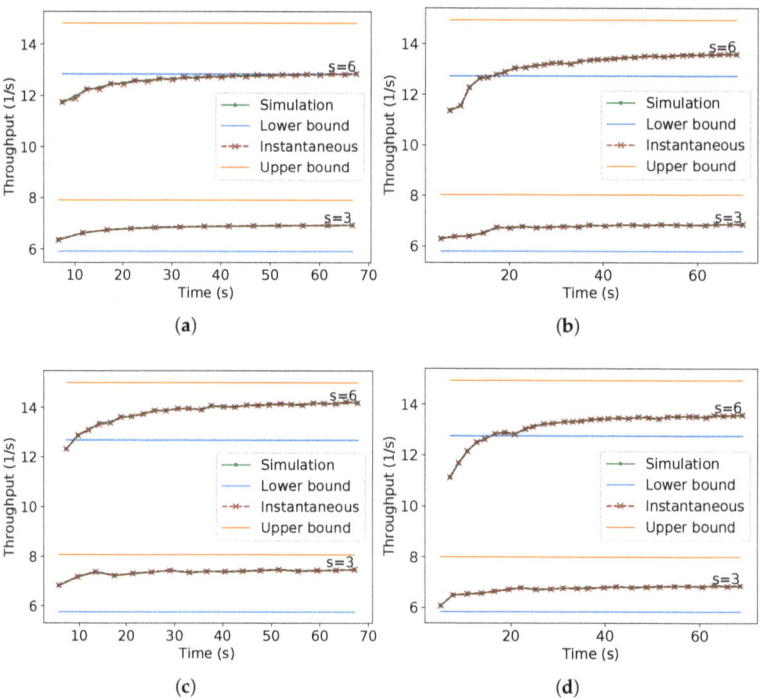

Figure 34. Comparison of simulation data with the asymptotic and instantaneous throughput for hexagonal packing with different values of s and θ. (**a**) $\theta = 0$; (**b**) $\theta = \pi/12$; (**c**) $\theta = \pi/6$; (**d**) $\theta = 5\pi/18$.

Additionally, note in Figure 34 that for any value of s or θ, as the time passes, the values of (9) asymptotically approach some value inside the bounds given by (11). Although the exact asymptotic value could not be given for the presented parameters, the experiments show that the bounds are correct. In the same manner, as occurred for parallel lanes, the higher throughput per time is reflected as a lower arrival time of the last robot per the number of robots, and it tends to infinity as the number of robots grows (Figure 35).

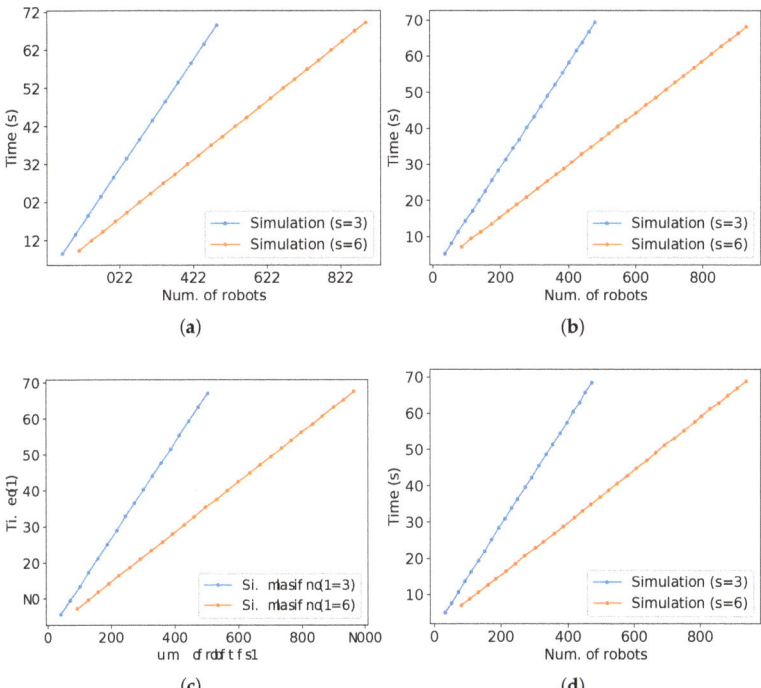

Figure 35. Time of arrival at the target of the last robot versus the number of robots for the same simulations in Figure 34. (**a**) $\theta = 0$; (**b**) $\theta = \pi/12$; (**c**) $\theta = \pi/6$; (**d**) $\theta = 5\pi/18$.

4.4. Touch and Run

For the touch and run strategy, the robots maintain the linear speed over the whole experiment, then turn at a fixed constant rotational speed $\omega = v/r$, for r obtained from (13), when they are next to the target centre by the distance d_r obtained from (12). After they arrive at the target region, when they are distant from the target centre by d_r, they leave the curved path, stop turning and follow the linear exiting lane. On that lane, to stabilise their path following, the robots follow the queue using a turning speed equal to $\gamma - \beta$, such that β is the angle of the exit lane and γ is the robot orientation angle, both in relation to the x-axis.

The speed of these experiments was $v = 0.1$ m/s because the robots utilised on Stage have a maximum turning speed of $\pi/2$ rad/s. Choosing a low linear speed implies a greater number of lanes K, as the turning speed $\omega = v/r$ and r vary over K and s. In addition, a low linear speed diminishes the time measurement error, since the positions of the robots are sampled at every 0.1 s by the Stage simulator. Their positions are not guaranteed to be obtained at the exact moment they are far from the target centre by d_r; thus, this also yields an error in time measurement for their arrival in the target area.

The value of s is in $\{3, 6\}$ m and all allowed K values are used for experimenting with the touch and run strategy with 200 robots. By (14), for the former s value, there is a maximum $K = 18$ and for the later, $K = 37$. However, as the maximum angular speed is limited, the allowed K values range for $s = 3$ m is reduced to $\{3, \ldots, 16\}$ and for $s = 6$ m, $\{3, \ldots, 33\}$. Figures 36–39 present screenshots from executions using some of these parameters. The circle in the middle of these figures is the target region, and the lines where the robots are over represent the curved trajectory they follow by the touch and run strategy.

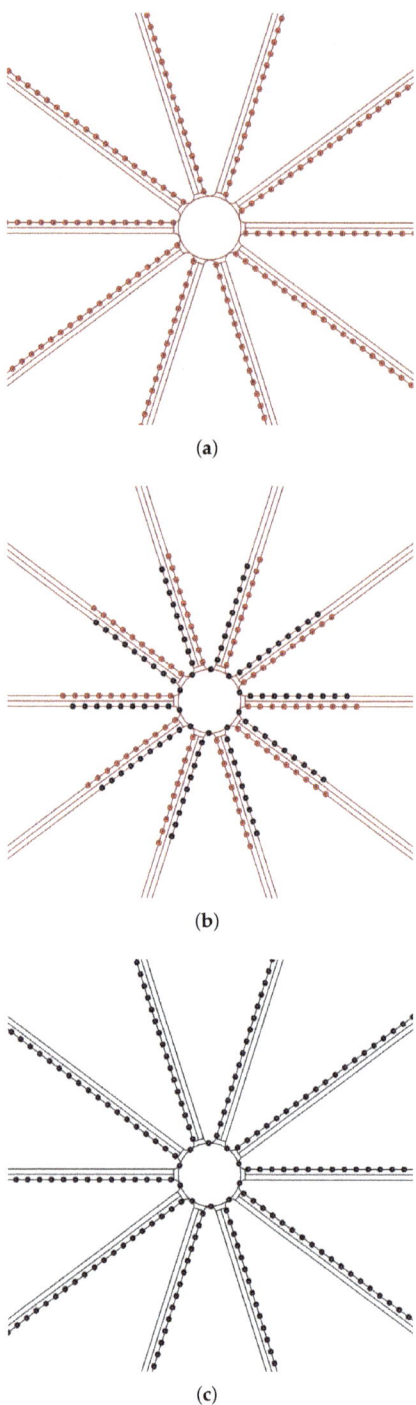

Figure 36. Simulation on Stage for the touch and run strategy using $s = 3$ m, $K = 10$ during $T = 228$ s at $v = 0.1$ m/s. Available on https://youtu.be/Z-ruOMYFyBU, accessed on 12 June 2022. (**a**) 0 s: beginning of the simulation; (**b**) After 114 s; (**c**) 228 s: ending of the simulation.

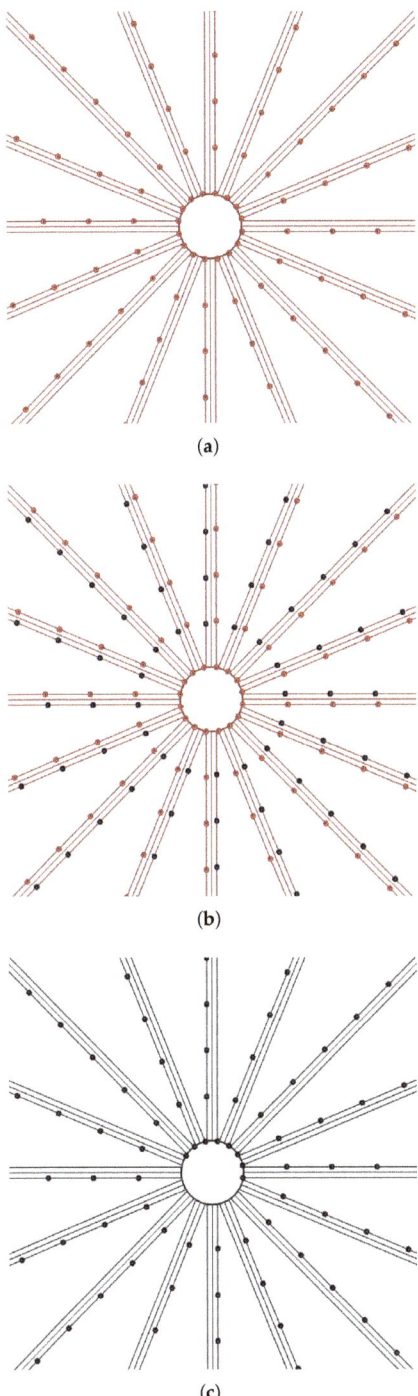

Figure 37. Simulation on Stage for the touch and run strategy using $s = 3$ m, $K = 16$ during $T = 523.1$ s at $v = 0.1$ m/s. Available on https://youtu.be/FvAqv0zD4_Y, accessed on 12 June 2022. (**a**) 0 s: beginning of the simulation; (**b**) After 261.6 s; (**c**) 523.1 s: ending of the simulation.

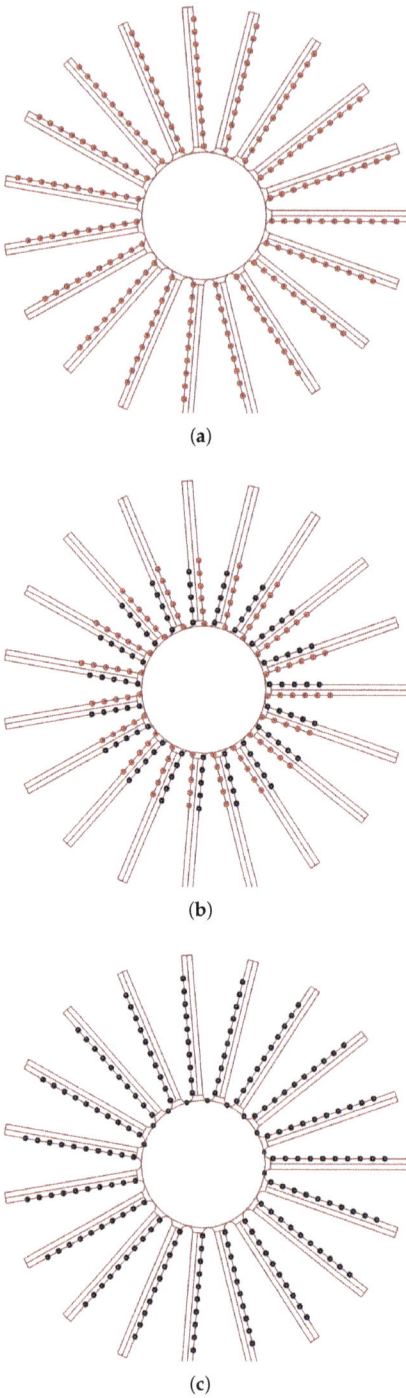

Figure 38. Simulation on Stage for the touch and run strategy using $s = 6$ m, $K = 19$ during $T = 127.4$ s at $v = 0.1$ m/s. Available on https://youtu.be/xJVoVCIjX5k, accessed on 12 June 2022. (**a**) 0 s: beginning of the simulation; (**b**) After 63.6 s; (**c**) 127.4 s: ending of the simulation.

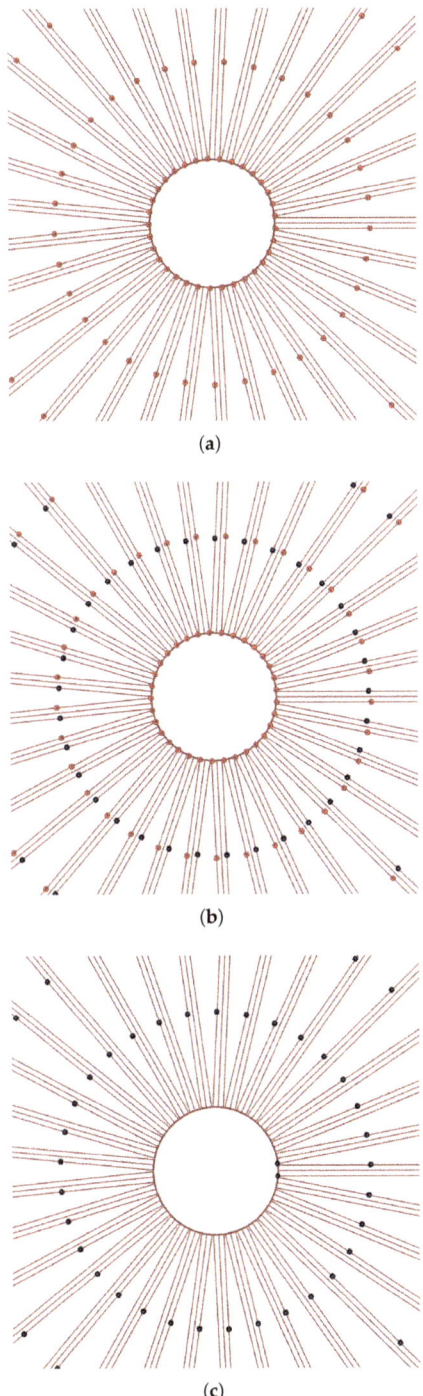

Figure 39. Simulation on Stage for the touch and run strategy using $s = 6$ m, $K = 33$ during $T = 548$ s at $v = 0.1$ m/s. Available on https://youtu.be/-xZz84npKV4, accessed on 12 June 2022. (**a**) 0 s: beginning of the simulation; (**b**) After 274 s; (**c**) 548 s: ending of the simulation.

Figure 40 presents the comparison of (16) and (19) for the throughput for a given time, the bound on its asymptotic value and the one obtained from Stage simulations. Although the total number of robots and the linear speed were fixed, the arrival times and the number of robots to reach the target change for each parameter used in this figure since the distance between the robots per lane varies and the number of robots simultaneously arriving is, in most cases, the number of lanes. In addition, the first two arrival times were not plotted because the first one is zero, yielding an indeterminate output by the throughput definition, and the second one is still too small in relation to the others, making the resultant throughput too high compared with the rest, thus producing an incomprehensible graph.

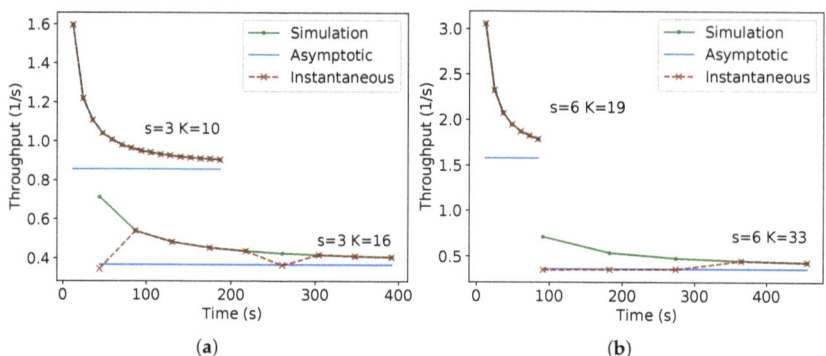

Figure 40. Throughput versus time comparison of the touch and run simulation on Stage with asymptotic values and the theoretical instantaneous equation for the throughput for different values of s and K. (**a**) $s = 3$ m and $K \in \{10, 16\}$; (**b**) $s = 6$ m and $K \in \{19, 33\}$.

Observe that the values from (16) are almost equal to the values from simulation, except for some points. They are different because of the floating-point error in the divisions and trigonometric functions before the use of floor function used on (16)—already mentioned in the introduction of Section 4—as well as the time measurement errors for the arrival of the robots on the target area as explained at the beginning of this section. As expected, the values of (16) tend to come nearer to the asymptotic value given by (19). Differently from the previous strategies, notice that, for small values of T, the throughput is higher than for larger ones because, for a fixed K, (16) is decreasing for T. As occurred for the previous strategies, higher throughput per time is reflected as a lower arrival time of the last robot per number of robots, which tends to infinity as the number of robots grows (Figure 41).

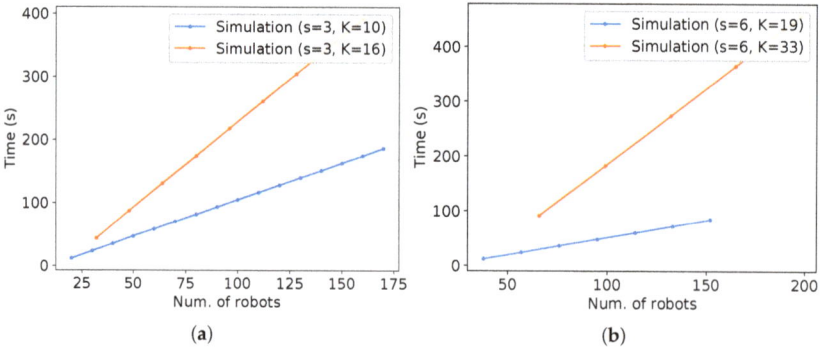

Figure 41. Time of arrival at the target of the last robot versus number of robots for the same simulations in Figure 40. (**a**) $s = 3$ m and $K \in \{10, 16\}$; (**b**) $s = 6$ m and $K \in \{19, 33\}$.

Figure 42 shows a comparison of the throughput at the end of the experiment—that is, for 200 robots and considering the difference between the time to reach the target region spent by the last robot and the first—and the asymptotic throughput obtained by (19) for all the possible number of lanes (K) for the used parameters and $s \in \{3, 6\}$ m. The simulation values tend to come close to the asymptotic value, confirming the theoretical results.

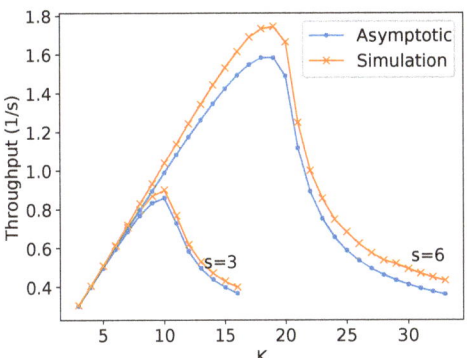

Figure 42. Throughput versus the number of lanes comparison of the simulation on Stage and asymptotic throughput for $s \in \{3, 6\}$ m.

4.5. Comparison between Hexagonal Packing and Parallel Lanes

As discussed in Section 3.3.4, it is observed that the parallel lane strategy has a higher throughput than hexagonal packing for values of $u = s/d$ from 0.5 to a value of about 0.85 and for high values of T, despite the parallel lanes having lower asymptotic throughput for other values of u. In order to validate this observation, experiments on Stage were performed for these strategies using $T = 10,000$ s, $v = 0.1$ m/s, $d = 1$ m and s ranging from 0.4 to 0.95 m in increments of 0.05 m. The best hexagonal packing angle θ was computed for hexagonal packing using the same method mentioned at the end of the theoretical section, i.e., the maximum throughput was searched using 1000 evenly spaced points between $[0, \pi/3)$ to find the best θ; then, it was compared with the result for $\pi/6$.

Figure 43 presents the results from the experiments with Stage and the theoretical results shown earlier. The functions f_h and f_p are the same presented in Figure 17. The labels "Simulation hex." and "Simulation par." stand for the throughput resultant from the experiments with hexagonal packing and parallel lanes strategies, respectively. The throughput improvement for the values of $u = s/d$ where the parallel lanes strategy overcomes the hexagonal packing is mainly caused by the square packing being more effective than hexagonal packing for fitting the robots inside the circle over the time for those values. To illustrate this, Figure 44 illustrates the execution for $v = 0.1$ m/s, $d = 1$ m and $s \in \{0.5, 0.85\}$ m. The robots run from right to left at a constant linear speed $v = 0.1$ m/s. The grey squares are highlighted—which measure 1×1 m^2—to help estimate the time needed for about eight robots to arrive in the target region. This figure shows that the square packing fits more robots than hexagonal packing over time in these cases.

Observe in these figures that when the robots are arranged in squares, more robots arrive per unit of time than using hexagonal packing. To help visualise this, heed that in Figure 44a, there are $N = 9$ robots in black, occupying a rectangle including the circular target area with a width of approximately $W \approx 4.5$ m (this distance can be roughly measured by the grey squares, counting from the two last black robots on the right side to the first one in the left side). As $v = 0.1$ m/s was assumed, the throughput in this case is approximately $\frac{N-1}{\frac{W}{v}} \approx \frac{(9-1)0.1}{4.5} \approx 0.178$ s^{-1}. Making similar calculations, Figure 44b–d have the approximate throughputs $\frac{(8-1)0.1}{3} \approx 0.233$ s^{-1}, $\frac{(8-1)0.1}{4} = 0.175$ s^{-1} and $\frac{(8-1)0.1}{3} \approx 0.233$ s^{-1}, respectively. The results from the parallel lanes in this illustration—about 0.233 for both values of s—surpass the values for the hexagonal packing.

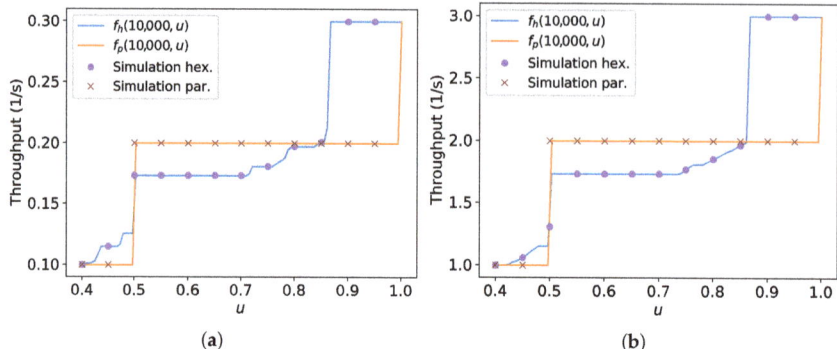

Figure 43. Throughput versus ratio $u = s/d$ comparing hexagonal packing and parallel lanes strategies for $v \in \{0.1, 1\}$ m/s, including results from Stage simulations. (**a**) $v = 0.1$ m/s; (**b**) $v = 1$ m/s.

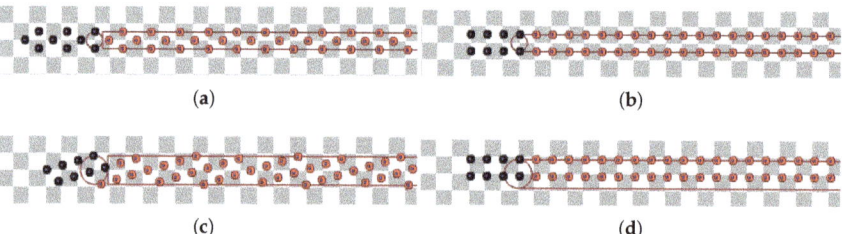

Figure 44. Screenshots of the Stage simulation for hexagonal packing and parallel lanes strategy for $d = 1$ m and $s \in \{0.5, 0.85\}$ m. (**a**) Hexagonal packing with best θ for $s = 0.5$ m. Available on https://youtu.be/IZBnFHLKXUA, accessed on 12 June 2022. (**b**) Parallel lanes for $s = 0.5$ m. Available on https://youtu.be/YYv1dJFkdPA, accessed on 12 June 2022. (**c**) Hexagonal packing with best θ for $s = 0.85$ m. Available on https://youtu.be/r9X0fsnngm0, accessed on 12 June 2022. (**d**) Parallel lanes for $s = 0.85$ m. Available on https://youtu.be/0cx-bHPIong, accessed on 12 June 2022.

5. Conclusions

A novel metric was proposed for measuring the effectiveness of algorithms to minimise congestion in a swarm of robots trying to reach the same goal: the common target area throughput. In addition, the asymptotic throughput for the common target area was defined as the throughput when the time tends to infinity.

Assuming that the robots move at constant maximum speed and the distance between each other is as close as possible to a fixed value, it was shown how to calculate the maximum throughput for different theoretical strategies to arrive at the common circular target region: (i) making parallel queues to reach the target region, (ii) using a corridor with robots in hexagonal packing to enter in the region, and (iii) following curved trajectories to touch the region boundary. Based on these calculations solely, it was possible to compare which strategy is better.

Due to their aim of maximising the target area throughput, these strategies were used as inspiration for new algorithms using artificial potential fields in [38]. Thus, for common target area congestion in robotic swarms, the throughput is well suited for comparing algorithms due to its abstraction of the rate of the target area access as the number of robots grows, whether the closed throughput equation is derived or not.

The key contribution of this work is a fundamental theoretical study of congestion in swarm robotics, which already served as inspiration to create new algorithms. However, future work could extend this theoretical study further, by considering varying linear speed and distances between the robots.

Supplementary Materials: The following supporting information can be downloaded at: https://www.mdpi.com/article/10.3390/math10142482/s1. References [39–41] are cited in the supplementary materials.

Author Contributions: Conceptualization, X.D.; Formal analysis, Y.T.d.P. and X.D.; Investigation, Y.T.d.P. and X.D.; Methodology, Y.T.d.P. and X.D.; Software, Y.T.d.P. and X.D.; Supervision, L.S.M.; Writing—original draft, X.D.; Writing—review & editing, Y.T.d.P. and L.S.M. All authors have read and agreed to the published version of the manuscript.

Funding: This research was funded by Lancaster University.

Acknowledgments: Yuri Tavares dos Passos gratefully acknowledges the Faculty of Science and Technology at Lancaster University for the scholarship and the Universidade Federal do Recôncavo da Bahia for granting the leave of absence to finish his PhD.

Conflicts of Interest: The authors declare no conflict of interest.

References

1. Navarro, I.; Matía, F. An introduction to swarm robotics. *Int. Sch. Res. Not.* **2013**, *2013*, 608164. [CrossRef]
2. Garnier, S.; Gautrais, J.; Theraulaz, G. The biological principles of swarm intelligence. *Swarm Intell.* **2007**, *1*, 3–31. [CrossRef]
3. Giavitto, J.L.; Dulman, S.O.; Spicher, A.; Viroli, M. *Proceedings of the Spatial Computing Workshop (SCW 2013) Colocated with AAMAS (W09)*; Version 1; IFAMAAS (International Foundation for Autonomous Agents and Multiagent Systems): Saint Paul, MN, USA, 2013. Available online: https://hal.archives-ouvertes.fr/hal-00821901 (accessed on 12 June 2022).
4. Giordano, J.P.; Wu, A.S.; Pherwani, A.; Mathias, H.D. Comparison of Desynchronization Methods for a Decentralized Swarm on a Logistical Resupply Problem. In Proceedings of the 20th International Conference on Autonomous Agents and MultiAgent Systems, AAMAS'21, Online, 3–7 May 2021; International Foundation for Autonomous Agents and Multiagent Systems: Richland, SC, USA, 2021; pp. 1510–1511.
5. Cohen, S.; Agmon, N. Spatial Consensus-Prevention in Robotic Swarms. In Proceedings of the 20th International Conference on Autonomous Agents and MultiAgent Systems, AAMAS'21, Online, 3–7 May 2021; International Foundation for Autonomous Agents and Multiagent Systems: Richland, SC, USA, 2021; pp. 359–367.
6. Cicerone, S.; Di Fonso, A.; Di Stefano, G.; Navarra, A. MOBLOT: Molecular Oblivious Robots. In Proceedings of the 20th International Conference on Autonomous Agents and MultiAgent Systems, AAMAS'21, Online, 3–7 May 2021; International Foundation for Autonomous Agents and Multiagent Systems: Richland, SC, USA, 2021; pp. 350–358.
7. Yang, B.; Ma, C.; Xia, X. Drone Formation Control via Belief-Correlated Imitation Learning. In Proceedings of the 20th International Conference on Autonomous Agents and MultiAgent Systems, AAMAS'21, Online, 3–7 May 2021; International Foundation for Autonomous Agents and Multiagent Systems: Richland, SC, USA, 2021; pp. 1407–1415.
8. Marcolino, L.S.; Chaimowicz, L. No Robot Left behind: Coordination to Overcome Local Minima in Swarm Navigation. In Proceedings of the 2008 IEEE International Conference on Robotics and Automation, Pasadena, CA, USA, 19–23 May 2008; pp. 1904–1909. [CrossRef]
9. Ducatelle, F.; Caro, G.A.D.; Pinciroli, C.; Mondada, F.; Gambardella, L. Communication Assisted Navigation in Robotic Swarms: Self-Organization and Cooperation. In Proceedings of the 2011 IEEE/RSJ International Conference on Intelligent Robots and Systems, San Francisco, CA, USA, 25–30 September 2011; pp. 4981–4988. [CrossRef]
10. Kato, S.; Nishiyama, S.; Takeno, J. Coordinating Mobile Robots By Applying Traffic Rules. In Proceedings of the IEEE/RSJ International Conference on Intelligent Robots and Systems, Raleigh, NC, USA, 7–10 July 1992; pp. 1535–1541. [CrossRef]
11. Caloud, P.; Choi, W.; Latombe, J.C.; Le Pape, C.; Yim, M. Indoor automation with many mobile robots. In Proceedings of the IEEE International Workshop on Intelligent Robots and Systems, Ibaraki, Japan, 3–6 July 1990; pp. 67–72. [CrossRef]
12. Grossman, D. Traffic control of multiple robot vehicles. *IEEE J. Robot. Autom.* **1988**, *4*, 491–497. [CrossRef]
13. Carlino, D.; Boyles, S.D.; Stone, P. Auction-Based Autonomous Intersection Management. In Proceedings of the 16th International IEEE Conference on Intelligent Transportation Systems (ITSC 2013), The Hague, The Netherlands, 6–9 October 2013; pp. 529–534. [CrossRef]
14. Sharon, G.; Stone, P. A Protocol for Mixed Autonomous and Human-Operated Vehicles at Intersections. In *Autonomous Agents and Multiagent Systems—AAMAS 2017 Workshops, Best Papers*; Lecture Notes in Artificial Intelligence; Sukthankar, G., Rodriguez-Aguilar, J.A., Eds.; Springer International Publishing: New York, NY, USA, 2017; Volume 10642, pp. 151–167. [CrossRef]
15. Cui, J.; Macke, W.; Yedidsion, H.; Goyal, A.; Urieli, D.; Stone, P. Scalable Multiagent Driving Policies for Reducing Traffic Congestion. In Proceedings of the 20th International Conference on Autonomous Agents and Multiagent Systems, AAMAS '21, Online, 3–7 May 2021; International Foundation for Autonomous Agents and Multiagent Systems: Richland, SC, USA, 2021; pp. 386–394.
16. Choudhury, S.; Solovey, K.; Kochenderfer, M.J.; Pavone, M. Efficient Large-Scale Multi-Drone Delivery using Transit Networks. *J. Artif. Intell. Res.* **2021**, *70*, 757–788. [CrossRef]
17. Shahar, T.; Shekhar, S.; Atzmon, D.; Saffidine, A.; Juba, B.; Stern, R. Safe Multi-Agent Pathfinding with Time Uncertainty. *J. Artif. Int. Res.* **2021**, *70*, 923–954. [CrossRef]

18. Xia, G.; Sun, X.; Xia, X. Distributed Swarm Control Algorithm of Multiple Unmanned Surface Vehicles Based on Grouping Method. *J. Mar. Sci. Eng.* **2021**, *9*, 1324. [CrossRef]
19. Sahin, E. Swarm Robotics: From Sources of Inspiration to Domains of Application. In *International Workshop on Swarm Robotics*; Sahin, E., Spears, W.M., Eds.; Springer: Berlin/Heidelberg, Germany, 2005; pp. 10–20. [CrossRef]
20. Sahin, E.; Girgin, S.; Bayindir, L.; Turgut, A.E. Swarm Robotics. In *Swarm Intelligence: Introduction and Applications*; Springer: Berlin/Heidelberg, Germany, 2008; pp. 87–100. [CrossRef]
21. Barca, J.C.; Sekercioglu, Y.A. Swarm robotics reviewed. *Robotica* **2013**, *31*, 345–359. [CrossRef]
22. Brambilla, M.; Ferrante, E.; Birattari, M.; Dorigo, M. Swarm robotics: A review from the swarm engineering perspective. *Swarm Intell.* **2013**, *7*, 1–41. [CrossRef]
23. Bayındır, L. A review of swarm robotics tasks. *Neurocomputing* **2016**, *172*, 292–321. [CrossRef]
24. Chung, S.; Paranjape, A.A.; Dames, P.; Shen, S.; Kumar, V. A Survey on Aerial Swarm Robotics. *IEEE Trans. Robot.* **2018**, *34*, 837–855. [CrossRef]
25. Hoy, M.; Matveev, A.S.; Savkin, A.V. Algorithms for Collision-Free Navigation of Mobile Robots in Complex Cluttered Environments: A Survey. *Robotica* **2015**, *33*, 463–497. [CrossRef]
26. Marcolino, L.S.; Passos, Y.T.; Souza, A.A.F.d.; Rodrigues, A.d.S.; Chaimowicz, L. Avoiding Target Congestion on the Navigation of Robotic Swarms. *Auton. Robot.* **2017**, *41*, 1297–1320. [CrossRef]
27. van den Berg, J.; Guy, S.J.; Lin, M.; Manocha, D. Reciprocal n-Body Collision Avoidance. In *Robotics Research*; Pradalier, C., Siegwart, R., Hirzinger, G., Eds.; Springer: Berlin/Heidelberg, Germany, 2011; pp. 3–19. [CrossRef]
28. Lima, D.A.; Oliveira, G.M. A cellular automata ant memory model of foraging in a swarm of robots. *Appl. Math. Model.* **2017**, *47*, 551–572. [CrossRef]
29. Varghese, B.; McKee, G. A mathematical model, implementation and study of a swarm system. *Robot. Auton. Syst.* **2010**, *58*, 287–294. [CrossRef]
30. Li, Y.; Chen, X. Modeling and Simulation of Swarms for Collecting Objects. *Robotica* **2006**, *24*, 315–324. [CrossRef]
31. Taylor-King, J.P.; Franz, B.; Yates, C.A.; Erban, R. Mathematical modelling of turning delays in swarm robotics. *IMA J. Appl. Math.* **2015**, *80*, 1454–1474. [CrossRef]
32. Galstyan, A.; Hogg, T.; Lerman, K. Modeling and mathematical analysis of swarms of microscopic robots. In Proceedings of the 2005 IEEE Swarm Intelligence Symposium, Pasadena, CA, USA, 8–10 June 2005; pp. 201–208. [CrossRef]
33. Khaluf, Y.; Dorigo, M. Modeling Robot Swarms Using Integrals of Birth-Death Processes. *ACM Trans. Auton. Adapt. Syst.* **2016**, *11*, 1–16 [CrossRef]
34. Mannone, M.; Seidita, V.; Chella, A. Categories, Quantum Computing, and Swarm Robotics: A Case Study. *Mathematics* **2022**, *10*, 372. [CrossRef]
35. Daduna, H.; Pestien, V.; Ramakrishnan, S. Asymptotic Throughput in Discrete-Time Cyclic Networks with Queue-Length-Dependent Service Rates. *Stoch. Model.* **2003**, *19*, 483–506. [CrossRef]
36. Hockney, R.W. The communication challenge for MPP: Intel Paragon and Meiko CS-2. *Parallel Comput.* **1994**, *20*, 389–398. [CrossRef]
37. Gerkey, B.P.; Vaughan, R.T.; Howard, A. The Player/Stage Project: Tools for Multi-Robot and Distributed Sensor Systems. In Proceedings of the 11th International Conference on Advanced Robotics, Coimbra, Portugal, 30 June–3 July 2003; pp. 317–323.
38. Passos, Y.T.; Duquesne, X.; Marcolino, L.S. Congestion control algorithms for robotic swarms with a common target based on the throughput of the target area. *arXiv* **2022**, arXiv:2201.09337.
39. Chang, H.C.; Wang, L.C. A Simple Proof of Thue's Theorem on Circle Packing. *arXiv* **2010**, arXiv:math.MG/1009.4322.
40. Red Blob Games. Hexagonal Grids. 2021. Available online: https://www.redblobgames.com/grids/hexagons/ (accessed on 16 November 2021).
41. Graham, R.L.; Knuth, D.E.; Patashnik, O. *Concrete Mathematics: A Foundation for Computer Science*, 2nd ed.; Addison-Wesley Professional: Reading, MA, USA, 1994.

Article

Learning to Utilize Curiosity: A New Approach of Automatic Curriculum Learning for Deep RL

Zeyang Lin, Jun Lai *, Xiliang Chen *, Lei Cao and Jun Wang

Command Control Engineering College, Army Engineering University of PLA, Nanjing 210007, China; hunterlzy@aeu.edu.cn (Z.L.); feiyuewuxian2018@aeu.edu.cn (L.C.); wangjun920811@aeu.edu.cn (J.W.)
* Correspondence: zhangk@aeu.edu.cn (J.L.); lgd_chenxiliang@aeu.edu.cn (X.C.)

Abstract: In recent years, reinforcement learning algorithms based on automatic curriculum learning have been increasingly applied to multi-agent system problems. However, in the sparse reward environment, the reinforcement learning agents get almost no feedback from the environment during the whole training process, which leads to a decrease in the convergence speed and learning efficiency of the curriculum reinforcement learning algorithm. Based on the automatic curriculum learning algorithm, this paper proposes a curriculum reinforcement learning method based on the curiosity model (CMCL). The method divides the curriculum sorting criteria into temporal-difference error and curiosity reward, uses the K-fold cross validation method to evaluate the difficulty priority of task samples, uses the Intrinsic Curiosity Module (ICM) to evaluate the curiosity priority of the task samples, and uses the curriculum factor to adjust the learning probability of the task samples. This study compares the CMCL algorithm with other baseline algorithms in cooperative-competitive environments, and the experimental simulation results show that the CMCL method can improve the training performance and robustness of multi-agent deep reinforcement learning algorithms.

Keywords: deep reinforcement learning; automatic curriculum learning; curiosity; sparse reward

MSC: 68T07

Citation: Lin, Z.; Lai, J.; Chen, X.; Cao, L.; Wang, J. Learning to Utilize Curiosity: A New Approach of Automatic Curriculum Learning for Deep RL. *Mathematics* **2022**, *10*, 2523. https://doi.org/10.3390/math10142523

Academic Editors: Jiangping Hu and Zhinan Peng

Received: 4 July 2022
Accepted: 15 July 2022
Published: 20 July 2022

Copyright: © 2022 by the authors. Licensee MDPI, Basel, Switzerland. This article is an open access article distributed under the terms and conditions of the Creative Commons Attribution (CC BY) license (https://creativecommons.org/licenses/by/4.0/).

1. Introduction

Deep reinforcement learning [1] combines the perception ability of deep learning with the decision-making ability of reinforcement learning, and has been widely used in the processing of complex decision-making tasks [2], such as Atari games [3], complex robot action control [4,5], and the application of AlphaGo intelligence [6]. In 2015, Hinton, Bengio and Lecun, famous experts in the field of machine learning, published a review paper on deep learning in *Nature*, which considered deep reinforcement learning as an important development direction of deep learning [7].

However, there is a significant problem in the application of deep reinforcement learning algorithms in multi-agent systems [8]. With the increase in the number of agents and the increase in the complexity of the environment, the coordination and cooperation between agents becomes more difficult, which can easily cause a situation where the Reinforcement Learning (RL) algorithm does not converge or even cannot be trained [9,10].

Curriculum learning [11], as a hot field of current artificial intelligence research, was proposed by Bengio et al. at the International Conference on Machine Learning (ICML) in 2009. Bengio et al. pointed out that the curriculum learning method can be regarded as a special kind of continuous optimization method, which can start with smoother (i.e., simpler) optimization problems and gradually add rougher (i.e., more difficult) non-convex optimization problems, and finally optimize the target task. In curriculum reinforcement learning algorithms [12], manually set the tasks of different difficulty levels, and gradually add more difficult tasks to the simple reinforcement learning tasks, so that the knowledge of the source tasks can be reused in the process of learning difficult tasks, thereby accelerating the convergence of model to the optimal policy.

The above-mentioned predefined curriculum learning methods need to be manually set in advance in the process of task generation and sorting, so the quality of the generated curriculums will be directly affected by the experience of experts. However, the learning method of pre-defined curriculums requires manual curriculum difficulty assessment and sorting, and lacks task versatility. The current curriculum learning field gradually adopts automatic curriculum learning (ACL) instead of predefined curriculum learning to train reinforcement learning agents.

Before the agent learns the whole task, the difficulty of the experience samples in the experience replay buffer is evaluated and sorted, and the experience samples are learned in order from easy to difficult, so automatic curriculum learning [13] can realize the learning of difficult tasks, shorten the training time, and improve the training performance of task learning.

Traditional automatic curriculum learning often uses the temporal-difference error method to evaluate and sort the difficulty of task samples, that is, to obtain the optimal policy by maximizing the external reward value that appears in the process of interacting with the environment, but in the reward sparsity environment, the agent is difficult to obtain environmental reward feedback in long-lasting time steps. The lack of reward signals will affect the iteration and update of the agent's action policy, so it is hard for the agent to learn an effective policy.

To solve the above problems, this paper proposes a curriculum learning method based on curiosity module (CMCL), adding curiosity intrinsic reward in curriculum sorting criteria, the curiosity reward value of the experience samples was evaluated to obtain the curiosity priority, and the curriculum sequence of the experience samples was sorted together with the temporal-difference error, and the selection progress of the curriculum difficulty was adjusted by setting the curriculum difficulty factor, so as to enhance the exploration and training performance of the curriculum reinforcement learning algorithm for the environment. The experimental results of two tasks in multi-agent particle environment show that the CMCL method proposed in this paper can greatly improve the processing performance of multi-agent tasks in sparse reward environments compared with the three baseline algorithms.

The contributions of this paper are as follows:

(1) This paper proposes a curriculum reinforcement learning method based on the curiosity module. By adding curiosity priority to the curriculum sorting criteria, it can enhance the exploratory and robustness of reinforcement learning agents and avoid the appearance of turn-in-place agent;
(2) This paper introduces a curriculum difficulty factor in the process of selecting the curriculum difficulty of the model, and dynamically adjusts the difficulty of the currently selected curriculum through the curriculum difficulty factor, so as to realize automatic curriculum learning from easy to difficult priority experience.

The rest of this paper is organized as follows. Section 2 introduces related work, Section 3 introduces the MADDPG algorithm and the theory of automatic curriculum learning, Section 4 introduces the CMCL algorithm in detail, Section 5 presents experimental results and analyzes them, Section 6 presents discussion and Section 7 draws some conclusions.

2. Related Work

How to reasonably arrange the sequence of curriculums and select curriculums in the process of curriculum learning is the main research problem of current automatic curriculum reinforcement learning research. Carlos Florensa et al. [14] used generative networks to propose tasks that the agent needs to implement to automatically generate curriculums capable of learning many types of tasks without requiring prior knowledge. Ren et al. [15] proposed an automatic curriculum reinforcement learning method that uses a priority curriculum sorting method to extract experience samples from the experience replay buffer to achieve automatic curriculum learning. Jiayu Chen et al. [16] used the perspective of variational inference to automatically generate training curriculums for the task environment and the number of agents from two aspects of task expansion and agent expansion, which can be used to solve cooperative multi-agent reinforcement learning problems in difficult environments.

Curiosity-driven agent exploration is an important approach in reward function design for reinforcement learning. In supervised learning, curiosity is used to alleviate the problem of imbalanced representation and distributional bias among data [17,18]. Pathak et al. [19] used curiosity as an intrinsic reward value for agents, which can encourage the agent to explore new environmental states. Our method is derived from the curiosity mechanism of the human brain [20]. Curiosity is used as a reference standard for automatic curriculum learning's curriculum sorting, which can complement the priority experience replay algorithm (PER). The selection probability of novel samples is increased in the samples to balance the exploration of the uncertain state in the process of environmental exploration of multi-agent system.

The most important works related to our method include the self-adaptive priority correction algorithm proposed by Hongjie Zhang et al. [21], the High-Value Prioritized Experience Replay proposed by Xi Cao et al. [22], and the Curriculum Guided Hindsight Experience Replay proposed by Meng Fang et al. [4]. Hongjie Zhang et al. predicted the sum of the real Temporal-Difference error of all samples in the experience replay, and corrected it by an importance weight. Xi Cao et al. designed a priority experience replay method based on the combination of temporal-difference error and value for the sparse reward environment, Meng Fang et al. applied the curiosity mechanism to the Hindsight experience replay algorithm (HER), and learned successful experience from failure through the HER mechanism. Our method provides a further improvement on the basis of the above methods. As one of the curriculum sorting standards in the priority experience replay algorithm, the curiosity mechanism can compensate for the exploratory and randomness of the agent in the sparse reward environment, thereby improving the training performance and robustness of the algorithm.

3. Basic Concepts

This chapter will sequentially introduce some important concepts of Deep Reinforcement Learning, Multi-Agent Deep Deterministic Policy Gradient algorithms (MADDPG), and Automatic Curriculum Learning (ACL).

3.1. Deep Reinforcement Learning

Reinforcement learning [23] consists of two parts: agents and environment. To maximize agents' total reward value, the agents observe the initial state in the environment, take actions from an action set, and the environment accepts the action and gives the agents a reward. This process can be modeled as a Markov decision quintuple (S, A, R, P, γ), where S represents the state space, A represents the action space, R represents the reward function, P represents the state transition function, and γ represents the discount factor. The schematic diagram of reinforcement learning is shown in Figure 1.

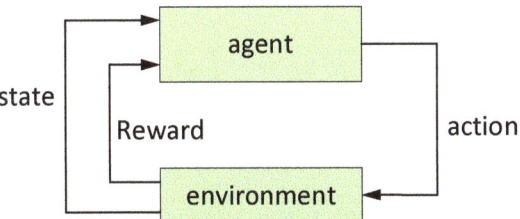

Figure 1. Schematic diagram of reinforcement learning.

Deep reinforcement learning approximates policy function and value function through a deep learning multi-layer neural network, thereby solving the high-dimensional mapping problem caused by continuous high-dimensional state-action pairs [24]. The goal of agents is to maximize expected reward $J(\pi_\theta) = E_{\tau \sim \pi_\theta}[R(\tau)]$ by continuously optimizing the policy π_θ, then the optimal policy is

$$\pi_\theta^* = \underset{\pi_\theta}{argmax} E_{\tau \sim \pi_\theta} \left(\sum_{t=0}^{\infty} \gamma^t r_t \right) \tag{1}$$

where r_t represents the reward of agents at time t.

Deep reinforcement learning algorithms can be divided into following three categories [25], deep reinforcement learning based on value function, deep reinforcement learning based on policy gradient, and deep reinforcement learning based on the actor-critic (AC) framework. The DRL algorithm based on the structure of the AC framework uses the error of the value function to guide the policy update and improve the performance of the algorithm training. The policy π_θ is updated by policy gradient $\nabla_\theta J(\pi_\theta)$ of expected reward, the formula is as follows:

$$\nabla_\theta J(\pi_\theta) = E_{\tau \sim \pi_\theta} \left[\sum_{t=0}^{T} \nabla_\theta \log \pi_\theta(a|s) R(\tau) \right] \tag{2}$$

where $\pi_\theta(a|s)$ represents the actor Function and $R(\tau)$ represents the critic Function.

3.2. MADDPG Algorithm

Multi-Agent Deep Deterministic Policy Gradient algorithm [26] (MADDPG) is an improved Multi-Agent Reinforcement Learning algorithm based on the AC network framework, which can be considered as an extended application of the DDPG algorithm in a multi-agent environment. To solve the problem of non-stationarity in Multi-agent Training Process [27], the MADDPG pioneered the principle of centralized training and distributed execution (CTDE), that is, in the training stage, the MADDPG algorithm allows the agents to obtain global information during learning, only local information is used in the decision execution. The AC training framework can be seen as an actor network for policy exploration, critic network as an evaluator to evaluate the policy, and obtain the current optimal policy. The algorithm structure consists of actor network, critic network, target actor network and target critic network. The training framework of the MADDPG algorithm is shown in Figure 2. The MADDPG algorithm stores experience tuples through the experience replay mechanism:

$$D_i = (o_1, \cdots, o_N, a_1, \cdots, a_N, r_1, \cdots, r_N, o_1', \cdots, o_N') \tag{3}$$

During the training process, experience tuples are stored in batches in the experience replay buffer, and the experience replay buffer extracts small samples of experience in stages and inputs them into the neural network for model training. This experience replay mechanism can reduce the degree of association between experience tuples, thus improving the neural network training efficiency. The MADDPG algorithm updates the action network of agents using the stochastic gradient descent method. The formula is as follows:

$$\nabla_{\theta^\pi} J = \frac{1}{K} \sum_{j=1}^{K} \nabla_{\theta^\pi} \pi(o, \theta^\pi) \nabla_a Q(s, a_1, a_2, \ldots, a_N, \theta^Q) \tag{4}$$

In the formula, o and a_i represent the observation value and action of the ith agent respectively; $\pi(o, \theta^\pi)$ represents the action of agent i obtained by inputting the observation value into actor network.

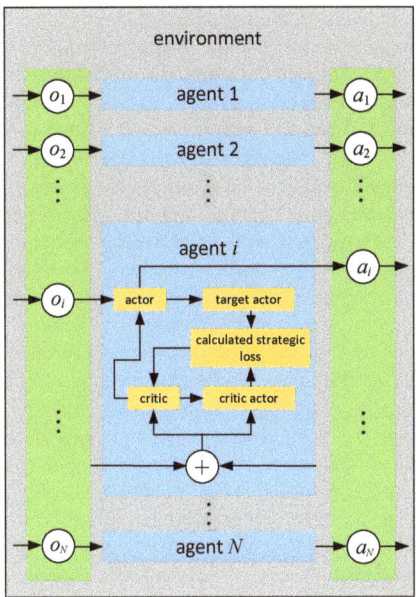

Figure 2. MADDPG algorithm training framework diagram.

The critic network of agents is iteratively updated as follows to minimize the loss function:

$$L = \frac{1}{K} \sum_{j=1}^{K} (y_j - Q(s_j, a_1, a_2, \ldots, a_N, \theta^Q))^2 \quad (5)$$

In the formula, the function y represents the cumulative average reward of agent i in the target actor network.

The network parameters of target actor network and target critic network are replicated and updated in stages:

$$\theta_i' = \tau \theta_i + (1 - \tau) \theta_i' \quad (6)$$

In the formula, τ represents the control parameter of the network parameter updating frequency, which can stabilize the parameter network update process. θ_i' represents the target network parameter of the ith agent, and θ_i represents the initial network parameter of the ith agent.

In view of the good stability and convergence of the MADDPG algorithm, it can be applied to various task scenarios such as cooperative, competitive and hybrid. The innovation and experimental verification of the algorithm in this paper are partly based on the MADDPG algorithm and its accompanying multi-agent particle environment (MPE).

3.3. Automatic Curriculum Learning

End-to-end deep reinforcement learning methods have led to breakthroughs in board games, real-time policy games, and path planning problems. However, reinforcement learning agents still face difficulties and challenges when dealing with many application scenarios [13]. The reason is that agents need to fully interact with the environment to obtain enough information to continuously modify its own policy, but the environment itself has the problems of reward sparseness, partial observability, delayed reward, and too high dimension of action space, which leads to the problem that the training time of the agent is too long or even unable to converge when dealing with difficult tasks.

In response to the above problems, Curriculum Learning (CL) can utilize knowledge from source tasks to speed up the learning of complex target tasks, thus improving the training performance of reinforcement learning agents on fixed task sets [28]. As an important paradigm in the field of machine learning, curriculum learning can imitate the human learning sequence from easy to difficult. In the initial stage of reinforcement learning, the curriculum learning algorithm trains the model in a simple simulation environment (fewer obstacles and more reward values), and as the training progresses, the simulation environment is gradually added with more and more difficult (sparse reward values and more obstacles), and finally, the algorithm is validated in a full simulation environment.

Most traditional curriculum learning methods use predefined methods [13], that is, using expert experience to evaluate the difficulty of task curriculums and formulate curriculum plans from the perspectives of the number of agents, initial state distribution, reward function, goals, environment distribution, opponent policy, etc., such as tasks with a higher number of agents and more obstacles are generally considered more difficult training environments. Because the predefined curriculum learning method requires manual assessment and sorting of curriculum difficulties and lacks task versatility, the current curriculum learning field gradually adopts automatic curriculum learning instead of predefined curriculum learning to train reinforcement learning agents [29].

The current automatic curriculum learning process can be divided into curriculum sorting stage and curriculum selection stage [30]. The main idea is to construct a task curriculum sampler $q(n, \phi)$ based on the experience replay buffer, which can evaluate the difficulty of the transitions in the experience replay buffer and sort them from easy to difficult, and then the task $M(n, \phi)$ that is currently most suitable for agent training is extracted in real time from the experience replay to maximize the cumulative reward value of the reinforcement learning agent $J(\theta)$, ϕ represents the environmental factor variables that affect the difficulty of task curriculum.

To prove that curriculum updating can increase the cumulative reward value of agents in the process of automatic curriculum learning, in this paper, the proof is performed as follows from the perspective of mathematics.

Proof. For a given number n of agents, $J(\theta)$ can be simplified as follows:

$$\begin{aligned} J &= E_{\phi \sim p}[V(\phi, \pi)] = E_{\phi \sim q}\left[\frac{p(\phi)}{q(\phi)} V(\phi, \pi)\right] \\ &= E_{\phi \sim q}\left[V(\phi, \pi) + \left(\frac{p(\phi)}{q(\phi)} - 1\right) V(\phi, \pi)\right] \\ &\geq \underbrace{E_{\phi \sim q}[V(\phi, \pi)]}_{J_1:\text{policy update}} + \underbrace{E_{\phi \sim q}\left[V(\phi, \pi) \log \frac{p(\phi)}{q(\phi)}\right]}_{J_2:\text{curriculum update}} \end{aligned} \quad (7)$$

□

In the formula, $p(\phi)$ represents the uniform distribution of ϕ in the range of possible values. For all ϕ, the inequality is due to $x - 1 \geq \log x$, the equal sign of the inequality holds if and only if $p(\phi) = q(\phi)$.

Through the simplification of the above equation, the cumulative reward value $J(\theta)$ can be composed of the policy update reward J_1 and the curriculum update reward J_2. The policy update reward J_1 represents that reinforcement learning agents update their own policy functions iteratively to maximize their reward value obtained from the environment, and the curriculum update reward J_2 represents the task curriculum sampling sorting and adjustment through the task curriculum sampler $q(n, \phi)$, which can improve the agent's ability to explore environment and the training performance of the model to maximize agents' cumulative reward value.

In traditional automatic curriculum learning algorithms, the ordering of task curriculums often takes the environmental reward value of agents as the reference standard, that is, it adjusts its own action policy according to the external reward value. However, in sparse reward environments, it is difficult for an agent to obtain positive or negative rewards from the environment during most of the exploration process. Under the framework of the traditional automatic curriculum learning algorithm, selecting the task curriculum from low to high according to temporal-difference error can easily lead to overfitting of the model training, and agents stay in circles in the environment, making it difficult to train a good policy.

4. Curriculum Reinforcement Learning Based on Curiosity Model

This paper proposes a general automatic curriculum learning framework—curiosity module-based curriculum learning for deep RL (CMCL), which is divided into two stages: curriculum sorting and curriculum selection. For all reinforcement learning tasks, suppose $D = \{d_1, d_2, \cdots, d_j, \cdots, d_K\}$ represents the experience sample set in experience replay buffer, and the task curriculum sampler $q(n, \phi)$ is used to operate on experience sample set D. The first stage is to evaluate and sort the difficulty of the samples in experience sample set to generate a curriculum learning plan; the second stage selects curriculums according to the set ability evaluation rules according to the curriculum plan.

The core of the curriculum difficulty sorting is to define the difficulty of the task samples. To convert the task samples in the experience replay into a curriculum sequence, a curriculum index function (CI) needs to be defined to calculate the priority p_{d_j} of task sample d_j.

Definition 1. *Curriculum Index Function (CI).*

The function $CI(d_j) \to \mathbb{R}$ is used to define the curriculum sequence of the task sample d_j in the experience replay D. For the task sample d_i and d_j, if $CI(d_i) < CI(d_j)$, the curriculum sequence of task sample d_i is before the task sample d_j.

$$CI(d_j) = KP(c_j, \lambda) + \eta CP(d_j) \tag{8}$$

In this paper, the curriculum sequence function is divided into two parts: $KP()$ and $CP()$. $KP()$ represents K-fold-priority function, $CP()$ represents curiosity-priority function, and c_j represents the K-fold teacher model score of task sample d_i, λ represents the curriculum learning factor, η represents the hyperparameter, which is used to control the efficiency and exploration of sample learning.

4.1. K-Fold Priority Experience Replay

In this paper, the absolute value of the temporal-difference error of the neural network is used as a reference standard for the curriculum sequence function $CI(d_j)$, and the difficult task is defined as the task with a large weight correction value for the current neural network model. The reason is that tasks with large temporal-difference error may have an adverse effect on the improvement of training model ability. For example, 1. The random noise during the model training process is prone to data deviation, thereby affecting the training accuracy of model; 2. In the stochastic gradient descent process of deep neural network training, tasks with large temporal-difference error often require a small update step size to obtain a better model convergence effect.

In this paper, the K-fold cross-validation method is used to evaluate the difficulty of the samples in the experience replay buffer, and experience replay D is divided into K equal parts $\left\{\widetilde{D}_i : i = 1, 2, \ldots, K\right\}$, and trained separately to obtain K teachers Model network $\theta = \{\theta_1, \theta_2, \cdots, \theta_K\}$, since the experience replay D is divided, the obtained K teacher model networks are independent of each other. The training formula of the teacher model network is as follows:

$$\widetilde{\theta}_i = \underset{\widetilde{\theta}_i}{\arg\min} \sum_{d_j \in \widetilde{D}_i} L(d_j, \widetilde{\theta}_i) \\ i = 1, 2, \ldots, K \tag{9}$$

where L represents the loss function of the temporal difference error.

The K teacher models obtained are cross-validated. For example, if sample d_j belongs to teacher model i, then the sample d_j is scored on the $K-1$ teacher models other than its own teacher model i. The scoring process can be expressed as follows:

$$c_{ji} = (y - Q_{teacher}^{\pi}(s, a_1, a_2, \ldots, a_N))^2$$
$$y = r_j + \gamma Q^{\pi'}(s', a'_1, a'_2, \ldots, a'_N)\Big|_{a'_v = \pi'_v(o_v)} \quad (10)$$

In the formula, c_{ji} represents the difficulty score of the teacher model i to the sample d_j, $Q_{teacher}^{\pi}$ represents the Q value obtained by inputting the state value s and the action value a into the value function network, and $Q^{\pi'}$ represents the Q value obtained after state s and the action values a are input into the policy function network, γ represents the discount factor, and the final difficulty score of the task sample d_j is the sum of the difficulty scores of all other teacher models:

$$c_j = \sum_{i \in (1,\ldots,K), i \neq k} c_{ji} \quad (11)$$

Definition 2. *K-Fold Priority Function (KP).*

The function $KP(c_j, \lambda) \to [0, 1]$ is used to define the K-fold priority of task sample d_j in experience replay D, c_j represents the final difficulty score of task sample d_j after K-fold cross-validation, λ represents the curriculum learning currently selected task curriculum difficulty factor. The K-fold priority function $KP(c_j, \lambda)$ is expressed as follows:

$$KP(c_j, \lambda) = \begin{cases} e^{c_j - \lambda} & , c_j \leq \lambda \\ \frac{1}{\log(1-\lambda)} \log(c_j - 2\lambda + 1), & \lambda < c_j < 2\lambda \\ 0 & , c_j \geq 2\lambda \end{cases} \quad (12)$$

where 1. $KP(cj, \lambda)$ is monotonically decreasing when $c_j > \lambda$; 2. $KP(cj, \lambda)$ is monotonically increasing when $c_j < \lambda$; 3. $KP(cj, \lambda)$ is the maximum value when $c_j = \lambda$.

The K-fold priority function outputs a scalar with a value range of $[0, 1]$ by inputting the difficulty score c_j of the task sample and the curriculum factor λ, thereby reflecting the sample priority of the task sample in the dimension of temporal-difference error. As the curriculum learning progresses, the curriculum factor λ can be gradually increased, thereby increasing the priority of the task curriculum with higher difficulty score c_j. Since the selection probability of task samples is proportional to the K-fold priority, agents can frequently select empirical samples which fit the current model capabilities. The graph of the K-fold priority function is shown in Figure 3, where $\lambda = 0.6$ is shown in the figure.

The framework of the K-Fold Cross-Validation method is shown in Figure 4.

4.2. Curiosity Exploration Rewards

In the K-fold priority function $KP(c_j, \lambda)$, we use the temporal-difference error as the reference standard for prioritization, which can improve the utilization efficiency of task samples and the robustness of training. However, in a multi-agent system, the traditional reinforcement learning algorithm uses extrinsic reward to guide agents to adjust their own policy. The agents take actions in environment to interact with the environment. When the policy is correct, it will get a positive reward value, otherwise it will get a negative reward value. This extrinsic reward method can achieve good performance in most RL environments, but in a sparse reward value environment, agents do not obtain immediate reward value most of the time they explore in the environment, and then agents are impossible to adjust their own policy according to their reward value, which will greatly reduce their convergence speed and training efficiency of the algorithm.

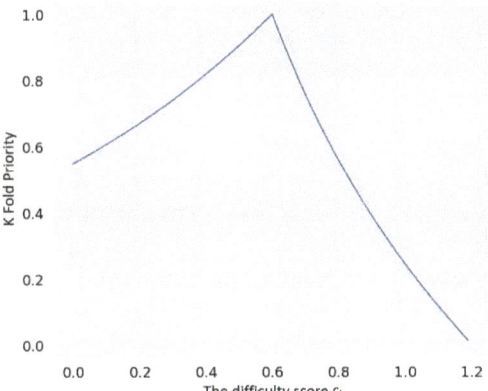

Figure 3. K-Fold priority function.

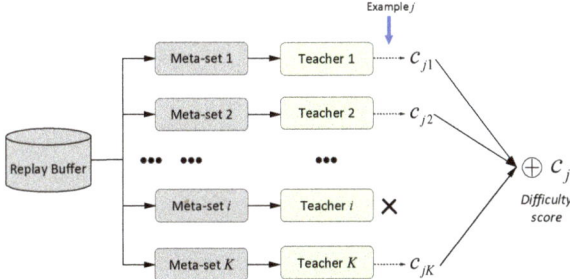

Figure 4. K-Fold Priority Cross Validation framework diagram.

Inspired by the theory of intrinsic motivation, based on the curiosity exploration mechanism [11], this paper uses the curiosity exploration reward as one of the reference standards of curriculum sequence function $CI(d_j)$ to enhance the agent's exploration of environment and avoid the over-fitting phenomenon of "turning in place" of agents.

The basic principle of curiosity exploration mechanism is that when the next state is inconsistent with the predicted state of policy network, the intrinsic reward of curiosity is generated. The greater the difference between actual state and predicted state, the greater the value of curiosity reward.

This curiosity-based mechanism is called the Intrinsic Curiosity Module (ICM), and the curiosity reward value is calculated through two sub-module networks. The first sub-module uses a feature convolutional neural network to extract the eigenvalues of the state s_t in experience samples, and encoded as $\phi(s_t)$, the second sub-module contains a forward neural network θ_F and an inverse dynamic network θ_I. The evaluation mechanism of curiosity reward value is shown in Figure 5.

In the ICM mechanism, the inverse dynamic network θ_I can estimate action value a_t through function g:

$$\hat{a}_t = g(s_t, s_{t+1}; \theta_I) \tag{13}$$

In the formula, a_t represents the actual action taken from state s_t to state s_{t+1}, $\hat{a}_t$ represents the estimated action of a_t, (s_t, a_t, r, s_{t+1}) experience tuple is obtained from the experience replay D, and the network parameters of reverse dynamic network θ_I are optimized by the following expressions:

$$\min_{\theta_I} L_I(\hat{a}_t, a_t) \tag{14}$$

where L_I represents the loss function between the predicted action value $\hat{a}_t$ and the actual action value a_t. The maximum likelihood estimates of the parameters θ_I of the inverse dynamic network can be obtained by minimizing L_I.

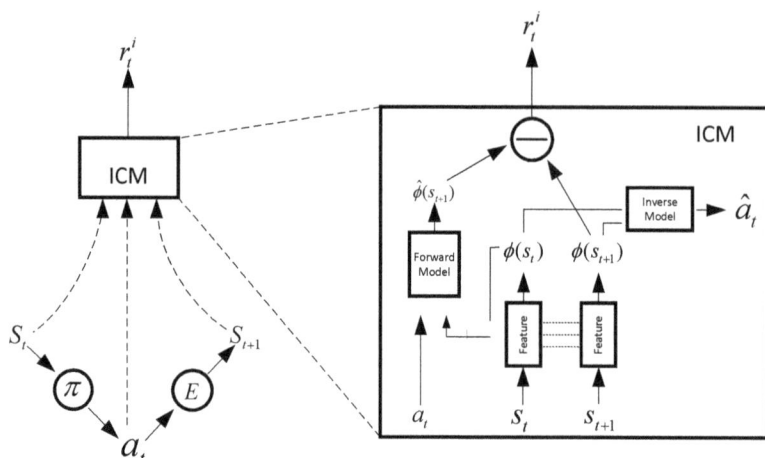

Figure 5. Curiosity reward evaluation mechanism.

For the forward neural network θ_F, the estimated state value a_t at the next time step $t+1$ can be obtained by inputting action value a_t and eigenvalue $\hat{\phi}(s_{t+1})$ of the state s_t.

$$\hat{\phi}(s_{t+1}) = f(\phi(s_t), a_t; \theta_F) \tag{15}$$

where the forward neural network parameter θ_F is optimized by the following loss function:

$$L_F(\phi(s_t), \hat{\phi}(s_{t+1})) = \frac{1}{2}\|\hat{\phi}(s_{t+1}) - \phi(s_{t+1})\|_2^2 \tag{16}$$

Then the overall optimization function learned by reinforcement learning agents is

$$\min_{\theta_P, \theta_I, \theta_F} \left[-\lambda E_{\pi(s_t;\theta_P)} \left[\sum_t r_t \right] + (1-\beta)L_I + \beta L_F \right] \tag{17}$$

In the formula, $0 \leq \beta \leq 1$ represents the weight parameter between the inverse dynamic network and the forward neural network, $\lambda > 0$ represents the weight parameter between the intrinsic curiosity reward value and the gradient descent loss function, and the available curiosity reward value is as follows:

$$r_t^i = \frac{1}{2}\|\hat{\phi}(s_{t+1}) - \phi(s_{t+1})\|_2^2 \tag{18}$$

Definition 3. *Curiosity Priority Function (CP).*

Function $CP(r_t^i(d_j)) \to [0, 1]$ is used to define the curiosity priority of task sample d_j in experience replay D, $r_t^i(d_j)$ represents the curiosity reward value of the task sample d_j. The curiosity-priority function expression of $CP(r_t^i(d_j))$ is as follows:

$$CP(r_t^i(d_j)) = -e^{\left(-\frac{(r_t^i(d_j))^2}{10}\right)} + 1 \tag{19}$$

where $CP(r_t^i(d_j))$ is a monotonically increasing function of $r_t^i(d_j)$.

From the above, the curriculum sequence function $CI(d_j) = KP(c_j, \lambda) + \eta CP(d_j)$ can be obtained, that is, the priority of each experience sample c_j in the experience replay D, then the sampling probability of each experience sample c_j is as follows:

$$P(d_j) = \frac{p_{d_j}^a}{\sum p_{d_j}^a} \qquad (20)$$

In the formula, p_{d_j} represents the priority of the task sample d_j, and a represents the use degree of the priority p_{d_j}.

4.3. Algorithm Framework and Pseudocode

The CMCL algorithm proposed in this paper combines the K-fold priority function and the curiosity priority function in the curriculum sorting stage, so as to use temporal-difference error and curiosity reward to jointly sort curriculums. Adjusting the curriculum factor, the K-fold priority selection of task samples can be controlled to ensure that agents frequently select samples that are most suitable for the current training difficulty, and to improve the exploration of the environment by agents. The basic framework of the CMCL algorithm is shown in Figure 6, and Algorithm 1 describes the training process of the CMCL algorithm.

Algorithm 1: CMCL algorithm.

Input: experience replay buffer D, curriculum factor λ, curriculum stride μ, balance weight η, curriculum sequence vector $ci = [ci_1, ci_2, \cdots, ci_N]$
Output: The final policy π_θ
for episode = 1 to *max_episode* **do**
  Initialize a random process N for reinforcement learning action exploration
  Receive initial state s_0
  for $t = 1$ to *max_episode_length* **do**
    In state s_t, the agents select action a through policy network $\pi_\theta(s_t)$
    Obtain the reward r given by environment E
    Store (s_t, a, s_{t+1}, r) in experience replay buffer D
    $s_t \leftarrow s_{t+1}$
    The experience samples in D are sampled for K-level teacher model training $\{\tilde{\theta}_i : i = 1, 2, \ldots, K\}$
$$\tilde{\theta}_i = \underset{\tilde{\theta}_i}{\operatorname{argmin}} \sum_{d_j \in \tilde{D}_i} L(d_j, \tilde{\theta}_i)$$
    The score of experience sample d_j is evaluated by cross validation $c_j = \sum_{i \in (1,\ldots,N), i \neq k} c_{ji}$
    The K-fold priority $kp_j = KP(c_j, \lambda)$ can be obtained according to Equation (12)
    Calculate the curiosity reward $r_t^i = \frac{1}{2} \|\hat{\phi}(s_{t+1}) - \phi(s_{t+1})\|_2^2$
    The curiosity priority $cp_j = CP(r_t^i(d_j))$ can be obtained according to Equation (19)
    Update curriculum sequence function ci_j by $ci(d_j) = kp(c_j, \lambda) + \eta cp(d_j)$
    for agent $v = 1$ to N_agent **do**
      Sample a minibatch of transitions (s_t, a, s_{t+1}, r) from D according to the priority sampling probability
$$P(d_j) = \frac{p_{d_j}^a}{\sum p_{d_j}^a}$$
      The neural network parameter θ was updated by gradient descent algorithm
    end for
  Adjust curriculum factor λ based on current model capabilities $\lambda = \lambda + \mu$
  end for
end for

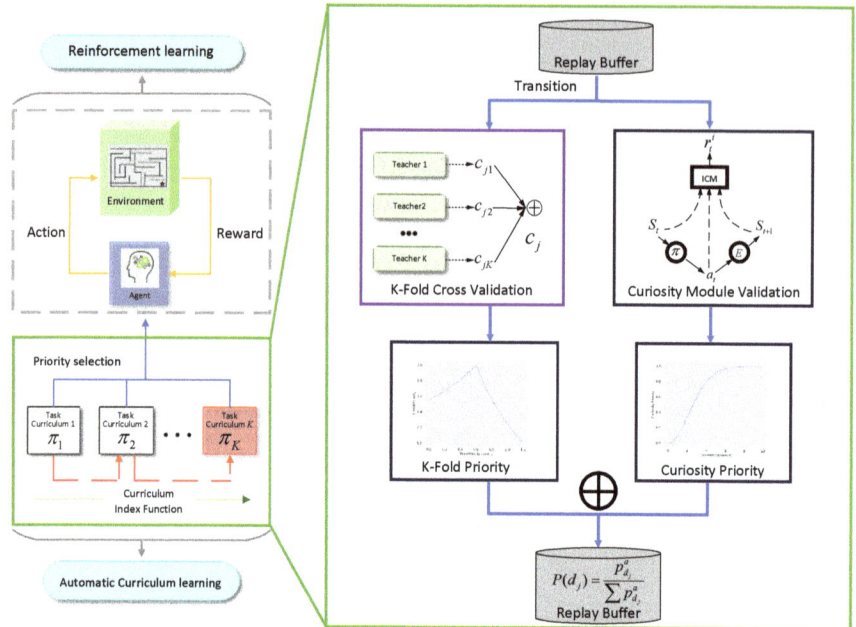

Figure 6. Framework diagram of curriculum reinforcement learning algorithm based on curiosity module.

5. Experiment

In this paper, the simulation verification of the CMCL algorithm is carried out in Multi-Agent Particle Environment [26] (MPE), and the multi-agent cooperative task and the competitive task are used as the target tasks. Based on the environment, a sparse reward value scenario is constructed to test the performance of the CMCL algorithm in teamwork and policy confrontation respectively. Each set of experiments is carried out in the experimental environment of Ubuntu18.04.3 + OpenAI + PyTorch, and adopts the hardware conditions of Intel Corei7-9700K + 64G + GeForceRTX2080. In our environment, the CMCL algorithm is compared with various baseline algorithms to demonstrate the effectiveness and feasibility of the CMCL algorithm. The key hyperparameters set for the RL training process are listed in Table 1. The state value and action value of the agents are input at the input end of the neural network, and the target Q value of the agents is obtained through the calculation of the neural network. The loss function is obtained by subtracting the original Q value, and the original Q value function is updated. Finally, the reinforcement learning algorithm is applied to the deep learning structure.

Table 1. Parameter setting of the DRL process.

Parameters	Values
Discount factor	0.99
Size of RNN hidden layers	64
Size of replay buffer	5000
Exploration	0.1
Initial curriculum factor λ	0.1
Batch size of replay buffer	128
Learning rate of actor network	0.001
Learning rate of critic network	0.001
Update rate of target network	0.01

5.1. Experimental Environment

5.1.1. Cooperative Experiment

The multi-agent cooperation experiment adopts the cooperative navigation experiment in the MPE environment. As shown in the Figure 7, N agents and N landmarks are randomly generated in a square two-dimensional plane with side length 1. The plane is surrounded by walls, and the agents can observe landmarks, but cannot observe the walls, and their missions are to reach landmarks in as few steps as possible and avoid collisions with other agents.

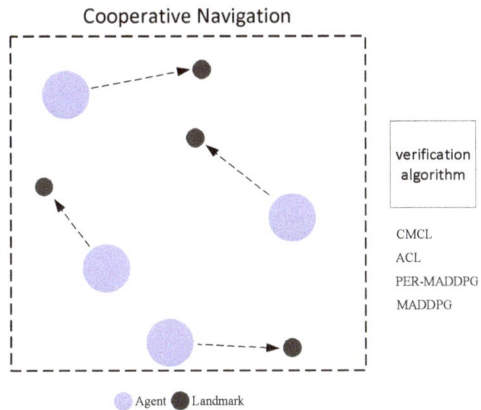

Figure 7. Cooperative navigation experimental environment.

Combined with the size of the two-dimensional plane, it is stipulated that when an agent enters an area with a radius of 0.1 around a landmark, the landmark is considered covered by the agent, and the cooperative navigation task is considered successful only when all landmarks are uniquely covered.

In the reward value setting of the experimental environment, to construct a sparse reward value scene, we cancel the dense reward function set according to the distance between the agent and the landmark in the original MPE environment. Therefore, the reward value obtained by each agent at each time step consists of only two parts, including 1. When there is a collision between the agents or the agent hits a wall, the environment gives a negative reward value, that is, agent collision reward value C_1; 2. When the agent covers the landmark, the environment gives a positive reward value, that is, the agent covers the landmark reward value C_2.

The agent collision reward value is as follows:

$$C_1 = \begin{cases} -1, if\ collided \\ 0, if\ not\ collided \end{cases} \quad (21)$$

The agent coverage landmark reward value is as follows:

$$C_2 = \begin{cases} +4, if\ covered \\ 0, if\ not\ coverd \end{cases} \quad (22)$$

As shown in Figure 7, in the $N = 4$ environment, the CMCL, ACL, PER-MADDPG, and MADDPG algorithms are used to control the movement of the agent. To prevent the agent from spinning in place or meaningless exploration, the episode duration is set to 30 steps, that is, when the agent finishes exploring after 30 steps, the environment is initialized to start a new episode of exploration.

Figure 8 shows the average reward value graph and the coverage graph obtained by the four algorithms after 20,000 episodes of training in the cooperative navigation environment. Figure 9 shows the bar graph of the average reward value of the four algorithms in 20,000 episodes, that is, the quotient of the total reward value obtained by the four algorithms in the whole training session and the number of sessions. As can be seen from the curve in Figure 8, at the beginning of the algorithm training, the agent is prone to colliding with other agents or with the wall. As the training progresses, agents gradually learn the policy of cooperatively covering landmarks. The curve of the CMCL algorithm oscillates slightly in the early training process, and gradually smooths in the later stage, and can obtain higher reward values and landmark coverage than other baseline algorithms, showing better training performance. Figure 10 shows the rendering of the agent training in cooperative navigation environment after the CMCL algorithm has been trained for 12,500 episodes. From the rendering, it can be seen that the agents can successfully approach and cover landmarks in the environment.

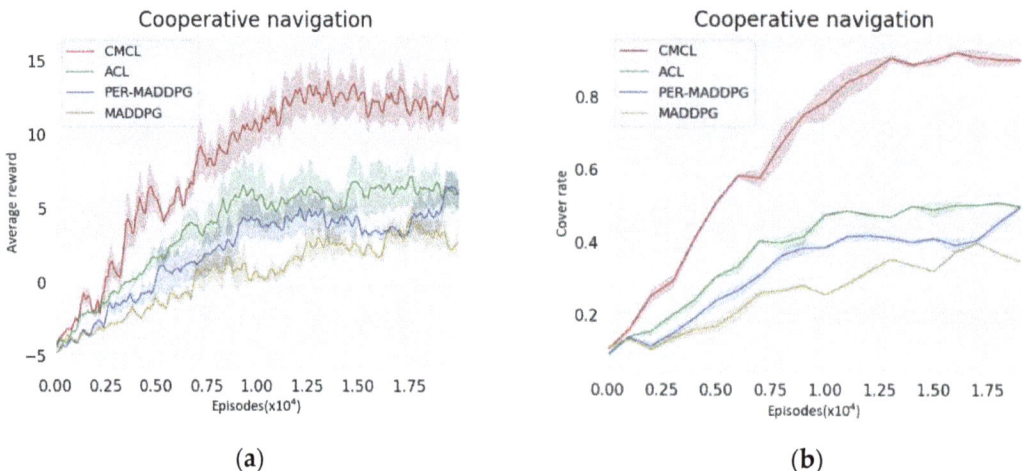

Figure 8. Representation diagram of agents in cooperative environment. (**a**) Average reward in cooperative environment; (**b**) landmark cover rate in cooperative environment.

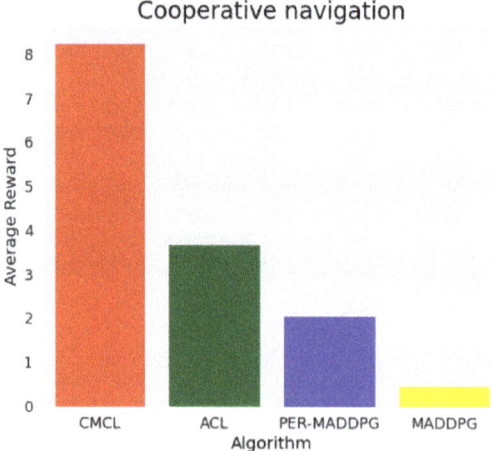

Figure 9. Bar chart of average reward value in cooperative environment.

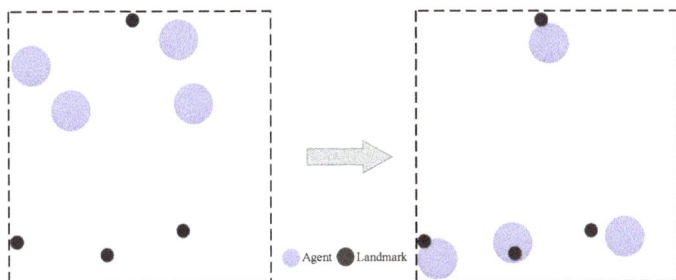

Figure 10. Diagram of the training effect of CMCL algorithm in cooperative environment.

5.1.2. Competition Experiment

In a cooperative training environment, agents share the observed value of the environment to maximize the total reward value, but in a multi-agent competition task, as training progresses, the policies of their opponents are constantly improved, resulting in the continuous fluctuation of the cumulative reward value. In addition to cooperating with other agents, the agent also needs to make policy corrections for the opponent's policy.

The multi-agent competition experiment uses the predator-prey experiment in the MPE environment. On a two-dimensional plane with side length 1, m predators and n prey are randomly generated, as well as three randomly generated obstacles, whose area is relatively large, which can prevent the intelligent body from observing and moving. The goal of predators is to capture prey as quickly as possible through team cooperation. During this process, the predators and the prey move randomly, and the prey move twice as fast as the predators. During the predation process, all predators form a team to hunt down the prey, and the capture is considered successful when the distance between the predator and the prey is less than the pursuit radius.

To construct the sparse reward scene of the predator-prey environment, the dense reward function set according to the distance between predator and prey is canceled. Therefore, the reward value obtained by the predator agent at each time step consists of two parts: 1. When the predator encounters the prey, it will receive a positive reward value, that is, the capture reward value D_1; 2. To prevent agents from escaping the boundary, when agent hits the wall, it will receive a negative reward value, that is, the collision reward value D_2.

The capture reward is as follows:

$$D_1 = \begin{cases} +5, if\ captured \\ 0, if\ not\ captured \end{cases} \qquad (23)$$

This represents that when the predator captures the prey, it gets a positive large reward value, while the prey gets a large negative reward value.

The collision boundary rewards are as follows:

$$D_2 = \begin{cases} -1, if\ collided \\ 0, if\ not\ collided \end{cases} \qquad (24)$$

This represents that the predator and prey get a negative reward when they collide with the boundary.

As shown in Figure 11, the CMCL, ACL, PER-MADDPG, and MADDPG algorithms are used to control the movements of predators and prey, respectively. To prevent the agent from spinning in place or performing meaningless exploration, the episode duration is set to 30 steps, which means the agent finishes the exploration after 30 steps and initializes the environment to restart the exploration.

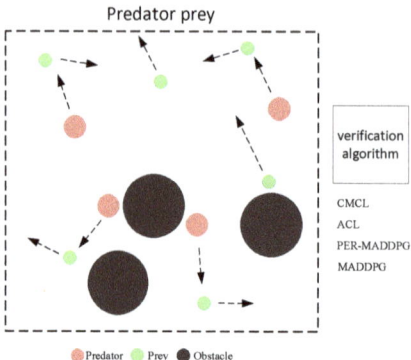

Figure 11. Schematic diagram of adversarial environment.

As shown in Figure 12, the predator agents are controlled by the CMCL, ACL, PER-MADDPG, and MADDPG algorithms respectively, and the prey agents are controlled by the MADDPG algorithm. The bar chart and the error band chart of the average reward value obtained after 20,000 episodes of training indicates that the average reward value in the bar chart is the quotient of the total reward value obtained during the whole training of the four algorithms and the number of episodes. As can be seen in the figure, as training progresses, predator agents controlled by the four algorithms gradually learn the cooperative hunting policy, which tends to stabilize after 10,000 episodes. Throughout the training process, the average reward value of the CMCL algorithm is generally higher than that of other baseline algorithms and is significantly higher than that of the other three algorithms after 10,000 episodes.

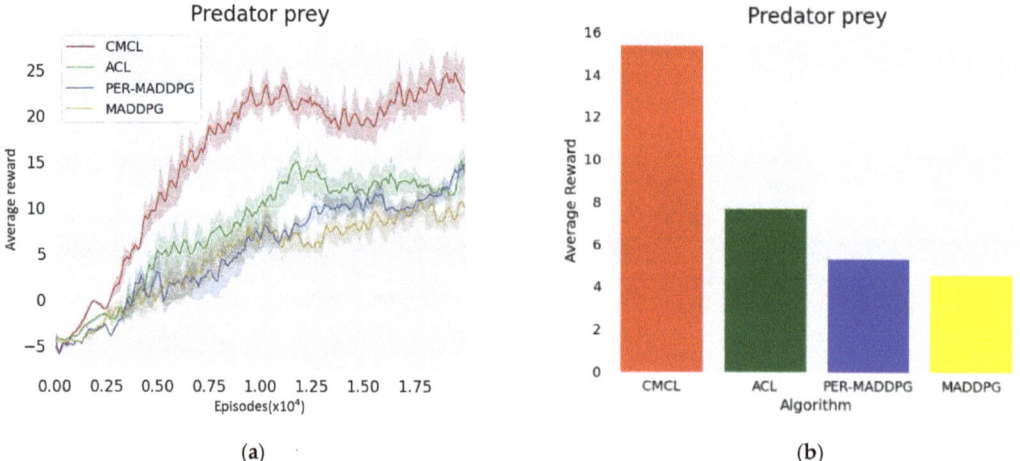

Figure 12. Representation diaram of agents in cooperative environment. (**a**) Episode reward achieved in adversarial environment; (**b**) the average reward obtained by the four algorithms in the adversarial environment.

Figure 13a shows the win rate charts obtained by both agents in each round under the condition that the predator agents adopt the CMCL algorithm and the prey agents adopt the ACL algorithm. Figure 13b shows the win rate charts obtained by both agents in each round under the condition that the predator agent adopts the CMCL algorithm and the prey agent adopts the PER-MADDPG algorithm. It can be seen from the figure that when the predator agents controlled by the CMCL algorithm fight against the prey agents controlled by the ACL algorithm, the two sides won and lost in the early stage. However, after a certain training period (5000 rounds), the predator agents controlled by the CMCL algorithm gain a significant advantage. Predator agents controlled by the CMCL algorithm can gain obvious advantages in a short period of time against the prey agents controlled by the PER-MADDPG algorithm, and the winning rate is above 0.85.

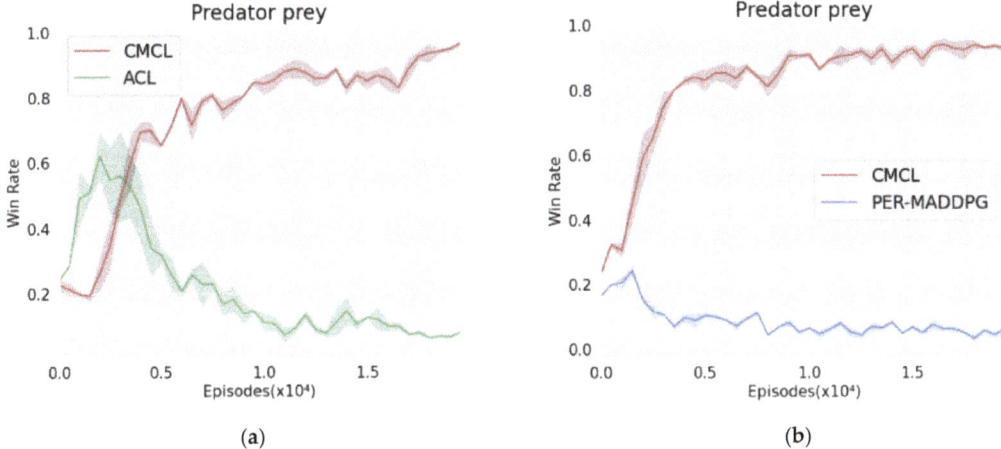

Figure 13. The win rate of predator and prey using two algorithms respectively in the adversarial environment. (**a**) CMCL vs. ACL; (**b**) CMCL vs. PER-MADDPG.

Figure 14 shows the training effect diagram of the CMCL algorithm obtained after 10,000 episodes of training in a competitive environment. It can be seen from the effect diagram that the predator agent can learn the batch-hunting policy, that is, to round up the prey agents in two batches by rational use of terrain obstacles. It can be seen that the CMCL algorithm can achieve better training performance than other baseline algorithms in the multi-agent competitive environment.

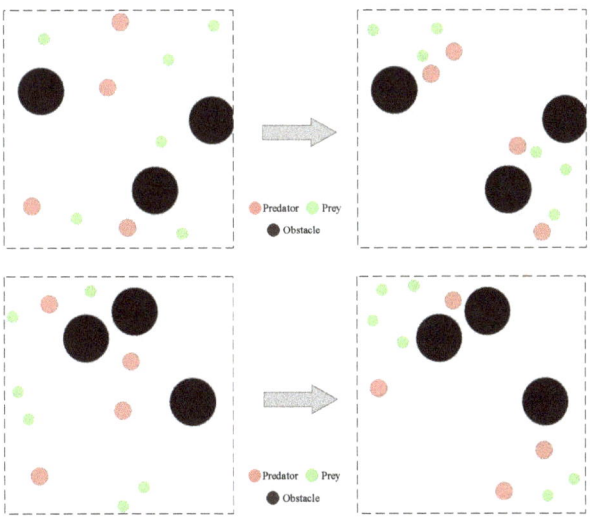

Figure 14. Training effect diagram of CMCL algorithm in competitive environment.

6. Discussion

On the basis of the analysis of the above two experimental environments, the overall performance of our proposed CMCL algorithm is better than that of the other three baseline algorithms, and the following experimental results can be obtained.

In the cooperative environment, the average reward value and landmark coverage of the CMCL algorithm are better than those of the ACL, PER-MADDPG and MADDPG algorithms. Combined with the screenshots of the actual performance of the agents in the experimental simulation environment, CMCL algorithm training in the cooperative environment can be performed. The agents can learn to execute policies dispersedly and cooperatively cover landmarks, avoiding collisions between agents or between agents and the wall.

In the competitive environment experiment, the average reward value of the CMCL algorithm is better than those of the ACL, PER-MADDPG and MADDPG algorithms, and when the predator agent controlled by the CMCL algorithm is confronted with the prey agent controlled by the ACL algorithm and the PER-MADDPG algorithm, after a period of training, a good win rate can be obtained. Combined with the actual performance screenshots of the agents in the experimental simulation environment, it can be concluded that the predator agents trained by CMCL algorithm in the competitive environment can learn to cooperate to surround the prey agents and group the prey agents to carry out the hunting strategy, and avoid the collision between agents or between agents and walls.

The current CMCL algorithm can achieve good training performance in the sparse reward value environment, but there are still two limitations:

1. The dimension explosion problem. A large number of agents in the reinforcement learning environment due to the excessively large state space and the action space, it is easy for the algorithm to fail to converge due to the explosion of dimensions.
2. The problem of reliability distribution. When multiple agents are trained in a reinforcement learning environment, the effective exploration of the environment by the agents can easily be affected due to the uneven distribution of reward functions, especially when multiple players are trained. This problem is more obvious.

7. Conclusions

To solve the problem that the training efficiency of the automatic curriculum reinforcement learning algorithm is not high in the scenario of sparse reward value, this paper adds a curiosity module on the basis of automatic curriculum learning, and uses the curiosity reward value and the temporal-difference error as the reference standard for curriculum sorting. The ICM module is used to evaluate the priority of curiosity, the curriculum factor is designed to control the selection of curriculum difficulty, and an automatic curriculum reinforcement learning algorithm based on the curiosity module is proposed, and the availability and superiority of the algorithm in sparse reward scenarios are verified by simulation experiments in cooperative and competitive environments. With the increase in the number of agents in multi-agent reinforcement learning, the input nodes of the neural network and the complexity of the neural network grow linearly, which can easily cause the problem of dimension explosion in the training process, which makes the algorithm difficult to converge. Methods that can be adopted include compression of state space and share parameters between agents. In the future, based on automatic curriculum reinforcement learning, further research will be conducted on how to reduce the time complexity of multi-agent reinforcement learning training under large-scale number conditions.

The main abbreviations are listed in Table 2.

Table 2. Main abbreviations.

Abbreviation	Explanation
RL	Reinforcement Learning
ACL	Automatic Curriculum Learning
DL	Deep Learning
MADDPG	Multi-Agent Deep Deterministic Policy Gradient
PER	Prioritized Experience Replay
ICM	Intrinsic Curiosity Module
MPE	Multi-Agent Particle Environment
CMCL	Curiosity Module-based Curriculum Learning
ICML	International Conference on Machine Learning
AC	Actor-Critic
KP	K-Fold Priority
CP	Curiosity Priority
CI	Curriculum Index
CTDE	Centralized Training and Distributed Execution

Author Contributions: Methodology, Z.L.; Software Z.L., J.L. and X.C.; Validation, Z.L., J.L. and X.C.; writing—original draft, Z.L., L.C. and J.W.; writing—review and editing, X.C. and J.L. All authors have read and agreed to the published version of the manuscript.

Funding: This research was partially supported by the National Natural Science Foundation of China (No. 61806221).

Institutional Review Board Statement: Not applicable.

Informed Consent Statement: Not applicable.

Data Availability Statement: The data used to support the findings of this study are available from the website https://github.com/openai/multiagent-particle-envs (accessed on 13 May 2022).

Conflicts of Interest: The authors declare no conflict of interest.

References

1. Sutton, R.S.; Barto, A.G. *Reinforcement Learning: An Introduction*; MIT Press: Cambridge, MA, USA, 2018.
2. Mnih, V.; Badia, A.P.; Mirza, M. Asynchronous methods for deep reinforcement learning. In Proceedings of the International Conference on Machine Learning (ICML), New York, NY, USA, 18–20 December 2016; pp. 1928–1937.
3. Foglino, F.; Christakou, C.C.; Gutierrez, R.L. Curriculum learning for cumulative return maximization. *arXiv* **2019**, arXiv:1906.06178.
4. Fang, M.; Zhou, T.; Du, Y. Curriculum-guided hindsight experience replay. *Adv. Neu. Infor. Pro. Sys.* **2019**, *19*, 12602–12613.
5. Gu, S.; Holly, E.; Lillicrap, T. Deep reinforcement learning for robotic manipulation with asynchronous off-policy updates. In Proceedings of the 2017 IEEE International Conference on Robotics and Automation (ICRA), Singapore, 29 May–3 June 2017; pp. 3389–3396.
6. Silver, D.; Huang, A.; Maddison, C.J. Mastering the game of Go with deep neural networks and tree search. *Nature* **2016**, *529*, 484–489. [CrossRef] [PubMed]
7. Lecun, Y.; Bengio, Y.; Hinton, G. Deep learning. *Nature* **2015**, *521*, 436–440. [CrossRef] [PubMed]
8. Singh, A.; Jain, T.; Sukhbaatar, S. Individualized controlled continuous communication model for multiagent cooperative and competitive tasks. In Proceedings of the International Conference on Learning Representations (ICLR), New Orleans, LA, USA, 6–9 May 2019; pp. 154–160.
9. Yang, Y.; Luo, R.; Li, M. Mean field multi-agent reinforcement learning. In Proceedings of the International Conference on Machine Learning (ICML), Stockholm, Sweden, 10–15 July 2018; pp. 5571–5580.
10. Liu, Q.; Cui, C.; Fan, Q. Self-Adaptive Constrained Multi-Objective Differential Evolution Algorithm Based on the State–Action–Reward–State–Action Method. *Mathematics* **2022**, *10*, 813. [CrossRef]
11. Bengio, Y.; Louradour, J.; Collobert, R. Curriculum learning. In Proceedings of the 26th Annual International Conference on Machine Learning (ICML), Quebec, MT, Canada, 14–18 June 2009; pp. 41–48.
12. Xue, H.; Hein, B.; Bakr, M. Using Deep Reinforcement Learning with Automatic Curriculum Learning for Mapless Navigation in Intralogistics. *Appl. Sci.* **2022**, *12*, 3153. [CrossRef]
13. Portelas, R.; Colas, C.; Weng, L. Automatic curriculum learning for deep rl: A short survey. *arXiv* **2020**, arXiv:2003.04664.

14. Florensa, C.; Held, D.; Geng, X. Automatic goal generation for reinforcement learning agents. In Proceedings of the International Conference on Machine Learning, Stockholm, Sweden, 10–15 July 2018; pp. 1515–1528.
15. Ren, Z.; Dong, D.; Li, H. Self-paced prioritized curriculum learning with coverage penalty in deep reinforcement learning. *IEEE Trans. Neu. Net. Learn. Syst.* **2018**, *29*, 2216–2226. [CrossRef]
16. Chen, J.; Zhang, Y.; Xu, Y. Variational Automatic Curriculum Learning for Sparse-Reward Cooperative Multi-Agent Problems. *Adv. Neu. Infor. Pro. Syst.* **2021**, *34*, 102–116.
17. Haibo, H.; Edwardo, A.G. Learning from imbalanced data. *IEEE Trans. Know. Data. Eng.* **2008**, *9*, 1263–1284. [CrossRef]
18. Geoffrey, E.; Hinton. To recognize shapes, first learn to generate images. *Pro. Bra. Res.* **2007**, *165*, 535–547.
19. Pathak, D.; Agrawal, P.; Efros, A.A. Curiosity-driven exploration by self-supervised prediction. In Proceedings of the International Conference on Machine Learning (ICML), Sydney, NSW, Australia, 6–11 August 2017; pp. 2778–2787.
20. Gruber, M.J.; Bernard, D.G.; Charan, R. States of curiosity modulate hippocampus-dependent learning via the dopaminergic circuit. *Neuron* **2014**, *84*, 486–496. [CrossRef]
21. Zhang, H.; Qu, C.; Zhang, J. Self-Adaptive Priority Correction for Prioritized Experience Replay. *Appl. Sci.* **2020**, *10*, 6925. [CrossRef]
22. Cao, X.; Wan, H.; Lin, Y. High-value prioritized experience replay for off-policy reinforcement learning. In Proceedings of the 2019 IEEE 31st International Conference on Tools with Artificial Intelligence (ICTAI), Portland, OR, USA, 4–6 November 2019; pp. 1510–1514.
23. Li, Y. Deep reinforcement learning: An overview. *arXiv* **2017**, arXiv:1701.07274.
24. Lv, K.; Pei, X.; Chen, C. A Safe and Efficient Lane Change Decision-Making Strategy of Autonomous Driving Based on Deep Reinforcement Learning. *Mathematics.* **2022**, *10*, 1551. [CrossRef]
25. Grondman, I.; Busoniu, L.; Lopes, G.A.D. A survey of actor-critic reinforcement learning: Standard and natural policy gradients. *IEEE Trans. Syst. Man. Cyber.* **2012**, *42*, 1291–1307. [CrossRef]
26. Lowe, R.; Wu, Y.I.; Tamar, A. Multi-agent actor-critic for mixed cooperative-competitive environments. *Adv. Neu. Infor. Pro. Syst.* **2017**, *30*, 133–160.
27. Lei, W.; Wen, H.; Wu, J. MADDPG-based security situational awareness for smart grid with intelligent edge. *Appl. Sci.* **2021**, *11*, 3101. [CrossRef]
28. Wang, X.; Chen, Y.; Zhu, W. A survey on curriculum learning. *IEEE Trans. Pat. Ana. Mac. Intel.* **2021**, *37*, 362–386. [CrossRef]
29. Parker-Holder, J.; Rajan, R.; Song, X. Automated Reinforcement Learning (AutoRL): A Survey and Open Problems. *arXiv* **2022**, arXiv:2201.03916. [CrossRef]
30. Kumar, M.; Packer, B.; Koller, D. Self-paced learning for latent variable models. *Adv. Neu. Infor. Pro. Syst.* **2010**, *23*, 154–160.

Article

Mixed-Delay-Dependent Augmented Functional for Synchronization of Uncertain Neutral-Type Neural Networks with Sampled-Data Control

Shuoting Wang [1] and Kaibo Shi [2,3,*]

1 School of Computer, Chengdu University, Chengdu 610106, China
2 School of Electronic Information and Electrical Engineering, Chengdu University, Chengdu 610106, China
3 Geomathematics Key Laboratory of Sichuan Province, Chengdu University of Technology, Chengdu 610059, China
* Correspondence: skbs111@163.com

Abstract: In this paper, the synchronization problem of uncertain neutral-type neural networks (NTNNs) with sampled-data control is investigated. First, a mixed-delay-dependent augmented Lyapunov–Krasovskii functional (LKF) is proposed, which not only considers the interaction between transmission delay and communication delay, but also takes the interconnected relationship between neutral delay and transmission delay into consideration. Then, a two-sided looped functional is also involved in the LKF, which effectively utilizes the information on the intervals $[t_k, t]$, $[t_k - \tau, t - \tau]$, $[t, t_{k+1}]$, $[t - \tau, t_{k+1} - \tau]$. Furthermore, based on the suitable LKF and a free-matrix-based integral inequality, two synchronization criteria via a sampled-data controller considering communication delay are derived in forms of linear matrix inequalities (LMIs). Finally, three numerical examples are carried out to confirm the validity of the proposed criteria.

Keywords: neutral-type neural networks; synchronization problem; mixed-delay-dependent functional; sampled-data control; communication delay

MSC: 93D20

Citation: Wang, S.; Shi, K. Mixed-Delay-Dependent Augmented Functional for Synchronization of Uncertain Neutral-Type Neural Networks with Sampled-Data Control. *Mathematics* **2023**, *11*, 872. https://doi.org/10.3390/math11040872

Academic Editor: António Lopes

Received: 13 January 2023
Revised: 5 February 2023
Accepted: 6 February 2023
Published: 8 February 2023

Copyright: © 2023 by the authors. Licensee MDPI, Basel, Switzerland. This article is an open access article distributed under the terms and conditions of the Creative Commons Attribution (CC BY) license (https://creativecommons.org/licenses/by/4.0/).

1. Introduction

Neural networks (NNs) are classes of mathematical models which simulate the neural processing mechanism in the human brain. Over the past several decades, NNs have attracted widespread attention due to their potential applications in many areas, such as image and signal processing [1], pattern recognition [2], optimization problems [3], parallel computation [4] and so on. In such systems, time delays may be generated due to the limited switching speeds of amplifiers and the inherent communication time among neurons. They may negatively impact the NNs and cause various undesired dynamical phenomena, such as oscillation or instability. Therefore, it is essential to consider time delays in the stability analysis of NNs.

Different forms of time delays have been conducted in the stability analysis of NNs, including variable delays [5], continuously distributed delays [6], and so on. In order to characterize the properties of neural reaction processes precisely, neutral-type time delays are involved in dynamical neural network models. In these models, the information about the derivative of the past state is considered. When both the current neuron state derivative and the past state derivative are involved in the NNs, neural network models are called neutral-type neural networks (NTNNs). Due to wide engineering applications in fields such as distributed networks [7], including lossless transmission lines [8] and heat exchangers [9], the stability and synchronization analysis of NTNNs having time delays and neutral delays has become an important research topic.

In recent years, various control methods have been brought to guarantee the synchronization of NTNNs. The sampled-data control method has been extensively applied due to its easy implementation [10,11]. This method can reduce network congestion and improve control efficiency. Three methods have been proposed to deal with the stability analysis of sampled-data control systems. The first one is the input delay method [12], where sampled-data systems are modeled as time-delay systems with a time-varying input. The second is the discrete-time method [13], in which sampled-data systems are transformed into discrete-time systems. However, this method suffers troubles when the sampling period is variable. The third method is the impulsive system method [14], which is often used to investigate systems with uncertain and bounded sampling intervals. Among these methods, the input delay method is widely used to investigate the synchronization of NTNNs with a sampled-data controller.

In addition, open communication networks are used in the control process. The communication delay is inevitably generated by the transmission of the signals from the sampler to the controller, and it may destabilize the sampled-data system. Therefore, it is crucial to design a sampled-data controller concerning communication delays. Recently, a new two-sided looped functional was proposed in [15], which can reduce conservativeness. However, only the information of intervals $[t_k, t]$, $[t, t_{k+1}]$ is considered, while the information on the intervals $[t_k - \tau_c, t - \tau_c]$, $[t - \tau_c, t_{k+1} - \tau_c)$ is ignored. In [16,17], the communication delays were considered in the design of the sampled-data controller, but some useful information about sample features is still lost. In [18,19], a mixed-delay-based LKF is proposed, and a less conservatism stability criterion is derived. Throughout analyzing these aspects, we find that a functional with more time-delay cross information can obtain better results than other functionals in the synchronization of NTNNs with sampled-data control.

Motivated by the above discussions, we further investigate the synchronization via sampled-data control for NTNNs with and without time-varying parameter uncertainties. The main contributions of this paper can be summarized as follows:

(i) A sampled-data controller is designed considering the communication delay τ_c. Such a method is easier to calculate and implement than the event-triggered communication scheme proposed in [20]. While guaranteeing system stability, this method can reduce network congestion and improve control efficiency.

(ii) A mixed-delay-dependent augmented LKF is constructed. The interconnected relationship between transmission delay and communication delay is taken into account, and the interaction between transmission delay and neutral delay is considered simultaneously. This interconnected relationship is utilized by introducing some integral and double integral terms associated with the mixed delay. Thus, the connection between those states is strengthened.

(iii) In order to reduce the conservatism of the synchronization criteria, a two-sided looped LKF is proposed, which utilizes the information of intervals $[t_k, t]$, $[t_k - \tau_c, t - \tau_c]$, $[t, t_{k+1})$, $[t - \tau_c, t_{k+1} - \tau_c)$. This functional can obtain better results for sampled-data synchronization problems in NTNNs. Two less conservative synchronization criteria are derived based on the LKF and a free-matrix-based integral inequality.

Notations: Throughout this paper, $\mathbb{N}_+$ represents the set of positive integers; $\mathbb{R}^n$ denotes the n-dimensional vector space; the superscript -1 and T stand for the inverse and the transpose, respectively; $P > 0$ ($P \geq 0$) means that P is a positive definite matrix; I and 0 denote the identity matrix and a zero matrix, respectively; $\text{diag}\{x_1, \ldots, x_n\}$ represents a diagonal matrix, in which its diagonal elements are $x_1, \ldots, x_n$, $\text{col}\{x_1, \ldots, x_n\} = [x_1^T, \ldots, x_n^T]^T$ and $\text{Sym}\{Z\} = Z + Z^T$; and the notation $*$ stands for the symmetric terms in a symmetric matrix.

2. Preliminaries

In this section, we consider the following NTNNs with time-varying parameter uncertainties:

$$\begin{aligned}\dot{y}(t) = &- (A + \Delta A(t))y(t) + (W_0 + \Delta W_0(t))g(y(t)) + (W_1 + \Delta W_1(t))g(y(t - \tau_2(t))) \\ &+ (W_2 + \Delta W_2(t))\dot{y}(t - \tau_1(t)) + J,\end{aligned} \quad (1)$$

where $y(\cdot) = \text{col}\{y_1(\cdot), y_2(\cdot), \cdots, y_n(\cdot)\} \in \mathbb{R}^n$ is the state vector with n neurons, $A = \text{diag}\{a_1, a_2, \cdots, a_n\}$ is a diagonal matrix with each $a_i > 0$ $(i = 1, 2, \cdots, n)$, $g(y(t)) = \text{col}\{g_1(y_1(t)), g_2(y_2(t)), \cdots, g_n(y_n(t))\} \in \mathbb{R}^n$ is the neural activation function indicating how the neuron responses to its input, W_0, W_1 and W_2 are the delayed interconnection weight matrices of appropriate dimensions, $\Delta A, \Delta W_0, \Delta W_1$ and ΔW_2 are parameter uncertainties, $J = \text{diag}\{J_1, J_2, \cdots, J_n\}$ is an external constant input vector.

$\tau_1(t)$ stands for the neutral-type time delay, $\tau_2(t)$ is the transmission delay accumulated during the transmission of information among neurons, which satisfy

$$0 \leq \tau_i(t) \leq \tau_i, \ \dot{\tau}_i(t) \leq \mu_i, \ i = 1, 2, \quad (2)$$

where τ_i and μ_i are known positive constants.

The neuron activation function $g(\cdot)$ is bounded by

$$k_i^- \leq \frac{g_i(u_1) - g_i(u_2)}{u_1 - u_2} \leq k_i^+, \ i = 1, 2, \cdots, n, \quad (3)$$

where $u_1, u_2 \in \mathbb{R}$, $u_1 \neq u_2$, k_i^- and k_i^+ are real scalars. For convenience, define $K^+ = \text{diag}\{k_1^+, k_2^+, \cdots, k_n^+\}$, $K^- = \text{diag}\{k_1^-, k_2^-, \cdots, k_n^-\}$.

$\Delta A(t)$, $\Delta W_0(t)$, $\Delta W_1(t)$, $\Delta W_2(t)$ are unknown matrices representing time-varying parameter uncertainties, which are assumed to be norm-bounded and satisfy

$$\begin{aligned}[\Delta A(t) \ \Delta W_0(t) \ \Delta W_1(t) \ \Delta W_2(t)] &= F\Sigma(t)[E_1 \ E_2 \ E_3 \ E_4], \\ \Sigma(t) &= \Delta(t)[I - G\Delta(t)]^{-1}, \ I - G^T G > 0,\end{aligned} \quad (4)$$

in which F and E_i $(i = 1, 2, 3, 4)$ are known constant matrices, $\Delta(t)$ is an unknown time-varying matrix satisfying $\Delta^T(t)\Delta(t) \leq I$.

Regarding system (1) as a master system, then, the corresponding slave system is described as follows:

$$\begin{aligned}\dot{z}(t) = &- (A + \Delta A(t))z(t) + (W_0 + \Delta W_0(t))g(z(t)) + (W_1 + \Delta W_1(t))g(z(t - \tau_2(t))) \\ &+ (W_2 + \Delta W_2(t))\dot{z}(t - \tau_1(t)) + u(t) + J,\end{aligned} \quad (5)$$

where $z(\cdot) = \text{col}\{z_1(\cdot), z_2(\cdot), \cdots, z_n(\cdot)\} \in \mathbb{R}^n$ is the neural state and $u(t)$ is the control input. The rest of the matrices and variables are defined in (1).

Define the error state as $e(t) = z(t) - y(t)$, and thus, the error system is obtained as follows:

$$\begin{aligned}\dot{e}(t) = &- (A + \Delta A(t))e(t) + (W_0 + \Delta W_0(t))f(e(t)) + (W_1 + \Delta W_1(t))f(e(t - \tau_2(t))) \\ &+ (W_2 + \Delta W_2(t))\dot{e}(t - \tau_1(t))) + u(t),\end{aligned} \quad (6)$$

where $f(e(t)) = g(z(t)) - g(y(t))$ satisfies

$$k_i^- \leq \frac{f_i(\beta)}{\beta} = \frac{g_i(e_i + y_i) - g_i(y_i)}{e_i} \leq k_i^+, \ f_i(0) = 0, \ i = 1, 2, \cdots, n, \quad (7)$$

where $\beta \in \mathbb{R}, \beta \neq 0$.

Denote the updating instant time of the zero-order-hold (ZOH) by t_k, $e(t_k)$ is the discrete measurements of $e(t)$ at the sampling instant t_k. For any integer $k \geq 0$, the sampling intervals are denoted by d_k, which satisfy

$$d_k = t_{k+1} - t_k, \ d_k \in (0, d_M], \quad (8)$$

where $d_M > 0$ stands for the upper bound of sampling intervals.

In practical systems, communication delays are inevitable during the transmission of signals from sampler to controller. Therefore, the communication delay τ_c is considered in the sampled-data controller. Then, the control input $u(t)$ is formulated as

$$u(t) = Ke(t_k - \tau_c), \ t \in [t_k, t_{k+1}), \tag{9}$$

where K is the controller gain matrix to be calculated.

For simplicity, we use η_k to represent $d_k + \tau_c$, where $\eta_k \in (\tau_c, \eta_M]$, $\eta_M = d_M + \tau_c$ equals to the upper bound of η_k.

Substituting the control input (9) into the error system (6), the corresponding error system can be reformulated as

$$\begin{aligned}\dot{e}(t) = &- (A + \Delta A(t))e(t) + (W_0 + \Delta W_0(t))f(e(t)) + (W_1 + \Delta W_1(t))f(e(t - \tau_2(t))) \\ &+ (W_2 + \Delta W_2(t))\dot{e}(t - \tau_1(t))) + Ke(t_k - \tau_c).\end{aligned} \tag{10}$$

In order to derive the stability criteria for the system (10), the following lemmas will be utilized.

Lemma 1 ([21]). *Let x be a differentiable function: $[\alpha, \beta] \to \mathbb{R}^n$. For symmetric matrices $R > 0$, and N_1, N_2, N_3, the following inequality holds:*

$$-\int_\alpha^\beta \dot{x}^T(s) R \dot{x}(s) ds \leq \zeta_i^T \Omega_i \zeta_i, \ i = 1, 2, \tag{11}$$

where

$$\Omega_i = (\beta - \alpha)\left(N_1 R^{-1} N_1^T + \frac{1}{3} N_2 R^{-1} N_2^T + \frac{1}{5} N_3 R^{-1} N_3^T\right) + \mathrm{Sym}\left\{N_1 \Pi_1 + N_2 \Pi_2 + N_3 \Pi_3^{(i)}\right\},$$

$$\Pi_1 = \bar{e}_1 - \bar{e}_2, \ \Pi_2 = \bar{e}_1 + \bar{e}_2 - 2\bar{e}_3,$$

$$\Pi_3^{(1)} = \bar{e}_1 - \bar{e}_2 - 6\bar{e}_3 + 6\bar{e}_4, \ \Pi_3^{(2)} = \bar{e}_1 - \bar{e}_2 + 6\bar{e}_3 - 6\bar{e}_4,$$

$$\zeta_1 = \mathrm{col}\left\{x(\beta), x(\alpha), \frac{1}{\beta - \alpha}\int_\alpha^\beta x(s) ds, \frac{2}{(\beta - \alpha)^2}\int_\alpha^\beta \int_\alpha^s x(u) du ds\right\},$$

$$\zeta_2 = \mathrm{col}\left\{x(\beta), x(\alpha), \frac{1}{\beta - \alpha}\int_\alpha^\beta x(s) ds, \frac{2}{(\beta - \alpha)^2}\int_\alpha^\beta \int_s^\beta x(u) du ds\right\},$$

$$\bar{e}_j = \begin{bmatrix} 0_{n \times (j-1)n} & I_n & 0_{n \times (4-j)n} \end{bmatrix}, \ j = 1, 2, \cdots, 4.$$

Lemma 2 ([22]). *Letting $I - G^T G > 0$, define the set $Y = \{\Delta(t) = \Sigma(t)[I - G\Sigma(t)]^{-1}, \Sigma^T(t)\Sigma(t) \leq I\}$ and for given matrices H, Q and R of appropriate dimensions with H symmetrical, then $H + Q\Delta(t)R + R^T\Delta^T(t)Q^T < 0$, if and only if there exists a scalar $\delta > 0$ such that*

$$\begin{bmatrix} H & R^T & \delta Q \\ * & -\delta I & \delta G \\ * & * & -\delta I \end{bmatrix} = H + \begin{bmatrix} \delta R \\ \delta^{-1} Q^T \end{bmatrix}^T \begin{bmatrix} I & -G \\ -G^T & I \end{bmatrix}^{-1} \begin{bmatrix} \delta R \\ \delta^{-1} Q^T \end{bmatrix} < 0. \tag{12}$$

3. Main Results

In this section, we will demonstrate the asymptotic stability of the NTNNs synchronization error system by constructing an augmented functional with mixed delays and a two-side looped functional. First, we take up the case $\Delta A(t) = \Delta W_0(t) = \Delta W_1(t) = \Delta W_2(t) = 0$.

Theorem 1. *Given scalars ϵ_1 and ϵ_2, if there exist symmetric matrices $P_i > 0$ ($i = 1, 2, 3$), $X_j > 0$ ($j = 1, 2, \cdots, 6$), $S_1 > 0$, $S_2 > 0$, $Z_1 > 0$, $Z_3 > 0$, $R_2 > 0$, $R_4 > 0$, Z_2, Q_3, Q_4, R_1, R_3, any matrices Q_1, Q_2, G, L, Y_k ($k = 1, 2, \cdots, 19$) and diagonal matrices $U \geq 0$, $V \geq 0$,*

$\Delta_l = diag\{\lambda_{l1}, \lambda_{l1}, \cdots, \lambda_{ln}\} \geq 0 \ (l = 1, 2)$, such that the followed linear matrix inequalities (LMIs) hold:

$$Z_3 + R_1 > 0, \ Z_2 + R_3 > 0, \ Z_2 + Z_3 > 0, \tag{13}$$

$$\begin{bmatrix} \Gamma_1 + d_k\Gamma_2 & \sqrt{d_k}\Theta_1 & \sqrt{\tau_c}\Theta_2 & \sqrt{\tau_2}\Theta_4 & \sqrt{\tau_1}\Pi^TY_{18} & \sqrt{\tau_{21}}\Pi^TY_{19} \\ * & -\chi_1 & 0 & 0 & 0 & 0 \\ * & * & -\chi_2 & 0 & 0 & 0 \\ * & * & * & -\chi_4 & 0 & 0 \\ * & * & * & * & -S_1 & 0 \\ * & * & * & * & * & -X_5 \end{bmatrix} < 0, \tag{14}$$

$$\begin{bmatrix} \Gamma_1 + d_k\Gamma_3 & \sqrt{d_k}\Theta_3 & \sqrt{\tau_c}\Theta_2 & \sqrt{\tau_2}\Theta_4 & \sqrt{\tau_1}\Pi^TY_{18} & \sqrt{\tau_{21}}\Pi^TY_{19} \\ * & -\chi_3 & 0 & 0 & 0 & 0 \\ * & * & -\chi_2 & 0 & 0 & 0 \\ * & * & * & -\chi_4 & 0 & 0 \\ * & * & * & * & -S_1 & 0 \\ * & * & * & * & * & -X_5 \end{bmatrix} < 0, \tag{15}$$

where

$$\Gamma_1 = \text{Sym}\Big\{\Omega_1^T P_1\Omega_2 + (r_{11} - K_pr_1)^T\Delta_1r_8 + (K_mr_1 - r_{11})^T\Delta_2r_8 + \Omega_9^T(Q_1\Omega_{10} + Q_2\Omega_{11})$$
$$+ \Omega_{20}^T\Omega_{21} + \Omega_{22}^TU\Omega_{23} + \Omega_{24}^TV\Omega_{25} + \Omega_9^T(Q_1\Omega_{10} + Q_2\Omega_{11}) + \text{Sym}\Big\{\sum_{i=1}^{19} Y_i\beth_i\Big\}\Big\}$$
$$+ \Omega_3^TP_2\Omega_3 - (1-u_2)\Omega_4^TP_2\Omega_4 + r_8^TP_3r_8 - (1-u_1)r_{26}^TP_3r_{26} + \Omega_5^T(X_1 + X_2 + X_3)\Omega_5$$
$$- \Omega_6^TX_1\Omega_6 - \Omega_7^TX_2\Omega_7 - \Omega_8^TX_3\Omega_8 + r_{24}^TX_4r_{24} - r_{11}^TX_4r_{11} + \tau_{21}r_8^TX_5r_8 + r_{11}^TX_6r_{11}$$
$$- r_2^TX_6r_2 + \tau_1r_8^TS_1r_8 + \tau_2r_8^TS_2r_8 + \tau_cr_8^TZ_1r_8 + d_Mr_8^TZ_2r_8 + \eta_Mr_8^TZ_3r_8,$$

$$\Gamma_2 = \text{Sym}\Big\{\Omega_{12}^T(Q_1\Omega_{10} + Q_2\Omega_{11}) + \Omega_{13}^TQ_1\Omega_{14} + \Omega_{17}^TQ_4\Omega_{18}\Big\} + \Omega_{11}^TQ_3\Omega_{11} + \Omega_{17}^TQ_4\Omega_{17}$$
$$+ r_8^TR_1r_8 + r_9^TR_3r_9,$$

$$\Gamma_3 = \text{Sym}\Big\{\Omega_{15}^T(Q_1\Omega_{10} + Q_2\Omega_{11}) + \Omega_{16}^TQ_1\Omega_{14} + \Omega_{17}^TQ_4\Omega_{19}\Big\} - \Omega_{11}^TQ_3\Omega_{11} - \Omega_{17}^TQ_4\Omega_{17}$$
$$+ r_8^TR_2r_8 + r_9^TR_4r_9,$$

$$\Gamma_4 = \tau_c\sum_{i=1}^{3}\frac{1}{2i-1}\Pi^TY_iZ_i^{-1}Y_i^T\Pi + \tau_c\Pi^TY_{14}Z_3^{-1}Y_{14}^T\Pi + \tau_2\sum_{i=1}^{2}\frac{1}{2i-1}\Pi^TY_{15+i}S_2^{-1}Y_{15+i}^T\Pi$$
$$+ \tau_1\Pi^TY_{18}S_1^{-1}Y_{18}^T\Pi + \tau_{21}\Pi^TY_{19}X_5^{-1}Y_{19}^T\Pi + \sum_{i=1}^{3}\frac{1}{2i-1}\Pi^TY_{3+i}R_2^{-1}Y_{3+i}^T\Pi$$
$$+ \sum_{i=1}^{2}\frac{1}{2i-1}\Pi^TY_{11+i}R_4^{-1}Y_{11+i}^T\Pi + \Pi^TY_{15}(Z_2 + Z_3)^{-1}Y_{15}^T\Pi,$$

$$\Gamma_5 = \tau_c\sum_{i=1}^{3}\frac{1}{2i-1}\Pi^TY_iZ_i^{-1}Y_i^T\Pi + \tau_c\Pi^TY_{14}Z_3^{-1}Y_{14}^T\Pi + \tau_2\sum_{i=1}^{2}\frac{1}{2i-1}\Pi^TY_{15+i}S_2^{-1}Y_{15+i}^T\Pi$$
$$+ \tau_1\Pi^TY_{18}S_1^{-1}Y_{18}^T\Pi + \tau_{21}\Pi^TY_{19}X_5^{-1}Y_{19}^T\Pi + \sum_{i=1}^{3}\frac{1}{2i-1}\Pi^TY_{6+i}(R_1 + Z_3)^{-1}Y_{6+i}^T\Pi$$
$$+ \sum_{i=1}^{2}\frac{1}{2i-1}\Pi^TY_{9+i}(R_3 + Z_2)^{-1}Y_{9+i}^T\Pi,$$

$\tau_{21} = \tau_2 - \tau_1$, $r_i = \begin{bmatrix} 0_{n \times (i-1)n} & I_n & 0_{n \times (26-i)n} \end{bmatrix}$, $i = 1, 2, \cdots, 26$,

$\Pi = \text{col}\{r_1, r_2, r_3, r_4, r_5, r_6, r_7, r_{11}, r_{15}, r_{16}, r_{17}, r_{18}, r_{19}, r_{20}, r_{21}, r_{22}, r_{23}, r_{24}\}$,

$\Theta_1 = \Pi^T(Y_4, Y_5, Y_6, Y_{12}, Y_{13}, Y_{15})$, $\Theta_2 = \Pi^T(Y_1, Y_2, Y_3, Y_{14})$,

$\Theta_3 = \Pi^T(Y_7, Y_8, Y_9, Y_{10}, Y_{11})$, $\Theta_4 = \Pi^T(Y_{16}, Y_{17})$,

$\chi_1 = \text{diag}\{R_2, 3R_2, 5R_2, R_4, 3R_4, Z_2 + Z_3\}$, $\chi_2 = \text{diag}\{Z_1, 3Z_1, 5Z_1, Z_3\}$,

$\chi_3 = \text{diag}\{R_1 + Z_3, 3(R_1 + Z_3), 5(R_1 + Z_3), R_3 + Z_2, 3(R_3 + Z_2)\}$, $\chi_4 = \text{diag}\{S_2, 3S_2\}$,

$\Omega_1 = \text{col}\{r_1, r_2, r_{24}, r_{11}, \tau_c r_{15} - \tau_2 r_{23}, \tau_c^2 r_{16}\}$,

$\Omega_2 = \text{col}\{r_8, r_9, r_{25}, r_{10}, r_{11} - r_2, \tau_c(r_{15} - r_2)\}$,

$\Omega_3 = \text{col}\{r_1, r_{13}\}$, $\Omega_4 = \text{col}\{r_{12}, r_{14}\}$, $\Omega_5 = \text{col}\{r_1, r_8\}$,

$\Omega_6 = \text{col}\{r_{11}, r_{10}\}$, $\Omega_7 = \text{col}\{r_2, r_9\}$, $\Omega_8 = \text{col}\{r_{24}, r_{25}\}$,

$\Omega_9 = \text{col}\{r_3 - r_1, r_1 - r_5, r_4 - r_2, r_2 - r_6\}$, $\Omega_{10} = \text{col}\{r_1 - r_3, r_1 - r_5, r_2 - r_4, r_2 - r_6\}$,

$\Omega_{11} = \text{col}\{r_3, r_4, r_5, r_6\}$, $\Omega_{12} = \text{col}\{r_8, 0, r_9, 0\}$, $\Omega_{13} = \text{col}\{r_1 - r_3, 0, r_2 - r_4, 0\}$,

$\Omega_{14} = \text{col}\{r_8, r_8, r_9, r_9\}$, $\Omega_{15} = \text{col}\{0, r_8, 0, r_9\}$, $\Omega_{16} = \text{col}\{0, r_1 - r_5, 0, r_2 - r_6\}$,

$\Omega_{17} = \text{col}\{r_{17}, r_{19}, r_{21}, r_{22}, r_{18}, r_{20}\}$, $\Omega_{18} = \text{col}(r_1 - r_{17}, 0, r_2 - r_{21}, 0, r_{17} - 2r_{18}, 0\}$,

$\Omega_{19} = \text{col}\{0, r_{19} - r_1, 0, r_{22} - r_2, 0, 2r_{20} - r_{19}\}$, $\Omega_{20} = \text{col}\{r_1 + \epsilon_1 r_4 + \epsilon_2 r_8\}$,

$\Omega_{21} = \text{col}\{-Gr_8 - GAr_1 + GW_0 r_{13} + GW_1 r_{14} + GW_2 r_{26} + Lr_4\}$,

$\Omega_{22} = r_{13} - K^- r_1$, $\Omega_{23} = K^+ r_1 - r_{13}$, $\Omega_{24} = r_{14} - K^- r_{12}$, $\Omega_{25} = K^+ r_{12} - r_{14}$,

$\beth_1 = r_1 - r_2$, $\beth_2 = r_1 + r_2 - 2r_{15}$, $\beth_3 = r_1 - r_2 - 6r_{15} + 12r_{16}$,

$\beth_4 = r_5 - r_1$, $\beth_5 = r_5 + r_1 - 2r_{19}$, $\beth_6 = r_5 - r_1 + 6r_{19} - 12r_{20}$,

$\beth_7 = r_1 - r_3$, $\beth_8 = r_1 + r_3 - 2r_{17}$, $\beth_9 = r_1 - r_3 - 6r_{17} + 12r_{18}$,

$\beth_{10} = r_2 - r_4$, $\beth_{11} = r_2 + r_4 - 2r_{21}$, $\beth_{12} = r_6 - r_2$, $\beth_{13} = r_6 + r_2 - 2r_{22}$, $\beth_{14} = r_3 - r_4$,

$\beth_{15} = r_4 - r_7$, $\beth_{16} = r_1 - r_{11}$, $\beth_{17} = r_1 + r_{11} - 2r_{23}$, $\beth_{18} = r_1 - r_{24}$, $\beth_{19} = r_{24} - r_{11}$.

Then, the slave system (5) can be synchronized with the master system (1). The gain matrix of the controller in (6) can be calculated by $K = G^{-1}L$.

Proof. Please see Appendix A. □

Remark 1. *The two-side looped functional $\mathcal{W}(t)$ was introduced in the LKF, which utilized the information on intervals $[t_k, t]$, $[t_k - \tau, t - \tau]$, $[t, t_{k+1})$, $[t - \tau, t_{k+1} - \tau)$. Notice that $\mathcal{W}_j(t)$ ($j = 1, 2, 3, 4, 5$) satisfy the requirement of the looped functional [23] as follows: $\mathcal{W}_j(t_k) = \mathcal{W}_j(t_{k+1}) = 0$.*

Remark 2. *Even though the existence of neutral delay $\tau_1(t)$ and transmission delay $\tau_2(t)$ were considered, the cross information associated with the states of the mixed delays were not exploited in [7,24,25]. In our work, the integral terms and double integral terms in $\mathcal{V}_4(t)$ can reflect the cross information associated with the time delays $\tau_i(t)$ ($i = 1, 2$). The derivative of LKF conducted not only depends on $\tau_i(t)$ ($i = 1, 2$), but also on the value τ_{21}. These integral terms strengthen the connection between those states and effectively reduce the conservatism when $\tau_1(t) \neq \tau_2(t)$.*

Remark 3. *The sampling periods d_k and the communication delay τ_c are coupled with the variable matrices in LMIs. We set the communication delay τ_c as a given scalar, and then use Algorithm 1 to derive the maximal sampling period d_M and the corresponding controller gain K.*

Algorithm 1 Find the maximum sampling period d_M and the controller gain matrix K.

Step 1: Input communication delay τ_c.
Step 2: Set the accuracy coefficient to $d_{ac} = 0.0001$, and initialize the search interval $[d_{\min}, d_{\max}]$ with $d_{\min} = 0$ and a large enough integer $d_{\max}$.
Step 3: By validating the feasibility of LMIs (14) and (15), determine whether the system has a sampling period given as $d_t = (d_{\min} + d_{\max})/2$.
Step 4: If (14) and (15) are feasible, calculate K_t by (A3), and set $d_{\min} = d_t$; else, set $d_{\max} = d_t$.
Step 5: If $|d_{\max} - d_{\min}| \leq d_{ac}$, record the maximum sampling period $d_M = d_{\min}$ and derive the corresponding controller gain $K = K_t$; else, repeat Step 3.
Step 6: End. Output d_M and K.

To demonstrate the validity of the mixed-delay-dependent terms in LKF, we remove them from Theorem 1, resulting in Corollary 1.

Corollary 1. *Given scalars ϵ_1 and ϵ_2, if there exist symmetric matrices $P_i > 0$ ($i = 1, 2, 3$), $X_j > 0$ ($j = 1, 2, \cdots, 6$), $S_1 > 0$, $S_2 > 0$, $Z_1 > 0$, $Z_3 > 0$, $R_2 > 0$, $R_4 > 0$, Z_2, Q_3, Q_4, R_1, R_3, any matrices Q_1, Q_2, G, L, Y_k ($k = 1, 2, \cdots, 18$) and diagonal matrices $U \geq 0$, $V \geq 0$, $\Delta_l = diag\{\lambda_{l1}, \lambda_{l1}, \cdots, \lambda_{ln}\} \geq 0$ ($l = 1, 2$), such that the followed linear matrix inequalities (LMIs) hold:*

$$Z_3 + R_1 > 0, \; Z_2 + R_3 > 0, \; Z_2 + Z_3 > 0, \tag{16}$$

$$\begin{bmatrix} \hat{\Gamma}_1 + d_k \Gamma_2 & \sqrt{d_k}\Theta_1 & \sqrt{\tau_c}\Theta_2 & \sqrt{\tau_2}\Theta_4 & \sqrt{\tau_1}\Pi^T Y_{18} \\ * & -\chi_1 & 0 & 0 & 0 \\ * & * & -\chi_2 & 0 & 0 \\ * & * & * & -\chi_4 & 0 \\ * & * & * & * & -S_1 \end{bmatrix} < 0, \tag{17}$$

$$\begin{bmatrix} \hat{\Gamma}_1 + d_k \Gamma_3 & \sqrt{d_k}\Theta_3 & \sqrt{\tau_c}\Theta_2 & \sqrt{\tau_2}\Theta_4 & \sqrt{\tau_1}\Pi^T Y_{18} \\ * & -\chi_3 & 0 & 0 & 0 \\ * & * & -\chi_2 & 0 & 0 \\ * & * & * & -\chi_4 & 0 \\ * & * & * & * & -S_1 \end{bmatrix} < 0, \tag{18}$$

where

$$\hat{\Gamma}_1 = \text{Sym}\left\{r_1^T P_1 r_8 + (r_{11} - K^- r_1)^T \Delta_1 r_8 + (K^+ r_1 - r_{11})^T \Delta_2 r_8 + \Omega_9^T (Q_1 \Omega_{10} + Q_2 \Omega_{11})\right.$$

$$+ \Omega_{20}^T \Omega_{21} + \Omega_{22}^T U \Omega_{23} + \Omega_{24}^T V \Omega_{25} + \Pi_9^T (Q_1 \Pi_{10} + Q_2 \Pi_{11}) + \text{Sym}\left\{\sum_{i=1}^{18} Y_i \beth_i\right\}\right\}$$

$$+ \Omega_3^T P_2 \Omega_3 - (1-u_2)\Omega_4^T P_2 \Omega_4 + r_8^T P_3 r_8 - (1-u_1)r_{26}^T P_3 r_{26} + \Omega_5^T (X_1 + X_2 + X_3)\Omega_5$$

$$- \Omega_6^T X_1 \Omega_6 - \Omega_7^T X_2 \Omega_7 - \Omega_8^T X_3 \Omega_8 + \tau_1 r_8^T S_1 r_8 + \tau_2 r_8^T S_2 r_8 + \tau_c r_8^T Z_1 r_8 + d_M r_8^T Z_2 r_8$$

$$+ \eta_M r_8^T Z_3 r_8,$$

and the rest of the vectors and matrices are defined in Theorem 1. Then, the slave system (5) can be synchronized with the master system (1). The gain matrix of the controller in (6) can be calculated by $K = G^{-1}L$.

Proof. Please see Appendix B. □

Remark 4. *In Corollary 1, we remove the mixed-delay terms in $\mathcal{V}_4(t)$, and change the vector $\varpi_1(t)$ to $e(t)$. The comparison between Theorem 1 and Corollary 1 demonstrates the usefulness of the mixed-delay terms. Example 1 in Section 4 shows numerical comparisons.*

Based on Theorem 1 and considering the parameter uncertainties, we have the following stability conditions.

Theorem 2. *Given scalars ϵ_1, ϵ_2, matrices J, $E_m > 0$ ($m = 1,2,3,4$), if there exist positive scalars $\delta_n > 0$ ($n = 1,2$), symmetric matrices $P_i > 0$ ($i = 1,2,3$), $X_j > 0$ ($j = 1,2,\cdots,6$), $S_1 > 0$, $S_2 > 0$, $Z_1 > 0$, $Z_3 > 0$, $R_2 > 0$, $R_4 > 0$, Z_2, Q_3, Q_4, R_1, R_3, any matrices Q_1, Q_2, G, L, Y_k ($k = 1,2,\cdots,19$) and diagonal matrices $U \geq 0$, $V \geq 0$, $\Delta_l = diag\{\lambda_{l1}, \lambda_{l1}, \cdots, \lambda_{ln}\} \geq 0$ ($l = 1,2$), such that the following LMIs hold:*

$$Z_3 + R_1 > 0, \ Z_2 + R_3 > 0, \ Z_2 + Z_3 > 0, \quad (19)$$

$$\begin{bmatrix} \Gamma_1 + d_k\Gamma_2 & \sqrt{d_k}\Theta_1 & \sqrt{\tau_c}\Theta_2 & \sqrt{\tau_2}\Theta_4 & \sqrt{\tau_1}\tilde{Y}_{18} & \sqrt{\tau_{21}}\tilde{Y}_{19} & \delta_1\Psi_1 & \Psi_2 \\ * & -\chi_1 & 0 & 0 & 0 & 0 & 0 & 0 \\ * & * & -\chi_2 & 0 & 0 & 0 & 0 & 0 \\ * & * & * & -\chi_4 & 0 & 0 & 0 & 0 \\ * & * & * & * & -S_1 & 0 & 0 & 0 \\ * & * & * & * & * & -X_5 & 0 & 0 \\ * & * & * & * & * & * & -\delta_1 I & \delta_1 J \\ * & * & * & * & * & * & * & -\delta_1 I \end{bmatrix} < 0, \quad (20)$$

$$\begin{bmatrix} \Gamma_1 + d_k\Gamma_3 & \sqrt{d_k}\Theta_3 & \sqrt{\tau_c}\Theta_2 & \sqrt{\tau_2}\Theta_4 & \sqrt{\tau_1}\tilde{Y}_{18} & \sqrt{\tau_{21}}\tilde{Y}_{19} & \delta_2\Psi_1 & \Psi_2 \\ * & -\chi_3 & 0 & 0 & 0 & 0 & 0 & 0 \\ * & * & -\chi_2 & 0 & 0 & 0 & 0 & 0 \\ * & * & * & -\chi_4 & 0 & 0 & 0 & 0 \\ * & * & * & * & -S_1 & 0 & 0 & 0 \\ * & * & * & * & * & -X_5 & 0 & 0 \\ * & * & * & * & * & * & -\delta_2 I & \delta_2 J \\ * & * & * & * & * & * & * & -\delta_2 I \end{bmatrix} < 0, \quad (21)$$

where Γ_i, Θ_i, χ_i ($i = 1,2,3,4$) are defined well in Theorem 1 and $\tilde{Y}_{18} = \Pi^T Y_{18}$, $\tilde{Y}_{19} = \Pi^T Y_{19}$, $\Psi_1 = col\{E_1^T, 0_{n\cdot 11n}^T, E_2^T, E_3^T, 0_{n\cdot 11n}^T, E_4^T\}$, $\Psi_2 = col\{GF, 0_{n\cdot 2n}^T, \epsilon_1 GF, 0_{n\cdot 3n}^T, \epsilon_2 GF, 0_{n\cdot 18n}^T\}$. Then, the slave system (5) can be synchronized with the master system (1). The gain matrix of the controller in (6) can be calculated by $K = G^{-1}L$.

Proof. Please see Appendix C. □

To demonstrate the validity of the mixed-delay-dependent terms in LKF, we remove them from Theorem 2, resulting in Corollary 2.

Corollary 2. *Given scalars ϵ_1, ϵ_2, matrices J, $E_m > 0$ ($m = 1,2,3,4$), if there exist positive scalars $\delta_n > 0$ ($n = 1,2$), symmetric matrices $P_i > 0$ ($i = 1,2,3$), $X_j > 0$ ($j = 1,2,\cdots,6$), $S_1 > 0$, $S_2 > 0$, $Z_1 > 0$, $Z_3 > 0$, $R_2 > 0$, $R_4 > 0$, Z_2, Q_3, Q_4, R_1, R_3, any matrices Q_1, Q_2, G, L, Y_k ($k = 1,2,\cdots,18$) and diagonal matrices $U \geq 0$, $V \geq 0$, $\Delta_l = diag\{\lambda_{l1}, \lambda_{l1}, \cdots, \lambda_{ln}\} \geq 0$ ($l = 1,2$), such that the following LMIs hold:*

$$Z_3 + R_1 > 0, \ Z_2 + R_3 > 0, \ Z_2 + Z_3 > 0, \quad (22)$$

$$\begin{bmatrix} \hat{\Gamma}_1 + d_k\Gamma_2 & \sqrt{d_k}\Theta_1 & \sqrt{\tau_c}\Theta_2 & \sqrt{\tau_2}\Theta_4 & \sqrt{\tau_1}\tilde{Y}_{18} & \delta_1\Psi_1 & \Psi_2 \\ * & -\chi_1 & 0 & 0 & 0 & 0 & 0 \\ * & * & -\chi_2 & 0 & 0 & 0 & 0 \\ * & * & * & -\chi_4 & 0 & 0 & 0 \\ * & * & * & * & -S_1 & 0 & 0 \\ * & * & * & * & * & -\delta_1 I & \delta_1 J \\ * & * & * & * & * & * & -\delta_1 I \end{bmatrix} < 0, \quad (23)$$

$$\begin{bmatrix} \hat{\Gamma}_1 + d_k\Gamma_3 & \sqrt{d_k}\Theta_3 & \sqrt{\tau_c}\Theta_2 & \sqrt{\tau_2}\Theta_4 & \sqrt{\tau_1}\tilde{Y}_{18} & \delta_2\Psi_1 & \Psi_2 \\ * & -\chi_3 & 0 & 0 & 0 & 0 & 0 \\ * & * & -\chi_2 & 0 & 0 & 0 & 0 \\ * & * & * & -\chi_4 & 0 & 0 & 0 \\ * & * & * & * & -S_1 & 0 & 0 \\ * & * & * & * & * & -\delta_2 I & \delta_2 J \\ * & * & * & * & * & * & -\delta_2 I \end{bmatrix} < 0, \quad (24)$$

where

$$\hat{\Gamma}_1 = \mathrm{Sym}\Big\{r_1^T P_1 r_8 + (r_{11} - K^- r_1)^T \Delta_1 r_8 + (K^+ r_1 - r_{11})^T \Delta_2 r_8 + \Omega_9^T(Q_1\Omega_{10} + Q_2\Omega_{11})$$
$$+ \Omega_{20}^T\Omega_{21} + \Omega_{22}^T U\Omega_{23} + \Omega_{24}^T V\Omega_{25} + \Pi_9^T(Q_1\Pi_{10} + Q_2\Pi_{11}) + \mathrm{Sym}\Big\{\sum_{i=1}^{18} Y_i \beth_i\Big\}\Big\}$$
$$+ \Omega_3^T P_2\Omega_3 - (1-u_2)\Omega_4^T P_2\Omega_4 + r_8^T P_3 r_8 - (1-u_1)r_{26}^T P_3 r_{26} + \Omega_5^T(X_1 + X_2 + X_3)\Omega_5$$
$$- \Omega_6^T X_1\Omega_6 - \Omega_7^T X_2\Omega_7 - \Omega_8^T X_3\Omega_8 + \tau_1 r_8^T S_1 r_8 + \tau_2 r_8^T S_2 r_8 + \tau_c r_8^T Z_1 r_8 + d_M r_8^T Z_2 r_8$$
$$+ \eta_M r_8^T Z_3 r_8,$$

and the rest of the vectors and matrices are defined in Theorem 2. Then, the slave system (5) can be synchronized with the master system (1). The gain matrix of the controller in (6) can be calculated by $K = G^{-1}L$.

Proof. Please see Appendix D. □

If the neutral delay is not taken into consideration, the following synchronization criterion is derived.

Corollary 3. *Given scalars ϵ_1 and ϵ_2, if there exist symmetric matrices $P_i > 0$ ($i = 1, 2$), $X_j > 0$ ($j = 2, 3$), $S_1 > 0$, $Z_1 > 0$, $Z_3 > 0$, $R_2 > 0$, $R_4 > 0$, Z_2, Q_3, Q_4, R_1, R_3, any matrices Q_1, Q_2, G, L, Y_k ($k = 1, 2, \cdots, 17$) and diagonal matrices $U \geq 0$, $V \geq 0$, $\Delta_l = \mathrm{diag}\{\lambda_{l1}, \lambda_{l1}, \cdots, \lambda_{ln}\} \geq 0$ ($l = 1, 2$), such that the following linear matrix inequalities (LMIs) hold:*

$$Z_3 + R_1 > 0, \; Z_2 + R_3 > 0, \; Z_2 + Z_3 > 0, \quad (25)$$

$$\begin{bmatrix} \tilde{\Gamma}_1 + d_k\tilde{\Gamma}_2 & \sqrt{d_k}\tilde{\Theta}_1 & \sqrt{\tau_c}\tilde{\Theta}_2 & \sqrt{\tau_2}\tilde{\Theta}_4 \\ * & -\chi_1 & 0 & 0 \\ * & * & -\chi_2 & 0 \\ * & * & * & -\chi_4 \end{bmatrix} < 0, \quad (26)$$

$$\begin{bmatrix} \tilde{\Gamma}_1 + d_k\tilde{\Gamma}_3 & \sqrt{d_k}\tilde{\Theta}_3 & \sqrt{\tau_c}\tilde{\Theta}_2 & \sqrt{\tau_2}\tilde{\Theta}_4 \\ * & -\chi_3 & 0 & 0 \\ * & * & -\chi_2 & 0 \\ * & * & * & -\chi_4 \end{bmatrix} < 0, \quad (27)$$

where

$$\tilde{\Gamma}_1 = \mathrm{Sym}\Big\{\tilde{\Omega}_1^T P_1 \tilde{\Omega}_2 + (\tilde{r}_{11} - K^-\tilde{r}_1)^T \Delta_1 \tilde{r}_8 + (K^+\tilde{r}_1 - \tilde{r}_{11})^T \Delta_2 \tilde{r}_8 + \tilde{\Omega}_9^T(Q_1\tilde{\Omega}_{10} + Q_2\tilde{\Omega}_{11})$$
$$+ \tilde{\Omega}_{20}^T\tilde{\Omega}_{21} + \tilde{\Omega}_{22}^T U\tilde{\Omega}_{23} + \tilde{\Omega}_{24}^T V\tilde{\Omega}_{25} + \tilde{\Omega}_9^T(Q_1\tilde{\Omega}_{10} + Q_2\tilde{\Omega}_{11}) + \mathrm{Sym}\Big\{\sum_{i=1}^{17} Y_i \beth_i\Big\}\Big\}$$
$$+ \tilde{\Omega}_3^T P_2\tilde{\Omega}_3 - (1-u_2)\tilde{\Omega}_4^T P_2\tilde{\Omega}_4 + \tilde{\Omega}_5^T(X_2 + X_3)\tilde{\Omega}_5 - \tilde{\Omega}_7^T X_2\tilde{\Omega}_7 - \tilde{\Omega}_8^T X_3\tilde{\Omega}_8$$
$$+ \tilde{r}_{11}^T X_6 \tilde{r}_{11} - \tilde{r}_2^T X_6 \tilde{r}_2 + \tau_2 \tilde{r}_8^T S_1 \tilde{r}_8 + \tau_c \tilde{r}_8^T Z_1 \tilde{r}_8 + d_M \tilde{r}_8^T Z_2 \tilde{r}_8 + \eta_M \tilde{r}_8^T Z_3 \tilde{r}_8,$$

$$\tilde{\Gamma}_2 = \mathrm{Sym}\{\tilde{\Omega}_{12}^T(Q_1\tilde{\Omega}_{10} + Q_2\tilde{\Omega}_{11}) + \tilde{\Omega}_{13}^T Q_1 \tilde{\Omega}_{14} + \tilde{\Omega}_{17}^T Q_4 \tilde{\Omega}_{18}\} + \tilde{\Omega}_{11}^T Q_3 \tilde{\Omega}_{11}$$
$$+ \tilde{\Omega}_{17}^T Q_4 \tilde{\Omega}_{17} + \tilde{r}_8^T R_1 \tilde{r}_8 + \tilde{r}_9^T R_3 \tilde{r}_9,$$
$$\tilde{\Gamma}_3 = \mathrm{Sym}\{\tilde{\Omega}_{15}^T(Q_1\tilde{\Omega}_{10} + Q_2\tilde{\Omega}_{11}) + \tilde{\Omega}_{16}^T Q_1 \tilde{\Omega}_{14} + \tilde{\Omega}_{17}^T Q_4 \tilde{\Omega}_{19}\} - \tilde{\Omega}_{11}^T Q_3 \tilde{\Omega}_{11}$$
$$- \tilde{\Omega}_{17}^T Q_4 \tilde{\Omega}_{17} + \tilde{r}_8^T R_2 \tilde{r}_8 + \tilde{r}_9^T R_4 \tilde{r}_9,$$
$$\tilde{r}_i = \begin{bmatrix} 0_{n \times (i-1)n} & I_n & 0_{n \times (23-i)n} \end{bmatrix}, i = 1, 2, \cdots, 23,$$
$$\tilde{\Pi} = \mathrm{col}\{r_1, r_2, r_3, r_4, r_5, r_6, r_7, r_{11}, r_{15}, r_{16}, r_{17}, r_{18}, r_{19}, r_{20}, r_{21}, r_{22}, r_{23}\},$$
$$\tilde{\Theta}_1 = \tilde{\Pi}^T(Y_4, Y_5, Y_6, Y_{12}, Y_{13}, Y_{15}), \tilde{\Theta}_2 = \tilde{\Pi}^T(Y_1, Y_2, Y_3, Y_{14}),$$
$$\tilde{\Theta}_3 = \tilde{\Pi}^T(Y_7, Y_8, Y_9, Y_{10}, Y_{11}), \tilde{\Theta}_4 = \tilde{\Pi}^T(Y_{16}, Y_{17}),$$
$$\chi_1 = \mathrm{diag}\{R_2, 3R_2, 5R_2, R_4, 3R_4, Z_2 + Z_3\}, \chi_2 = \mathrm{diag}\{Z_1, 3Z_1, 5Z_1, Z_3\},$$
$$\chi_3 = \mathrm{diag}\{R_1 + Z_3, 3(R_1 + Z_3), 5(R_1 + Z_3), R_3 + Z_2, 3(R_3 + Z_2)\}, \chi_4 = \mathrm{diag}\{S_2, 3S_2\},$$
$$\tilde{\Omega}_1 = \mathrm{col}\{\tilde{r}_1, \tilde{r}_2, \tilde{r}_{11}, \tau_c \tilde{r}_{15} - \tau_2 \tilde{r}_{23}, \tau_c^2 \tilde{r}_{16}\},$$
$$\tilde{\Omega}_2 = \mathrm{col}\{\tilde{r}_8, \tilde{r}_9, \tilde{r}_{10}, \tilde{r}_{11} - \tilde{r}_2, \tau_c(\tilde{r}_{15} - \tilde{r}_2)\},$$
$$\tilde{\Omega}_3 = \mathrm{col}\{\tilde{r}_1, \tilde{r}_{13}\}, \tilde{\Omega}_4 = \mathrm{col}\{\tilde{r}_{12}, \tilde{r}_{14}\}, \tilde{\Omega}_5 = \mathrm{col}\{\tilde{r}_1, \tilde{r}_8\},$$
$$\tilde{\Omega}_7 = \mathrm{col}\{\tilde{r}_2, \tilde{r}_9\}, \tilde{\Omega}_8 = \mathrm{col}\{\tilde{r}_{11}, \tilde{r}_{10}\},$$
$$\tilde{\Omega}_9 = \mathrm{col}\{\tilde{r}_3 - \tilde{r}_1, \tilde{r}_1 - \tilde{r}_5, \tilde{r}_4 - \tilde{r}_2, \tilde{r}_2 - \tilde{r}_6\}, \tilde{\Omega}_{10} = \mathrm{col}\{\tilde{r}_1 - \tilde{r}_3, \tilde{r}_1 - \tilde{r}_5, \tilde{r}_2 - \tilde{r}_4, \tilde{r}_2 - \tilde{r}_6\},$$
$$\tilde{\Omega}_{11} = \mathrm{col}\{\tilde{r}_3, \tilde{r}_4, \tilde{r}_5, \tilde{r}_6\}, \tilde{\Omega}_{12} = \mathrm{col}\{\tilde{r}_8, 0, \tilde{r}_9, 0\}, \tilde{\Omega}_{13} = \mathrm{col}\{\tilde{r}_1 - \tilde{r}_3, 0, \tilde{r}_2 - \tilde{r}_4, 0\},$$
$$\tilde{\Omega}_{14} = \mathrm{col}\{\tilde{r}_8, \tilde{r}_8, \tilde{r}_9, \tilde{r}_9\}, \tilde{\Omega}_{15} = \mathrm{col}\{0, \tilde{r}_8, 0, \tilde{r}_9\}, \tilde{\Omega}_{16} = \mathrm{col}\{0, \tilde{r}_1 - \tilde{r}_5, 0, \tilde{r}_2 - \tilde{r}_6\},$$
$$\tilde{\Omega}_{17} = \mathrm{col}\{\tilde{r}_{17}, \tilde{r}_{19}, \tilde{r}_{21}, \tilde{r}_{22}, \tilde{r}_{18}, \tilde{r}_{20}\}, \tilde{\Omega}_{18} = \mathrm{col}\{\tilde{r}_1 - \tilde{r}_{17}, 0, \tilde{r}_2 - \tilde{r}_{21}, 0, \tilde{r}_{17} - 2\tilde{r}_{18}, 0\},$$
$$\tilde{\Omega}_{19} = \mathrm{col}\{0, \tilde{r}_{19} - \tilde{r}_1, 0, \tilde{r}_{22} - \tilde{r}_2, 0, 2\tilde{r}_{20} - \tilde{r}_{19}\}, \tilde{\Omega}_{20} = \mathrm{col}\{\tilde{r}_1 + \epsilon_1 \tilde{r}_4 + \epsilon_2 \tilde{r}_8\},$$
$$\tilde{\Omega}_{21} = \mathrm{col}\{-G\tilde{r}_8 - GA\tilde{r}_1 + GW_0\tilde{r}_{13} + GW_1\tilde{r}_{14} + L\tilde{r}_4\},$$
$$\tilde{\Omega}_{22} = \tilde{r}_{13} - K^-\tilde{r}_1, \tilde{\Omega}_{23} = K^+\tilde{r}_1 - \tilde{r}_{13}, \tilde{\Omega}_{24} = \tilde{r}_{14} - K^-\tilde{r}_{12}, \tilde{\Omega}_{25} = K^+\tilde{r}_{12} - \tilde{r}_{14},$$
$$\beth_1 = \tilde{r}_1 - \tilde{r}_2, \beth_2 = \tilde{r}_1 + \tilde{r}_2 - 2\tilde{r}_{15}, \beth_3 = \tilde{r}_1 - \tilde{r}_2 - 6\tilde{r}_{15} + 12\tilde{r}_{16},$$
$$\beth_4 = \tilde{r}_5 - \tilde{r}_1, \beth_5 = \tilde{r}_5 + \tilde{r}_1 - 2\tilde{r}_{19}, \beth_6 = \tilde{r}_5 - \tilde{r}_1 + 6\tilde{r}_{19} - 12\tilde{r}_{20},$$
$$\beth_7 = \tilde{r}_1 - \tilde{r}_3, \beth_8 = \tilde{r}_1 + \tilde{r}_3 - 2\tilde{r}_{17}, \beth_9 = \tilde{r}_1 - \tilde{r}_3 - 6\tilde{r}_{17} + 12\tilde{r}_{18},$$
$$\beth_{10} = \tilde{r}_2 - \tilde{r}_4, \beth_{11} = \tilde{r}_2 + \tilde{r}_4 - 2\tilde{r}_{21}, \beth_{12} = \tilde{r}_6 - \tilde{r}_2, \beth_{13} = \tilde{r}_6 + \tilde{r}_2 - 2\tilde{r}_{22}, \beth_{14} = \tilde{r}_3 - \tilde{r}_4,$$
$$\beth_{15} = \tilde{r}_4 - \tilde{r}_7, \beth_{16} = \tilde{r}_1 - \tilde{r}_{11}, \beth_{17} = \tilde{r}_1 + \tilde{r}_{11} - 2\tilde{r}_{23}.$$

Then, the slave system can be synchronized with the master system. The gain matrix of the controller can be calculated by $K = G^{-1}L$.

Proof. Please see Appendix E. □

4. Illustrative Examples

In this section, three numerical examples will be presented to illustrate the validity of the derived criteria.

Example 1. *Consider the neutral-type neural networks (10) with the following parameters [19]:*

$$A = \begin{bmatrix} 3 & 0 \\ 0 & 3 \end{bmatrix}, W_0 = \begin{bmatrix} 0.5 & 0.3 \\ 0.3 & 0.5 \end{bmatrix}, W_1 = \begin{bmatrix} 0.2 & 0.1 \\ 0.1 & 0.2 \end{bmatrix}, W_2 = \begin{bmatrix} 0.15 & 0 \\ 0 & 0.15 \end{bmatrix},$$
$$K^- = \mathrm{diag}\{0,0\}, K^+ = \mathrm{diag}\{1,1\}, \Delta A(t) = 0, \Delta W_0(t) = 0, \Delta W_1(t) = 0, \Delta W_2(t) = 0.$$

We choose $\tau_1 = 0.5$, $\mu_1 = 0.9$, $\tau_2 = 2.0$, $\mu_2 = 0.5$, $\epsilon_1 = \epsilon_2 = 0.24$. For different communication delays τ_c, via solving the LMIs in Theorem 1 and Corollary 1, the maximal sampling period d_m computed by Algorithm 1 are listed in Table 1. It can be seen from Table 1 that as the communication delays τ_c increases, the maximum sampling period d_M decreases. Theorem 1 and

Corollary 1 are derived via a similar approach. When $\tau_c = 0.09$, the maximum sampling period d_M calculated by Corollary 1 is 1.2996 and by Theorem 1 is 1.3738. Theorem 1 provides a larger allowable sampling period than Corollary 1. Therefore, a mixed-delay-dependent LKF can lead to a less conservative result.

When $\tau_c = 0.01$, $d_M = 1.3881$, the corresponding controller gain matrix is obtained by Theorem 1 as follows:

$$K = \begin{bmatrix} 0.0653 & -0.0414 \\ -0.0414 & 0.0653 \end{bmatrix}.$$

Table 1. The maximum sampling period d_M for different τ_c (Example 1).

τ_c	0.01	0.03	0.05	0.07	0.09
Corollary 1	1.3766	1.3499	1.3296	1.3134	1.2996
Theorem 1	1.3881	1.3865	1.3820	1.3777	1.3738
Improvement	0.835%	4.9406%	3.9411%	4.8957%	5.7094%

Let $\tau_c = 0.01$, the largest sampling interval $d_M = 1.3881$, and the corresponding controller is obtained as

$$K = \begin{bmatrix} 0.0653 & -0.0414 \\ -0.0414 & 0.0653 \end{bmatrix}.$$

The neural activation function is taken as the form $f_i(x_i) = tanh(x_i)$ $(i = 1, 2)$, which satisfies the assumption (7). The time-delay $\tau_1(t) = 0.5sin^2(0.9t)$, and $\tau_2(t) = 2cos^2(0.5t)$. The initial condition of the master system and the slave system are taken as $x(0) = col\{-0.2, -0.3\}$, $y(0) = col\{0.3, 0.6\}$, respectively. The initial control input is $u(t) = 0$. Let the sampling period $d_k = d_M$, based on the above sample-data controller, the state response of the error system (6) is shown in Figure 1, and the control input (9) is shown in Figure 2. It can be seen from Figure 1 that the error state converges to zero. That is to say, the error system (6) is asymptotic stable, and the slave system (5) is synchronous with the master system (1). It verifies the effectiveness of our methods.

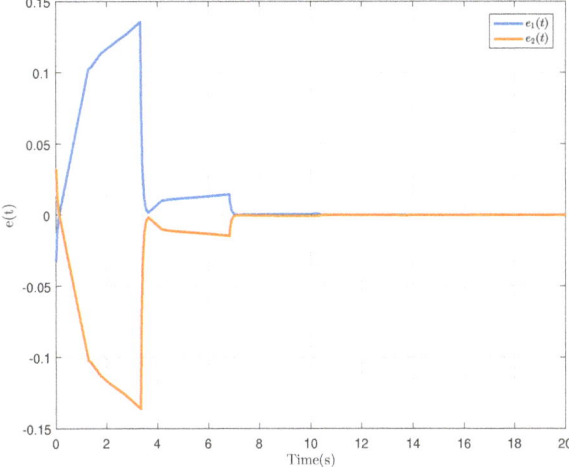

Figure 1. State responses of the error system.

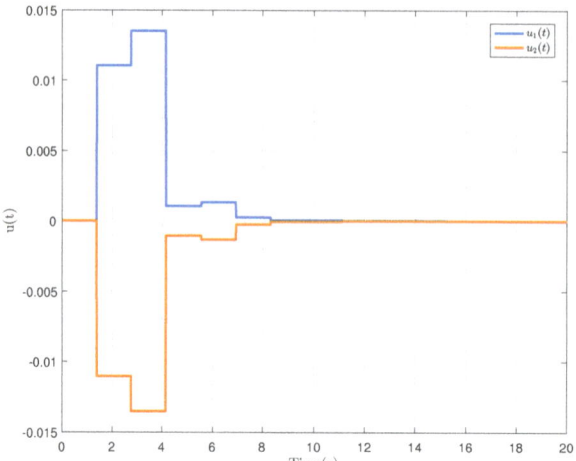

Figure 2. Control input $u(t)$.

Example 2. *Consider the neutral-type neural networks (10) with the following parameters [19]:*

$$A = \text{diag}\{3,3\},\ K^- = \text{diag}\{0,0\},\ K^+ = \text{diag}\{1,1\},$$
$$W_0 = \begin{bmatrix} 1 & 0.1 \\ 0.1 & 1 \end{bmatrix},\ W_1 = \begin{bmatrix} 0.4 & 0.1 \\ 0.04 & 0.1 \end{bmatrix},\ W_2 = \begin{bmatrix} 0.01 & 0 \\ 0 & 0.01 \end{bmatrix}.$$

In this example, the time-varying parameter uncertainties $\Delta A(t)$, $\Delta W_0(t)$, $\Delta W_1(t)$, $\Delta W_2(t)$ are defined as $[\Delta A(t)\ \Delta W_0(t)\ \Delta W_1(t)\ \Delta W_2(t)] = F\Sigma(t)[E_1\ E_2\ E_3\ E_4]$, where $E_1 = [0.1, 0.02]$, $E_2 = [-0.07, 0.1]$, $E_3 = [0.02, -0.02]$, $E_4 = [0.01, 0.02]$, $F = \text{diag}\{1,1\}$, $\Sigma(t) = \sin t$. We choose $\tau_1 = 0.5$, $\mu_1 = 0.9$, $\mu_2 = 0.5$, $\tau_2 = 2$.

The maximum sampling period d_M for different τ_c by Theorem 2 is listed in Table 2. By solving the LMIs (19)–(21), Theorem 2 provides a larger allowable upper bound of delays than those in Corollary 2. That is to say, the LKF containing a mixed delay part can lead to a less conservative result effectively.

Based on Theorem 2, when $\tau_c = 0.01$, the maximum sampling period $d_M = 26.2134$, the desired controller gain matrix can be calculated as

$$K = \begin{bmatrix} -0.4317 & 0.0106 \\ 0.0168 & -0.4401 \end{bmatrix}.$$

Table 2. The maximum sampling period d_M with parameter uncertainties (Example 2).

τ_c	0.01	0.03	0.05	0.07	0.09
Corollary 2	25.9631	24.1682	22.6097	21.2437	20.0398
Theorem 2	26.2134	25.2686	24.3970	23.4741	22.5952
Improvement	0.964%	4.5530%	7.9050%	10.4991%	12.7516%

Furthermore, if we choose the neural activation functions as $f_i(x_i) = \tanh(x_i)$ ($i = 1, 2$), $\tau_1(t) = 0.5\sin^2(0.9t)$, and $\tau_2(t) = 2\cos^2(0.5t)$, the initial condition $e(0) = \text{col}\{1, -1\}$. The response curves of the error system (6) with $u(t)$ are given in Figure 3, and the control input $u(t)$ is shown in Figure 4. Figure 3 shows that the NTNNs with parameter uncertainties are stable at their equilibrium points, which verifies that the slave system (5) is synchronous with the master system (1).

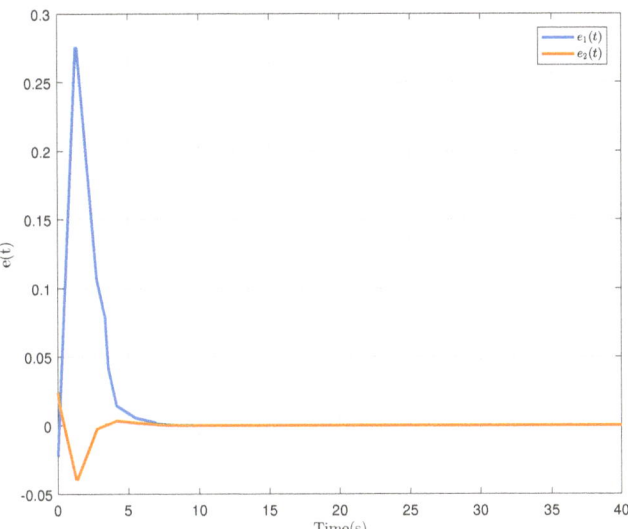

Figure 3. State responses of the error system.

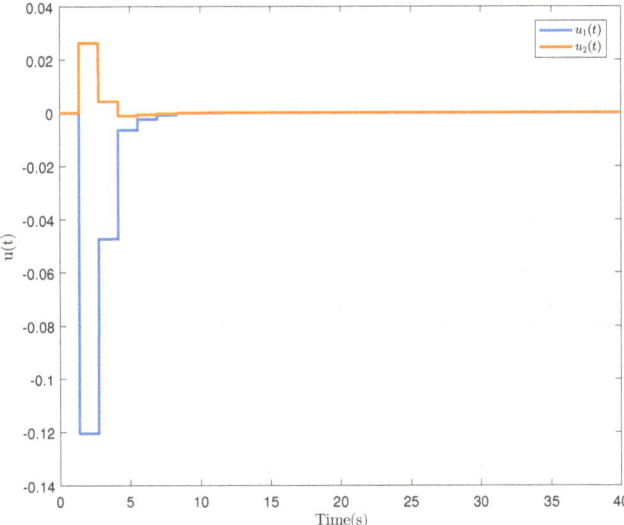

Figure 4. Control input $u(t)$.

Example 3. *Consider the neutral-type neural networks (10) with the following parameters [17]:*

$$A = \begin{bmatrix} 1 & 0 \\ 0 & 0.5 \end{bmatrix}, W_0 = \begin{bmatrix} 1.8 & -0.15 \\ -5.2 & 1.5 \end{bmatrix}, W_1 = \begin{bmatrix} 1.7 & -0.12 \\ -0.26 & -2.5 \end{bmatrix}, W_2 = \begin{bmatrix} 0 & 0 \\ 0 & 0 \end{bmatrix}.$$

The neural activation functions are taken as the form $f_i(x_i) = \tanh(x_i)$ $(i = 1, 2)$, which satisfies the assumption (7) with $K^- = \text{diag}(0,0)$, $K^+ = \text{diag}(1,1)$. The time-delay $\tau(t) = e^t/(e^t + 1)$. The initial condition of the master system and the slave system is taken as $x(0) = \text{col}(-0.2, -0.3)$, $y(0) = \text{col}(0.3, 0.6)$ and $u(t) = 0$. For given scalars $\tau_2 = 1$, $\mu_2 = 0.25$, via solving the LMIs in Corollary 3, the derived maximum sampling intervals of the NTNNs for different communication delays τ_c are obtained when $\epsilon_1 = \epsilon_2 = 0.24$, and $d_k = d_M$. The results are shown in Table 3.

For various τ_c, the maximum sampling period d_M by Corollary 3 in this paper and the related methods in [26,27] are listed in Table 3. It can be found that Corollary 3 provides a larger maximum sampling period compared with the results of the literature in Table 3. It shows that the proposed criteria are less conservative than the ones in the literature. That is to say, the LKF containing a mixed delay part can lead to a less conservative result.

Table 3. The maximum sampling period d_M for different τ_c (Example 2).

τ_c	0.01	0.03	0.05	0.07	0.09
[26]	0.23	0.19	0.16	0.13	0.10
[27]	0.24	0.20	0.17	0.13	0.10
Corollary 3	0.26	0.21	0.17	0.14	0.10

Based on Corollary 3, when $\tau_c = 0.01$, the maximum sampling period $d_M = 0.2621$, the desired controller gain matrix can be calculated as

$$K = \begin{bmatrix} -4.2096 & 0.0865 \\ 0.8097 & -4.5161 \end{bmatrix}.$$

Based on the above sample-data controller, the state response curves of the error system (6) and the control input (9) are shown in Figures 5 and 6, respectively. Figure 5 shows that the NNs are stable at their equilibrium points, which verifies that the error system is asymptotic stable. The slave system (5) is synchronous with the master system (1).

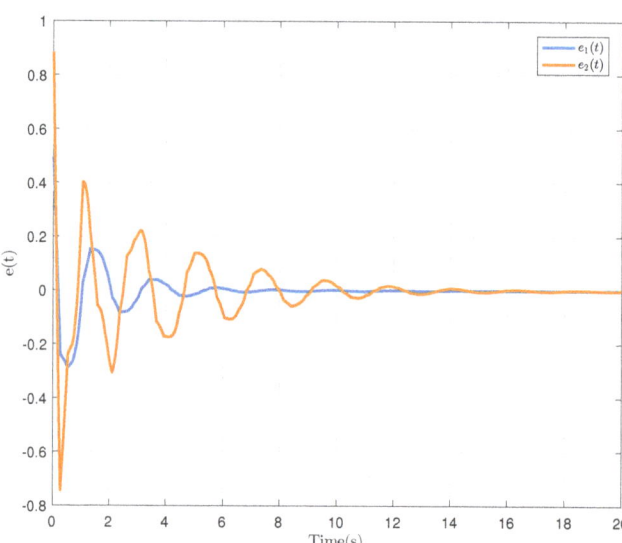

Figure 5. State responses of the error system.

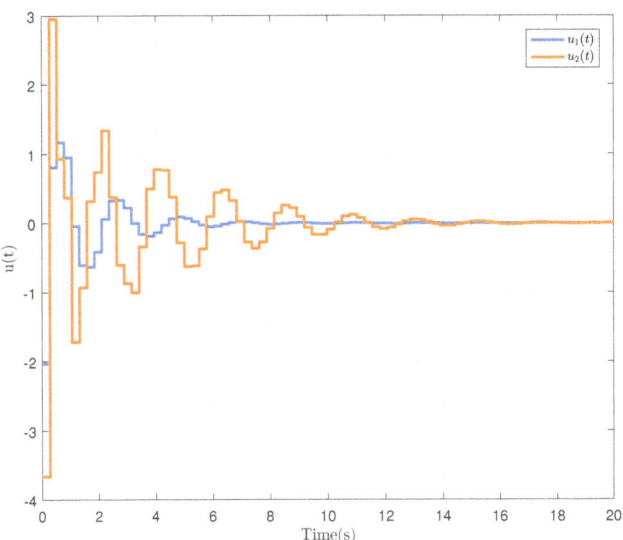

Figure 6. Control input $u(t)$.

5. Discussion

A function that considers the information of the intervals $[t_k, t]$, $[t_k - \tau, t - \tau]$, $[t, t_{k+1}]$ and $[t - \tau, t_{k+1} - \tau]$ is known as a two-sided looped functional. As more state information is taken into account, a less conservative stability criterion will be developed. The sample-data control approach can increase the control effectiveness while reducing network congestion. However, in practical implementations, communication delays are unavoidable when signals are sent from the sampler to the controller. Our limited knowledge indicates that there is no research that investigate NTNNs synchronization control using a sample-data controller considering communication delay. These factors prompt our investigation on NTNNs synchronization control.

In addition, systems with a single delay were the main subject of the research. Any form of time delay might be detrimental to the practical synchronization control of NTNNs. An important area for research is how the interaction between the delays affect each other. The connection between transmission delay and communication delay as well as the interconnectedness between neutral delay and transmission delay are all taken into account in this work. Three examples show that a mixed-delay-dependent LKF may lead to a less conservative criterion. Therefore, future research will concentrate on gathering more information of the delays and exploring ways to relax the restrictions placed on LKF.

In this article, the integral terms of the derivative of the LKF are estimated by using a free-matrix-based integral inequality. This approach produces a large number of free matrices, which makes the calculations more complex. Therefore, future study will concentrate on finding ways to improve the result by decreasing the LKF limitations and achieving a less conservative criterion.

6. Conclusions

In this study, we present a sampled-data synchronization scheme for uncertain NTNNs. A new LKF with a mixed-delay-dependent augmented part and a two-sided looped part is proposed. Benefiting from the LKF, two synchronization criteria are derived to guarantee the stability of the error systems, thereby allowing the slave system to synchronize with the master systems. Based on the criterion, a corresponding sampled-data controller scheme with a communication delay is designed. Finally, the validity of the proposed criteria is demonstrated through three numerical examples.

Author Contributions: Conceptualization, S.W. and K.S.; formal analysis, S.W. and K.S.; funding acquisition, K.S.; methodology, S.W. and K.S.; project administration, K.S.; software, S.W.; supervision, K.S.; validation, S.W.; visualization, S.W.; writing—original draft, S.W.; writing—review and editing, K.S. All authors have read and agreed to the published version of the manuscript.

Funding: This work was supported by the Opening Fund of Geomathematics Key Laboratory of Sichuan Province (scsxdz2018zd04 and scsxdz2020zd01) and Sichuan Science and Technology Program under Grant Nos. 21YYJC0469 and 23ZDYF0645.

Institutional Review Board Statement: Not applicable.

Informed Consent Statement: Not applicable.

Data Availability Statement: Not applicable.

Conflicts of Interest: The authors declare no conflict of interest.

Appendix A

Proof. We choose the following LKFs candidates as

$$V(t) = \sum_{i=1}^{5} \mathcal{V}_i(t) + \sum_{j=1}^{4} \mathcal{W}_j(t), \tag{A1}$$

where

$$\mathcal{V}_1(t) = \varpi_1^T(t) P_1 \varpi_1(t) + 2\sum_{i=1}^{n} \lambda_{1i} \int_0^{e_i(t)} (f_i(s) - k_i^- s) ds + 2\sum_{i=1}^{n} \lambda_{2i} \int_0^{e_i(t)} (k_i^+ s - f_i(s)) ds,$$

$$\mathcal{V}_2(t) = \int_{t-\tau_2(t)}^{t} \varpi_2^T(s) P_2 \varpi_2(s) ds + \int_{t-\tau_1(t)}^{t} \dot{e}^T(s) P_3 \dot{e}(s) ds,$$

$$\mathcal{V}_3(t) = \int_{t-\tau_1}^{t} \varpi_3^T(s) X_1 \varpi_3(s) ds + \int_{t-\tau_2}^{t} \varpi_3^T(s) X_2 \varpi_3(s) + \int_{t-\tau_c}^{t} \varpi_3^T(s) X_3 \varpi_3(s) ds,$$

$$\mathcal{V}_4(t) = \int_{t-\tau_2}^{t-\tau_1} e^T(s) X_4 e(s) ds + \int_{-\tau_2}^{-\tau_1} \int_{t+\theta}^{t} \dot{e}^T(u) X_5 \dot{e}(u) du d\theta + \int_{t-\tau_c}^{t-\tau_2} e^T(s) X_6 e(s) ds,$$

$$\mathcal{V}_5(t) = \int_{-\tau_1}^{0} \int_{t+\theta}^{t} \dot{e}^T(u) S_1 \dot{e}(u) du d\theta + \int_{-\tau_2}^{0} \int_{t+\theta}^{t} \dot{e}^T(u) S_2 \dot{e}(u) du d\theta + \int_{-\tau_c}^{0} \int_{t+\theta}^{t} \dot{e}^T(u) Z_1 \dot{e}(u) du d\theta$$
$$+ \int_{-\eta_M}^{-\tau_c} \int_{t+\theta}^{t} \dot{e}^T(u) Z_2 \dot{e}(u) du d\theta + \int_{-\eta_M}^{0} \int_{t+\theta}^{t} \dot{e}^T(u) Z_3 \dot{e}(u) du d\theta,$$

$$\mathcal{W}_1(t) = 2\varpi_4^T(t)(Q_1 \varpi_5(t) + Q_2 \varpi_6(t)),$$
$$\mathcal{W}_2(t) = (t_{k+1} - t)(t - t_k) \varpi_6^T(t) Q_3 \varpi_6(t),$$
$$\mathcal{W}_3(t) = (t_{k+1} - t)(t - t_k) \varpi_7(t)^T Q_4 \varpi_7(t),$$
$$\mathcal{W}_4(t) = (t_{k+1} - t) \int_{t_k}^{t} \dot{e}^T(s) R_1 \dot{e}(s) ds - (t - t_k) \int_{t}^{t_{k+1}} \dot{e}^T(s) R_2 \dot{e}(s) ds$$
$$+ (t_{k+1} - t) \int_{t_k - \tau_c}^{t - \tau_c} \dot{e}^T(s) R_3 \dot{e}(s) ds - (t - t_k) \int_{t-\tau_c}^{t_{k+1} - \tau_c} \dot{e}^T(s) R_4 \dot{e}(s) ds$$

Taking the derivatives of $V(t)$ along the trajectory of the error system (6) yields

$$\dot{\mathcal{V}}_1(t) = \xi^T(t)\text{Sym}\Big\{\Omega_1^T P_1\Omega_2 + (r_{11} - K_p r_1)^T \Delta_1 r_8 + (K_m r_1 - r_{11})^T \Delta_2 r_8\Big\}\xi(t),$$

$$\dot{\mathcal{V}}_2(t) \leq \xi^T(t)\Big(\Omega_3^T P_2\Omega_3 - (1-u_2)\Omega_4^T P_2\Omega_4 + r_8^T P_3 r_8 - (1-u_1)r_{26}^T P_3 r_{26}\Big)\xi(t),$$

$$\dot{\mathcal{V}}_3(t) \leq \xi^T(t)\Big(\Omega_5^T(X_1 + X_2 + X_3)\Omega_5 - \Omega_6^T X_1\Omega_6 - \Omega_7^T X_2\Omega_7 - \Omega_8^T X_3\Omega_8\Big)\xi(t),$$

$$\dot{\mathcal{V}}_4(t) = \xi^T(t)\Big\{r_{24}^T X_4 r_{24} - r_{11}^T X_4 r_{11} + \tau_{21} r_8^T X_5 r_8 + r_{11}^T X_6 r_{11} - r_2^T X_6 r_2\Big\}\xi(t) + J_0,$$

$$\dot{\mathcal{V}}_5(t) = \xi^T(t)\Big(\tau_1 r_8^T S_1 r_8 + \tau_2 r_8^T S_2 r_8 + \tau_c r_8^T Z_1 r_8 + d_M r_8^T Z_2 r_8 + \eta_M r_8^T Z_3 r_8\Big)\xi(t)$$
$$+ J_1 + J_2 + J_3 + J_4 + J_5,$$

$$\dot{\mathcal{W}}_1(t) = \xi^T(t)\text{Sym}\Big\{\Omega_9^T(Q_1\Omega_{10} + Q_2\Omega_{11}) + (t_{k+1}-t)(\Omega_{12}^T(Q_1\Omega_{10} + Q_2\Omega_{11}) + \Omega_{13}^T Q_1\Omega_{14})$$
$$+ (t-t_k)(\Omega_{15}^T(Q_1\Omega_{10} + Q_2\Omega_{11}) + \Omega_{16}^T Q_1\Omega_{14})\Big\}\xi(t),$$

$$\dot{\mathcal{W}}_2(t) = \xi^T(t)((t_{k+1}-t)-(t-t_k))\Omega_{11}^T Q_3\Omega_{11}\xi(t),$$

$$\dot{\mathcal{W}}_3(t) = \xi^T(t)\Big\{(t_{k+1}-t)\big[\text{Sym}\{\Omega_{17}^T Q_4\Omega_{18}\} + \Omega_{17}^T Q_4\Omega_{17}\big]$$
$$+ (t-t_k)\big[\text{Sym}\{\Omega_{17}^T Q_4\Omega_{19}\} - \Omega_{17}^T Q_4\Omega_{17}\big]\Big\}\xi(t),$$

$$\dot{\mathcal{W}}_4(t) = \xi^T(t)\Big\{(t_{k+1}-t)(r_8^T R_1 r_8 + r_9^T R_3 r_9) + (t-t_k)(r_8^T R_2 r_8 + r_9^T R_4 r_9)\Big\}\xi(t)$$
$$+ J_6 + J_7 + J_8 + J_9,$$

where

$$J_0 = -\int_{t-\tau_2}^{t-\tau_1}\dot{e}^T(s)X_5\dot{e}(s)ds,\ J_1 = -\int_{t-\tau_1}^{t}\dot{e}^T(s)S_1\dot{e}(s)ds,\ J_2 = -\int_{t-\tau_2}^{t}\dot{e}^T(s)S_2\dot{e}(s)ds,$$

$$J_3 = -\int_{t-\tau_c}^{t}\dot{e}^T(s)Z_1\dot{e}(s)ds,\ J_4 = -\int_{t-\eta_M}^{t-\tau_c}\dot{e}^T(s)Z_2\dot{e}(s)ds,\ J_5 = -\int_{t-\eta_M}^{t}\dot{e}^T(s)Z_3\dot{e}(s)ds,$$

$$J_6 = -\int_{t_k}^{t}\dot{e}^T(s)R_1\dot{e}(s)ds,\ J_7 = -\int_{t}^{t_{k+1}}\dot{e}^T(s)R_2\dot{e}(s)ds,\ J_8 = -\int_{t_k-\tau_c}^{t-\tau_c}\dot{e}^T(s)R_3\dot{e}(s)ds,$$

$$J_9 = -\int_{t-\tau_c}^{t_{k+1}-\tau_c}\dot{e}^T(s)R_4\dot{e}(s)ds.$$

From $\eta_M \geq \eta_k$ and $Z_2 + Z_3 > 0$, the integral quadratic terms can be rearranged as

$$\sum_{i=0}^{9}J_i \leq -\int_{t-\tau_1}^{t}\dot{e}^T(s)S_1\dot{e}(s)ds - \int_{t-\tau_2}^{t}\dot{e}^T(s)S_2\dot{e}(s)ds - \int_{t-\tau_c}^{t}\dot{e}^T(s)Z_1\dot{e}(s)ds - \int_{t}^{t_{k+1}}\dot{e}^T(s)R_2\dot{e}(s)ds$$
$$- \int_{t_k}^{t}\dot{e}^T(s)(R_1+Z_3)\dot{e}(s)ds - \int_{t_k-\tau_c}^{t-\tau_c}\dot{e}^T(s)(R_3+Z_2)\dot{e}(s)ds - \int_{t-\tau_c}^{t_{k+1}-\tau_c}\dot{e}^T(s)R_4\dot{e}(s)ds$$
$$- \int_{t_k-\tau_c}^{t_k}\dot{e}^T(s)Z_3\dot{e}(s)ds - \int_{t-\eta_k}^{t_k-\tau_c}\dot{e}^T(s)(Z_2+Z_3)\dot{e}(s)ds - \int_{t-\tau_2}^{t-\tau_1}\dot{e}^T(s)X_5\dot{e}(s)ds.$$

Using Lemma 1 to estimate J_i ($i = 0, 1, \cdots, 9$), we obtain

$$\sum_{i=0}^{9} J_i \leq \xi^{\mathrm{T}}(t)\Pi^{\mathrm{T}}\Big\{\tau_c \sum_{i=1}^{3}\frac{1}{2i-1}Y_i Z_1^{-1}Y_i^{\mathrm{T}}\Pi + \tau_c Y_{14} Z_3^{-1} Y_{14}^{\mathrm{T}}\Pi + \tau_2 \sum_{i=1}^{2}\frac{1}{2i-1}Y_{15+i}S_2^{-1}Y_{15+i}^{\mathrm{T}}\Pi$$

$$+ \tau_1 Y_{18} S_1^{-1} Y_{18}^{\mathrm{T}}\Pi + \tau_{21} Y_{19} X_5^{-1} Y_{19}^{\mathrm{T}}\Pi + (t - t_k)\Big[\sum_{i=1}^{3}\frac{1}{2i-1}Y_{6+i}(R_1 + Z_3)^{-1}Y_{6+i}^{\mathrm{T}}\Pi$$

$$+ \sum_{i=1}^{2}\frac{1}{2i-1}Y_{9+i}(R_3 + Z_2)^{-1}Y_{9+i}^{\mathrm{T}}\Pi\Big] + (t_{k+1} - t)\Big[\sum_{i=1}^{3}\frac{1}{2i-1}Y_{3+i}R_2^{-1}Y_{3+i}^{\mathrm{T}}\Pi$$

$$+ \sum_{i=1}^{2}\frac{1}{2i-1}Y_{11+i}R_4^{-1}Y_{11+i}^{\mathrm{T}}\Pi + Y_{15}(Z_2 + Z_3)^{-1}Y_{15}^{\mathrm{T}}\Pi\Big] + \mathrm{Sym}\Big\{\sum_{i=1}^{19}Y_i \beth_i\Big\}\Big\}\xi(t).$$

For given scalars ϵ_1, ϵ_2 and any matrix G with appropriate dimensions, the following equations hold:

$$\begin{aligned}
0 =& 2\big[e^{\mathrm{T}}(t) + \epsilon_1 e^{\mathrm{T}}(t_k - \tau_c) + \epsilon_2 \dot{e}^{\mathrm{T}}(t)\big] G\big[-\dot{e}(t) - Ae(t) + W_0 f(e(t)) \\
& + W_1 f(e(t - \tau_2(t))) + W_2 \dot{e}(t - \tau_1(t)) + Ke(t_k - \tau_c)\big] \\
=& 2\xi^{\mathrm{T}}(t)\Omega_{20}^{\mathrm{T}}\Omega_{21}\xi(t),
\end{aligned} \tag{A2}$$

where $L = GK$, and the controller gain matrix in (6) can be calculated by

$$K = G^{-1}L. \tag{A3}$$

To make use of the information of the activation function (7), we can obtain

$$0 \leq \sum_{i=1}^{n}\delta_{1i}\big[f_i(e_i(t)) - k_i^- e_i(t)\big]\big[k_i^+ e_i(t) - f_i(e_i(t))\big] = 2\xi^{\mathrm{T}}(t)\Omega_{22}^{\mathrm{T}}U\Omega_{23}\xi(t), \tag{A4}$$

$$\begin{aligned}
0 \leq & \sum_{i=1}^{n}\delta_{2i}\big[f_i(e_i(t - \tau_2(t))) - k_i^- e_i(t - \tau_2(t))\big]\big[k_i^+ e_i(t - \tau_2(t)) - f_i(e_i(t - \tau_2(t)))\big] \\
= & 2\xi^{\mathrm{T}}(t)\Omega_{24}^{\mathrm{T}}V\Omega_{25}\xi(t),
\end{aligned} \tag{A5}$$

where $U = \mathrm{diag}\{\delta_{11}, \delta_{12}, \cdots, \delta_{1n}\} \geq 0$, $V = \mathrm{diag}\{\delta_{21}, \delta_{22}, \cdots, \delta_{2n}\} \geq 0$.

Adding the right-hand sides of (A2)–(A5) into $\dot{V}(t)$, we can obtain

$$\dot{V}(t) \leq \xi^{\mathrm{T}}(t)\Big[\frac{t_{k+1} - t}{d_k}\Xi_1 + \frac{t - t_k}{d_k}\Xi_2\Big]\xi(t), \tag{A6}$$

where $\Xi_1 = \Gamma_1 + d_k\Gamma_2 + \Gamma_4$, $\Xi_2 = \Gamma_1 + d_k\Gamma_3 + \Gamma_5$, and Γ_j ($j = 1, 2, 3, 4, 5$) are defined in Theorem 1.

Note that the LMIs (14) and (15) are equal to $\Xi_1 < 0$ and $\Xi_2 < 0$ based on the Schur complement, respectively. That is to say, if the inequality conditions of the LMIs (14) and (15) hold, it can guarantee $\dot{V}(t) \leq -\sigma\|e(t)\|^2$ for a sufficient small scalar $\sigma > 0$. Then, the synchronization error system (6) is asymptotically stable, and the slave system is synchronized with the master system. This completes the proof. □

Appendix B

Proof. We choose the following LKFs candidates as

$$V(t) = \widehat{\mathcal{V}}_1(t) + \mathcal{V}_2(t) + \mathcal{V}_3(t) + \mathcal{V}_5(t) + \sum_{j=1}^{4}\mathcal{W}_j(t), \tag{A7}$$

where

$$\widehat{\mathcal{V}}_1(t) = e^\mathrm{T}(t)P_1 e(t) + 2\sum_{i=1}^{n} \lambda_{1i} \int_0^{e_i(t)} (f_i(s) - k_i^- s)ds + 2\sum_{i=1}^{n} \lambda_{2i} \int_0^{e_i(t)} (k_i^+ s - f_i(s))ds,$$

as a result,

$$\dot{\mathcal{V}}_1(t) = \xi^\mathrm{T}(t)\mathrm{Sym}\left\{r_1^\mathrm{T} P_1 r_8 + (r_{11} - K^- r_1)^\mathrm{T} \Delta_1 r_8 + (K^+ r_1 - r_{11})^\mathrm{T} \Delta_2 r_8\right\}\xi(t).$$

The rest of the process is similar to that in the proof of Theorem 1, and hence, we omit it here. □

Appendix C

Proof. Based on the LMI results in Theorem 1, replacing A, W_0, W_1 and W_2 with $A + F\Sigma(t)E_1$, $W_0 + F\Sigma(t)E_2$, $W_1 + F\Sigma(t)E_3$ and $W_2 + F\Sigma(t)E_4$, respectively, we can check that the derived inequalities are equivalent to the following terms:

$$\widehat{\Xi}_1 = \Xi_1 + \Psi_1 \Sigma(t)\Psi_2^\mathrm{T} + (\Psi_1 \Sigma(t)\Psi_2^\mathrm{T})^\mathrm{T} < 0, \tag{A8}$$

$$\widehat{\Xi}_2 = \Xi_2 + \Psi_1 \Sigma(t)\Psi_2^\mathrm{T} + (\Psi_1 \Sigma(t)\Psi_2^\mathrm{T})^\mathrm{T} < 0. \tag{A9}$$

Now by utilizing Lemma 2, there exist two positive scalars $\sigma_i > 0$ $(i = 1, 2)$, such that

$$\widehat{\Xi}_1 = \Xi_1 + \begin{bmatrix} \sigma_1^{-1}\Psi_1^\mathrm{T} \\ \sigma_1 \Psi_2^\mathrm{T} \end{bmatrix}^\mathrm{T} \begin{bmatrix} I & -J \\ -J^\mathrm{T} & I \end{bmatrix}^{-1} \begin{bmatrix} \sigma_1^{-1}\Psi_1^\mathrm{T} \\ \sigma_1 \Psi_2^\mathrm{T} \end{bmatrix} < 0, \tag{A10}$$

$$\widehat{\Xi}_2 = \Xi_2 + \begin{bmatrix} \sigma_2^{-1}\Psi_1^\mathrm{T} \\ \sigma_2 \Psi_2^\mathrm{T} \end{bmatrix}^\mathrm{T} \begin{bmatrix} I & -J \\ -J^\mathrm{T} & I \end{bmatrix}^{-1} \begin{bmatrix} \sigma_2^{-1}\Psi_1^\mathrm{T} \\ \sigma_2 \Psi_2^\mathrm{T} \end{bmatrix} < 0. \tag{A11}$$

Then by setting $\delta_n = \sigma_i^{-2}$ $(n = 1, 2)$, it is easy to verify that $\widehat{\Xi}_1 < 0$ and $\widehat{\Xi}_2 < 0$ are equal to the LMIs (20) and (21) based on the Schur complement. The rest of the process is similar to that in the proof of Theorem 1; hence, we omit it here. □

Appendix D

Proof. We choose the following LKFs candidates as

$$V(t) = \widehat{\mathcal{V}}_1(t) + \mathcal{V}_2(t) + \mathcal{V}_3(t) + \mathcal{V}_5(t) + \sum_{j=1}^{4} \mathcal{W}_j(t), \tag{A12}$$

where

$$\widehat{\mathcal{V}}_1(t) = e^\mathrm{T}(t)P_1 e(t) + 2\sum_{i=1}^{n} \lambda_{1i} \int_0^{e_i(t)} (f_i(s) - k_i^- s)ds + 2\sum_{i=1}^{n} \lambda_{2i} \int_0^{e_i(t)} (k_i^+ s - f_i(s))ds,$$

as a result,

$$\dot{\mathcal{V}}_1(t) = \xi^\mathrm{T}(t)\mathrm{Sym}\left\{r_1^\mathrm{T} P_1 r_8 + (r_{11} - K^- r_1)^\mathrm{T} \Delta_1 r_8 + (K^+ r_1 - r_{11})^\mathrm{T} \Delta_2 r_8\right\}\xi(t).$$

The rest of the process is similar to that in the proof of Theorem 1; hence, we omit it here. □

Appendix E

Proof. We choose the following LKFs candidates as

$$V(t) = \sum_{i=1}^{5} \widetilde{\mathcal{V}}_i(t) + \sum_{j=1}^{4} \mathcal{W}_j(t), \tag{A13}$$

where

$$\widetilde{\mathcal{V}}_1(t) = \widetilde{\omega}_1^T(t) P_1 \widetilde{\omega}_1(t) + 2 \sum_{i=1}^{n} \lambda_{1i} \int_0^{e_i(t)} \left(f_i(s) - k_i^- s \right) ds + 2 \sum_{i=1}^{n} \lambda_{2i} \int_0^{e_i(t)} \left(k_i^+ s - f_i(s) \right) ds,$$

$$\widetilde{\mathcal{V}}_2(t) = \int_{t-\tau_2(t)}^{t} \omega_2^T(s) P_2 \omega_2(s) ds,$$

$$\widetilde{\mathcal{V}}_3(t) = \int_{t-\tau_2}^{t} \omega_3^T(s) X_2 \omega_3(s) + \int_{t-\tau_c}^{t} \omega_3^T(s) X_3 \omega_3(s) ds,$$

$$\widetilde{\mathcal{V}}_4(t) = \int_{t-\tau_c}^{t-\tau_2} e^T(s) X_6 e(s) ds,$$

$$\widetilde{\mathcal{V}}_5(t) = \int_{-\tau_2}^{0} \int_{t+\theta}^{t} \dot{e}^T(u) S_1 \dot{e}(u) du d\theta + \int_{-\tau_c}^{0} \int_{t+\theta}^{t} \dot{e}^T(u) Z_1 \dot{e}(u) du d\theta$$
$$+ \int_{-\eta_M}^{-\tau_c} \int_{t+\theta}^{t} \dot{e}^T(u) Z_2 \dot{e}(u) du d\theta + \int_{-\eta_M}^{0} \int_{t+\theta}^{t} \dot{e}^T(u) Z_3 \dot{e}(u) du d\theta.$$

$$\widetilde{\omega}_1(t) = \mathrm{col}\left\{ e(t), e(t-\tau_c), e(t-\tau_2), \tau_c v_5(t) - \tau_2 v_{13}(t), \tau_c^2 v_6(t) \right\},$$

$$\widetilde{\xi}(t) = \mathrm{col}\{ e(t), e(t-\tau_c), e(t_k), e(t_k-\tau_c), e(t_{k+1}), e(t_{k+1}-\tau_c), e(t-\eta_k), \dot{e}(t), \dot{e}(t-\tau_c),$$
$$\dot{e}(t-\tau_2), e(t-\tau_2), e(t-\tau_2(t)), f(e(t)), f(e(t-\tau_2(t))), v_5(t), v_6(t), v_7(t),$$
$$v_8(t), v_9(t), v_{10}(t), v_{11}(t), v_{12}(t), v_{13}(t) \}.$$

The rest of the proof process is similar to that in the proof of Theorem 2; hence, we omit it here. □

References

1. Egmont-Petersen, M.; de Ridder, D.; Handels, H. Image processing with neural networks—A review. *Pattern Recognit.* **2002**, *35*, 2279–2301. [CrossRef]
2. Kothari, S.; Oh, H. Neural Networks for Pattern Recognition. *Adv. Comput.* **1993**, *37*, 119–166.
3. Mestari, M.; Benzirar, M.; Saber, N.; Khouil, M. Solving Nonlinear Equality Constrained Multiobjective Optimization Problems Using Neural Networks. *IEEE Trans. Neural Netw. Learn Syst.* **2015**, *26*, 2500–2520. [CrossRef]
4. Chen, Y.H.; Fang, S.C. Neurocomputing with time delay analysis for solving convex quadratic programming problems. *IEEE Trans. Neural Netw. Learn Syst.* **2000**, *11*, 230–240. [CrossRef]
5. Zhang, J. Globally exponential stability of neural networks with variable delays. *IEEE Trans. Circuits Syst. I Fundam. Theory Appl.* **2003**, *50*, 288–290. [CrossRef]
6. Lv, Y.; Lv, W.; Sun, J. Convergence dynamics of stochastic reaction-diffusion recurrent neural networks with continuously distributed delays. *Nonlinear Anal. Real World Appl.* **2008**, *9*, 1590–1606. [CrossRef]
7. Shi, K.; Zhu, H.; Zhong, S.; Zeng, Y.; Zhang, Y. New stability analysis for neutral type neural networks with discrete and distributed delays using a multiple integral approach. *J. Frankl. Inst.* **2015**, *352*, 155–176. [CrossRef]
8. Kolmanovskii, V.B.; Nosov, V.R. *Stability of Functional Differential Equations*; Academic: London, UK, 1986.
9. Kuang, Y. *Delay Differential Equations with Applications in Population Dynamics*; Academic: Boston, MA, USA, 1993.
10. Cai, X.; Shi, K.; She, K.; Zhong, S.; Tang, Y. Quantized Sampled-Data Control Tactic for T-S Fuzzy NCS Under Stochastic Cyber-Attacks and Its Application to Truck-Trailer System. *IEEE Trans. Veh. Technol.* **2022**, *71*, 7023–7032. [CrossRef]
11. Cai, X.; Shi, K.; She, K.; Zhong, S.; Soh, Y.; Yu, Y. Performance Error Estimation and Elastic Integral Event Triggering Mechanism Design for T-S Fuzzy Networked Control System Under DoS Attacks. *IEEE Trans. Fuzzy Syst.* **2022**, 1–12. [CrossRef]
12. Fridman, E.; Seuret, A.; Richard, J.P. Robust sampled-data stabilization of linear systems: An input delay approach. *Automatica* **2004**, *40*, 1441–1446. [CrossRef]
13. Fujioka, H. A Discrete-Time Approach to Stability Analysis of Systems With Aperiodic Sample-and-Hold Devices. *IEEE Trans. Autom. Control* **2009**, *54*, 2440–2445. [CrossRef]
14. Naghshtabrizi, P.; Hespanha, J.P.; Teel, A.R. Exponential stability of impulsive systems with application to uncertain sampled-data systems. *Syst. Control Lett.* **2008**, *57*, 378–385. [CrossRef]

15. Zeng, H.B.; Teo, K.; He, Y. A new looped-functional for stability analysis of sampled-data systems. *Automatica* **2017**, *82*, 328–331. [CrossRef]
16. Zeng, H.B.; Zhai, Z.L.; He, Y.; Teo, K.L.; Wang, W. New insights on stability of sampled-data systems with time-delay. *Appl. Math. Comput.* **2020**, *374*, 125041. [CrossRef]
17. Zeng, H.B.; Zhai, Z.L.; Yan, H.; Wang, W. A New Looped Functional to Synchronize Neural Networks With Sampled-Data Control. *IEEE Trans. Neural Netw. Learn Syst.* **2022**, *33*, 406–415. [CrossRef]
18. Zhang, Y.; He, Y.; Long, F.; Zhang, C.K. Mixed-Delay-Based Augmented Functional for Sampled-Data Synchronization of Delayed Neural Networks With Communication Delay. *IEEE Trans. Neural Netw. Learn Syst.* **2022**, 1–10. [CrossRef] [PubMed]
19. Zhang, G.; Wang, T.; Li, T.; Fei, S. Multiple integral Lyapunov approach to mixed-delay-dependent stability of neutral neural networks. *Neurocomputing* **2018**, *275*, 1782–1792. [CrossRef]
20. Zhang, H.; Ma, Q.; Lu, J.; Chu, Y.; Li, Y. Synchronization control of neutral-type neural networks with sampled-data via adaptive event-triggered communication scheme. *J. Frankl. Inst.* **2021**, *358*, 1999–2014. [CrossRef]
21. Zeng, H.B.; He, Y.; Wu, M.; She, J. New results on stability analysis for systems with discrete distributed delay. *Automatica* **2015**, *60*, 189–192. [CrossRef]
22. Li, T.; Guo Song, A.; Min Fei, S. Robust stability of stochastic Cohen–Grossberg neural networks with mixed time-varying delays. *Neurocomputing* **2009**, *73*, 542–551. [CrossRef]
23. Seuret, A. A novel stability analysis of linear systems under asynchronous samplings. *Automatica* **2012**, *48*, 177–182. [CrossRef]
24. Liu, P.L. Further improvement on delay-dependent robust stability criteria for neutral-type recurrent neural networks with time-varying delays. *ISA Trans.* **2015**, *55*, 92–99. [CrossRef]
25. Yin, C.; Cheng, Y.; Huang, X.; ming Zhong, S.; Li, Y.; Shi, K. Delay-partitioning approach design for stochastic stability analysis of uncertain neutral-type neural networks with Markovian jumping parameters. *Neurocomputing* **2016**, *207*, 437–449. [CrossRef]
26. Wei, W.; Zeng, H.B.; Teo, K.L. Free-matrix-based time-dependent discontinuous Lyapunov functional for synchronization of delayed neural networks with sampled-data control. *Chin. Phys. B* **2017**, *26*, 127–134.
27. Xiao, S.P.; Lian, H.H.; Teo, K.L.; Zeng, H.B.; Zhang, X.H. A new Lyapunov functional approach to sampled-data synchronization control for delayed neural networks. *J. Frankl. Inst.* **2018**, *355*, 8857–8873. [CrossRef]

Disclaimer/Publisher's Note: The statements, opinions and data contained in all publications are solely those of the individual author(s) and contributor(s) and not of MDPI and/or the editor(s). MDPI and/or the editor(s) disclaim responsibility for any injury to people or property resulting from any ideas, methods, instructions or products referred to in the content.

Review

On Model Identification Based Optimal Control and It's Applications to Multi-Agent Learning and Control

Rui Luo [1,2], Zhinan Peng [1] and Jiangping Hu [1,2,*]

[1] School of Automation Engineering, University of Electronic Science and Technology of China, Chengdu 611731, China
[2] Yangtze Delta Region Institute (Huzhou), University of Electronic Science and Technology of China, Huzhou 313001, China
* Correspondence: hujp@uestc.edu.cn

Abstract: This paper reviews recent progress in model identification-based learning and optimal control and its applications to multi-agent systems (MASs). First, a class of learning-based optimal control method, namely adaptive dynamic programming (ADP), is introduced, and the existing results using ADP methods to solve optimal control problems are reviewed. Then, this paper investigates various kinds of model identification methods and analyzes the feasibility of combining the model identification method with the ADP method to solve optimal control of unknown systems. In addition, this paper expounds the current applications of model identification-based ADP methods in the fields of single-agent systems (SASs) and MASs. Finally, some conclusions and some future directions are presented.

Keywords: model identification; optimal control; multi-agent systems; adaptive dynamic programming; reinforcement learning

MSC: 49L20; 93B30; 68T07

1. Introduction

In recent years, with the rapid development of communication and network technology, MASs have been deeply applied in many fields, such as transportation, industrial production, etc. Facing increasingly large-scale and complex systems, the integration solutions to single-agent systems (SASs) are often limited by various resources and conditions. The MASs can effectively improve the robustness, reliability, and flexibility of large-scale complex systems [1,2].

MASs are composed of multiple agents with particular capabilities of sensing, computation, communication and control, and agents can coordinate to complete some common tasks through local interactions among agents [3,4]. Compared with traditional SASs, MASs involve relatively simple agents and thus reduce costs while improving robustness. Meanwhile, distributed coordination mechanisms exerted on multiple agents can improve the operation efficiency and reduce resource consumption. MASs have been widely used in real applications, such as resource detection, safety monitoring, natural disaster preparedness, etc. In some scenarios, agents can replace humans to guarantee the safety of military or agricultural production. In industrial applications, using multiple agents instead of single-agent can reduce production costs. Especially via coordination, such as mobile multi-unmanned aerial vehicles (Multi-UAV) systems, multi-robot systems, and multi-agent supporting systems, agents can complete more complex and challenging tasks while safety and reliability can be guaranteed [5–7].

The concerns in system control have gradually shifted from stabilization and stability to high steady-state accuracy, rapidity, strong robustness, and anti-interference performances. In many engineering application fields, scientists and engineers usually not

Citation: Luo, R.; Peng, Z.; Hu, J. On Model Identification Based Optimal Control and It's Applications to Multi-Agent Learning and Control. *Mathematics* **2023**, *11*, 906. https://doi.org/10.3390/math11040906

Academic Editors: Aydin Azizi and Irina Bashkirtseva

Received: 20 December 2022
Revised: 17 January 2023
Accepted: 9 February 2023
Published: 10 February 2023

Copyright: © 2023 by the authors. Licensee MDPI, Basel, Switzerland. This article is an open access article distributed under the terms and conditions of the Creative Commons Attribution (CC BY) license (https://creativecommons.org/licenses/by/4.0/).

only want to ensure the stability of controllable systems, but also aim to optimize certain performances (energy consumption and cost) at the same time. In this way, considering optimization is a key topic with greater practical implications for MASs. That is, a group of autonomous agents set out to complete some difficult tasks while also optimizing their performance indices.

Recently, optimization and optimal control employing a preset performance criterion have become increasingly hot research topics in the system and control fields. By interacting with an environment, an agent or decision maker develops a strategy to maximize a long-term reward using reinforcement learning (RL), a goal-oriented learning technology, which has achieved great success in the field of artificial intelligence (AI) [8–10]. In this context, the ADP method with strong self-learning ability has become a promising intelligent optimization technology. At present, in the field of multi-agent optimal control, most existing ADP methods are partially model-dependent or completely model-dependent. Unfortunately, model uncertainties exist in most of actual control systems, which leads to inaccurate modeling. In order to solve this problem, model identification-based ADP methods have been developed to solve MAS optimal control problems.

Motivated by the observations mentioned above, this paper aims at giving a brief survey for important developments in model identification based optimal control and its applications to multi-agent learning and control. In particular, we mainly focus on adaptive dynamic programming based optimal control method, model identification method, and the combination of ADP and model identification for dealing with the kinds of control problems of unknown system dynamics.

2. Adaptive Dynamic Programming-Based Optimal Control Method

Adaptive Dynamic Programming (ADP) is a learning-based intelligent control method with capabilities of adaption and optimization, which has great potential in solving optimal control problems. This section mainly introduces the origin of ADP, its basic structures and the development in the field of optimal control of dynamical systems, respectively.

2.1. Basic Structures of ADP

ADP, as a fusion technology of AI and control theory, is based on the traditional optimal control theory and RL principle. ADP can effectively solve a series of complex optimal control problems by learning through the continuous interactions between the agent and the environment. It is noted that there are some synonyms for ADP, such as Approximate Dynamic Programming [11], Neuro-Dynamic Programming [12], Adaptive Critic Design [13].

In the early stage, ADP was mainly used in the fields of computer science and operational research [14] and then gradually integrated with RL technology to solve optimal control problems later. Theoretically, ADP borrows from the basic principle of RL. That is, an agent interacts with the environment and constantly adjusts its strategy to achieve the optimal cumulative feedback (return) to solve an optimal decision problem. In 1977, Werbos proposed four basic ADP structures [11,15]: Heuristic Dynamic Programming (HDP), Dual Heuristic Programming (DHP), Action Dependent HDP (ADHDP), and Action Dependent DHP (ADDHP). Generally speaking, these ADP structures mainly include an actor-critic framework with the use of neural network approximation structure, which significantly improves the online learning and adaptive abilities of ADP. The basic structure of ADP is given in Figure 1. The ADP method not only avoids the "dimensional disaster" problem in dynamic programming (DP) methods, but also provides an effective way to solve the decision control problem of complex nonlinear systems, which makes it become an important research direction in the fields of artificial intelligence and control theory [9,16].

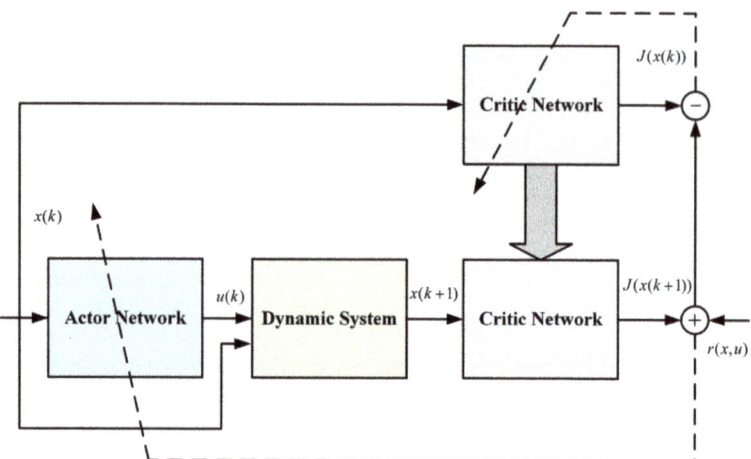

Figure 1. The basic structure of ADP.

2.2. Developments of ADP-Based Optimal Control

As an important optimal control method, ADP has been widely used in the field of optimal control. Particularly, many significant scientific research achievements have been made in early theoretical studies (including algorithm and convergence) [17]. In 2002, under the ADP framework proposed by Werbos, Murray et al. [8] firstly proposed an off-line iterative algorithm of the ADP strategy to solve an optimal control problem for nonlinear systems. At the same time, the authors offered rigorous proofs for the convergence of the iterative technique and the stability of the closed-loop system with an initial admissible control. This important theoretical result laid a solid theoretical foundation for the subsequent studies related to ADP.

The early groups engaged in ADP-related research mainly included Professor Frank L. Lewis' Team from the University of Texas at Arlington, Professor Zhongping Jiang's Team from New York University, Professor Huaguang Zhang's Team from Northeastern University, Professor Derong Liu's Team, etc. They have done much pioneering research in the field of optimal control based on ADP in the early stage. Frank L. Lewis [18] designed an ADP policy iterative algorithm to solve an input-constrained control problem for nonlinear systems. In [18], they introduced a special non-quadratic performance index function for the first time and proposed a Hamilton–Jacobi–Isaac (HJI) equation simultaneously. However, the limitation of this algorithm is that the controller design depended on the complete dynamics information of the system. To overcome this limitation, Vrabie [19] proposed a partially dynamics-dependent online optimal control algorithm based on a policy iteration, namely Integral Reinforcement Learning (IRL), for nonlinear systems with partially unknown dynamics. This algorithm parametrically represents the system's control strategy and performance using an actor-critic neural network framework, which makes the algorithm converge to the optimal control solution without requiring the system's internal dynamics, and guarantees the stability of the closed-loop system as well. After that, in order to solve a tracking control problem for partially unknown nonlinear systems, Hamidreza Modares [20] developed an IRL-based control method. The authors proposed an augmented system containing both error states and desired states, and used the augmented system to define a new non-quadratic discount performance index function.

In recent years, in order to improve the parameter updating efficiency of the actor-critic structure, Vamvoudakis [21] proposed an online policy iteration algorithm. In this algorithm, new parameter update laws were designed for the actor and critic networks, respectively, so that the two networks can realize online updates synchronously. In addition, Zhang [22] proposed a Greedy HDP iterative algorithm to solve a tracking control problem for discrete-time nonlinear systems by introducing a new tracking error performance index function. The above research results provided an essential theoretical basis for the developments of ADP methods.

In the following, we will describe the formulation of optimal control problems for two class of nonlinear dynamical systems, that is, discrete-time system and continuous-time system, respectively.

(1) For a continuous-time nonlinear system whose dynamics are modeled as follows

$$\dot{x}(t) = f(x) + g(x)u(t), \tag{1}$$

where $f(x)$ and $g(x)$ are the system matrices. $x(t) = [x_1(t), x_2(t), \cdots, x_n(t)] \in R^n$ denotes the system state, and $u(t) = [u_1(t), u_2(t), \cdots, u_m(t)] \in R^m$ is the control input. The objective is to find an optimal controller to stabilize the system (1) as well as minimize a pre-defined performance index function, which is given by

$$V(x(t), u(t)) = \int_t^\infty r(x(\tau), u(\tau)) d\tau, \tag{2}$$

where $r(x(t), u(t)) = x^\top(t) Q x(t) + u^\top(t) R u(t)$ represents the utility function, and Q and R are symmetric positive definite matrices with appropriate dimension. It is important to assume that the control input must be admissible such that a finite performance index function can be ensured.

The Hamiltonian of the system (1) is defined as

$$H(x(t), V_x(t), u(t)) = r(x(t), u(t)) + V_x^T(f(x(t)) + g(t)u(t)), \tag{3}$$

where $V_x = \partial V / \partial x$ is a partial derivative of x.

The optimal performance index function satisfies the continuous-time HJB (CT-HJB), i.e.,

$$0 = \min_{u(t)} \{ H(x(t), V_x^*(t), u(t)) \}. \tag{4}$$

By applying the stationarity condition, the ideal optimal control is then given by

$$u^*(t) = -\frac{1}{2} R^{-1} g(t)^\top \frac{\partial V^*(x(t))}{\partial x(t)}. \tag{5}$$

In order to obtain the optimal controller, it is necessary to solve the CT-HJB Equation (4). However, it is very difficult to solve (4) because it contains nonlinear and partial differential items, and requires knowledge of system dynamics model $g(x)$ (that is, it needs to be known in advance). Therefore, the CT-HJB is difficult to be solved directly.

(2) For a discrete-time nonlinear system, whose dynamics is given as follows

$$x(k+1) = f(x(k), u(k)), \tag{6}$$

where $x(k)$ is system state, $u(k)$ is control input, and $k = 0, 1, 2, \ldots$ denotes the sampling index. The goal is to design a controller $u(k)$ to minimize the following performance index function

$$J(x(k), u(k)) = \sum_{j=k}^\infty r(x(j), u(j)), \tag{7}$$

where $r(x(j), u(j))$ denotes the utility function. By using the performance index (7), the following Bellman Equation (nonlinear Lyapunov equation) can be obtained

$$J(x(k)) = r(x(k), u(k)) + J(x(k+1), u(k+1)). \tag{8}$$

According to the Bellman's principle of optimality, the optimal performance index function satisfies the following discrete-time Hamilton–Jacobi–Bellman (DT-HJB) equation

$$J^*(x(k)) = \min_{u(k)}\{r(x(k),u(k)) + J^*(x(k+1),u(k+1))\}. \tag{9}$$

Then, we can obtain the optimal controller as

$$u^*(k) = \arg\min_{u(k)}\{r(x(k),u(k)) + J^*(x(k+1),u(k+1))\}. \tag{10}$$

It is noted from above process that the optimal controller relies on the performance index at next time step $J^*(x(k+1), u(k+1))$. No matter how, the HJB equation is the key part for computing optimal control for both discrete-time and continuous-time nonlinear systems. Thus, it is important to obtain the approximate solution to the HJB equation. In the past decades, many researchers have made great efforts to propose all kinds of iterative algorithms to deal with this issue.

2.3. ADP-Based Approximate Solution to HJB Equations

In fact, most of the research results discussed above are mainly obtained for optimal control of nonlinear systems. Theoretically, the solutions to optimal control problems for nonlinear systems usually rely on Hamilton Jacobi Bellman (HJB) Equations [18]. However, it is very difficult to compute the analytical solutions to HJB equations in general, and thus numerous researches are essentially dedicated to approximate HJB equations. Till now, from the perspective of approximate solution methods, ADP-based algorithms can be divided into two categories: Value Iteration (VI) [23,24] and Policy Iteration (PI) [18,25].

Policy Iteration (PI):

Step 1: Initialization: Initial an admissible control $u^0(t)$;

Step 2: Policy evaluation: For a given iterative control strategy $u^k(t)$, the cost function can be updated according to the following rules:

$$0 = \min_{u(k)}\{H(x(t), V_x^k(t), u^k(t))\};$$

Step 3: Strategy improvement: the iterative control strategy is updated as follows:

$$u^{k+1}(t) = -\frac{\alpha}{2}R^{-1}g(t)^\top \frac{\partial V^k(x(t))}{\partial x(t)},$$

where k is the iterative index, the policy evaluation and policy improvement are updated alternately until the performance function and control policy converge to the optimal value. In addition, for the above PI iterative algorithm, the convergence of the algorithm has been proved.

Value Iteration (VI):

Step 1: Initialization: given an any control $u^0(t)$ and $V^0(t)$;

Step 2: Policy evaluation: the control policy can be updated according to the following rule:

$$u^k(t) = \min_{u(k)}\{H(x(t), V_x^k(t), u^k(t))\};$$

Step 3: Value improvement: the index function is updated according to the following Bellman equation:

$$V^{k=1}(x(t)) = r(x(t), u^k(t)) + V^{k+1}(x(t+1)),$$

where k is the iterative index and the policy evaluation and value improvement are updated alternately until the performance function and control converge to the optimal value.

The PI algorithm starts from an initial admissible control strategy and solves a series of HJB equations to obtain the optimal control strategy. In contrast, PI has a faster convergence rate than VI. The advantage of VI algorithm is that it does not require an initial admissible control. However, the iterative control during the iterative processing may not guarantee the stability of the closed-loop system. Al-Tamimi [23] presented a VI method (also known as greedy iterative ADP algorithm) for a discrete-time system and studied its convergence and stability under the approximation optimum controller. In [25], Liu et al. proposed a PI algorithm. Compared with other ADP algorithms, this paper presented a complete convergence analysis of the proposed PI algorithm for discrete-time nonlinear systems for the first time.

In recent decades, ADP methods have been widely concerned by academia and industry because of their theoretical research and practical application values. However, most ADP methods are partially model dependent or completely model dependent [26,27], so it is difficult to deal with the situation that accurate system information cannot be obtained. In most practical cases, the system model structure of the controlled object is unknown, or the model structure is known but the model parameters are unknown. Actually, the first consideration for the unknown model in the engineering field is to identify the model. Because accurate system models can reflect the system structure information, corresponding control strategies can then be better formulated.

From another perspective, in order to address the issue of unknown system dynamics, ADP can be divided into two main types: the indirect method and the direct method. In the direct method, the optimal control law is directly designed based on the measurable system data including the state information or input/output information without system identification process [28–30]. The indirect technique might be a significant new trend in the development of model-free optimal control, where the reconstructed system model is firstly established by approximate approaches such as neural networks (NNs) based identifiers. Then, an ADP algorithm is introduced to design an optimal controller for the approximate model. However, Modares et al. [31] have shown that the error of model identification directly affects the convergence effect of NN weights in the ADP algorithm. Therefore, the synthesis of model identification and ADP is an important trend and also a challenging issue, which has been widely attracted in this field very recently.

3. Model Identification

From the perspective of model structure, model identification methods can be divided into parametric model identification and non-parametric model identification, which will be introduced in the following, respectively.

3.1. Parametric Model Identification Method

A parametric model identification method needs to determine the model structure and order of the system in advance, and then estimates the unknown parameters of the system model. This method mainly includes the least squares method, the gradient method, the maximum likelihood estimation method, and expectation maximization method. The overview of the parametric model identification methods is illustrated in Figure 2.

Least squares methods have formed a complete theoretical system architecture and been widely applied in many model identification problems till now. Aiming at a parameter identification problem of linear-in-parameter systems with missing data, Ding et al. [32] developed an interval-varying auxiliary model based on the recursive least squares (AM-RLS) algorithm with the help of the auxiliary model identification idea. By introducing the forgetting factors, the parameter estimation accuracy and convergence rates can be improved. For the multivariable pseudo-linear autoregressive moving average (ARMA) systems, Ding et al. [33] proposed a decomposition-based least squares iterative identification algorithm. The key in the proposed algorithm is to transform the original system to a hierarchical identification model using a data filtering technique. The model was then divided into three subsystems, with each subsystem being identified separately. The proposed approach involves less processing effort than least squares-based iterative techniques. For the identification of bilinear forms, Camelia [34] proposed a recursive least-squares for bilinear forms (RLS-BF) algorithm. Two variations of the RLS-BF algorithm based on the dichotomous coordinate descent (DCD) approach were presented to lower the computing complexity of the process. Meanwhile, a regularized version of the RLS-BF method was created to increase the resilience of the RLS-BF technique in noisy situations.

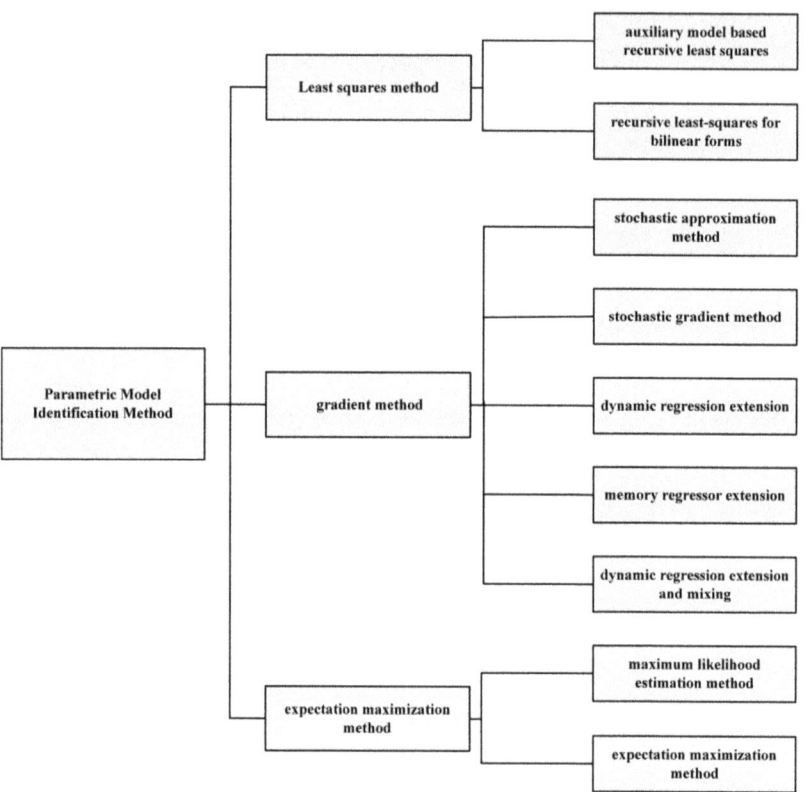

Figure 2. Overview of Parametric Model Identification Methods.

Essentially, a gradient method is an iterative algorithm. Compared with the recursive least squares, it has a slower convergence rate and lager error variance of parameter estimation. However, the computation of each step in the recursive process of gradient methods is smaller. According to the different search steps, the gradient method can be divided into the stochastic approximation method and the stochastic gradient method. There are two commonly used stochastic approximation methods, the Robbins-Monro algorithm, and the Kiefer-Wolfpwitz algorithm. However, because of the slow convergence rates of these two algorithms near the extreme points, they have not received widespread attention.

On the basis of stochastic approximation method, the stochastic gradient method adjusts the search step and accelerates the convergence rate. Recently, this method is widely used in the identification of various systems. For multivariate output-error systems, Wu [35] developed an auxiliary model based stochastic gradient (AM-SG) method and a coupled AM-SG algorithm, which ensured the parameter estimation error converged to zero under the persistence excitation (PE) condition. For the bilinear system with white noise, Ding [36] introduced a stochastic gradient (SG) technique and a gradient-based iterative approach for estimating system parameters with known input-output data using an auxiliary model. Experimental results show that the proposed gradient-based iterative algorithm has higher estimation accuracy than the auxiliary model based stochastic gradient.

In recent years, a new class of algorithms has been derived in the field of adaptive control based on the gradient method. An important concern in developing parameter identification and adaptive control schemes is transforming the original system model to a linear regressor Equation (LRE), in which the unknown parameters are linearly related to the measurable data. Then the unknown parameter estimation problem of the original system is transformed to solving the LRE, which derives a series of parameter identification methods based on the LRE of the original system.

The classical LRE can be expressed as

$$y = \phi^T \theta,$$

where $y \in R$ and $\phi(t) \in R^q$ are measurable signals. $\theta \in R^q$ is an unknown constant signal. Herein, $\phi(t)$ is also called the regression vector. Generally, we can use the least square method [37] or the gradient method [38] to solve the unknown parameters of the original system LRE. The gradient-descent based adaptive law is designed as

$$\dot{\hat{\theta}}(t) = \alpha \phi(t)[y(t) - \phi^T(t)\hat{\theta}(t)],$$

where $\hat{\theta}$ is the estimation of θ, $\alpha > 0$ presents adaptive learning gain. The idea of these two methods is to generate a linear time-varying (LTV) dynamic equation, known as the parameter error Equation (PEE) that can describe the estimation error, and then design the parameter estimator based on the PEE. However, the fundamental disadvantage of these techniques is that parameter estimation convergence is dependent on the PE condition of the regression vector.

Mathematically, the PE condition means that there exist some constants $t > 0$ and $\Delta > 0$ such that

$$\int_t^{t+t} \phi(s)\phi^T(s)ds \geq \Delta I$$

for any time t. That is, the input signal should excite all kinds of system modality so that the measurable signal contains enough information about the system, and then the convergence of parameter estimation can be guaranteed. In practice, input signals need to be designed to satisfy the PE condition. However, this is seldom practicable and difficult to verify online. Even if the input signal meets the PE criteria, the adaptive control's parameter convergence is largely reliant on the PE intensity, which leads to a slower convergence rate.

Moreover, the transient performance of these two methods is highly unpredictable and can only guarantee weak (vector norm) monotonicity of the estimation errors. Unfortunately, poor transient estimation error performance (such as significant overshoot and slow convergence in the first few seconds) may severely degrade the estimation response, resulting in identification and adaptive control instability. Therefore, engineering applications increasingly need fast, accurate, and robust parameter estimation method to maintain the security and reliability of control systems.

To improve the parameter convergence of the gradient method, most ideas are to convert the LRE of the original system into an alternative LRE to generate a new PEE with stronger convergence properties. By introducing multiple linear filter operators to apply on the LRE of the original system, Lion [39] piled up the filtered signals to generate an extended LRE. Then, a gradient estimator based on the extended LRE was proposed. The way of developing the extended LRE is called dynamic regression extension (DRE). Compared with the classical gradient estimation method, the parameter convergence rate of the DRE-based gradient estimator can be made arbitrarily fast by increasing the adaptive gain. Kreisselmeier [40] also proposed a filter method, namely memory regressor extension (MRE), to design new LREs. Unlike DRE, Kreisselmeier only applied one linear filter operator to $\phi(s)\phi^T(s)$. In fact, DRE can be transformed into MRE by rationally choosing the filter operator in the DRE algorithm. That is, MRE is a particular case of DRE. Except the advantages of the DRE-based gradient estimator, the MRE-based gradient estimator has better estimation performance than traditional gradient estimators for systems which do not satisfy the PE conditions.

To improve the transient performance of parameter identification, some researchers advocated combining the tracking error in direct adaptive control and the identification error in indirect control to form a new PEE. Then, parameter estimation algorithms based on tracking and identification errors were successively proposed [41–43]. Duarte et al. [41] used such an approach for model reference adaptive control (MRAC) of linear time-invariant (LTI) systems and gave the name composite adaptive control. In [42], position tracking control of robot manipulators was considered with composite adaptive control. Panteley [43] applied the composite adaptive control algorithm to the adaptive control of a class of nonlinear systems with measurable states, and relaxed a rather restrictive–detectability assumption in the stability proof. Later, Lavretsky [44] applied the work of Panteley to linear systems.

The above two types of parameter estimation frameworks lay the foundation for LRE parameter estimation. Five new adaptive control methods have gradually evolved in recent years based on these two types of original system LRE parameter estimation frameworks.

For the adaptive control of linear LTI systems, Chowdhary [45] used recorded and current data concurrently to estimate unknown parameters when designing composite adaptive law. This technique is named concurrent learning. Notably, the technique does not rely on the PE condition but guarantees the global exponential stability (GES) of the closed-loop system under an interval excitation (IE) condition. Compared with the traditional PE condition, the IE condition focuses on the evolution of integrals within an interval which is strictly weaker than the PE condition.

Cho [46] and Roy [47] designed a new composite estimator by constructing residual signals. Smilar with Chowdhary, the proposed algorithm used an "offline data selection method". That is, the incoming data are first accumulated to build the information matrix. A composite estimator is designed by the full rank information matrix after sufficient but not persistent excitation.

In [48–50], a variant algorithm of MRE is proposed, which selects the filter operator as a pure integral form. Actually, this improvement leads to a positive semi-definite open-loop integral in the parameter estimator, which affects the noise sensitivity and high-gain adaptive alertness of the parameter estimator. It will make the algorithm difficult to apply in practical engineering.

Adetola [51] proposed a finite-time parameter estimation algorithm for nonlinear systems. This algorithm combines the pure integrator based MRE technique with the "initialization" process proposed in [48], and the unknown parameters of the original system can be estimated in finite time under the condition that the regression vector satisfies IE.

Aranovskiy [52] proposed a modified algorithm for DRE and named it "DRE and mixing" (DREM). The DREM algorithm adds a key mixing step to DRE and decouples vector PEEs into scalar PEEs. The scalar PEE ensures the monotonicity of each element in the parameter estimation error, which is stronger than the norm monotonicity of the traditional parameter error vector. It means the parameter estimator designed based on the scalar PEE has stronger transient stability. At the same time, the algorithm guarantees the parameter convergence and proposes a new parameter convergence condition that does not depend on the PE condition.

The least squares method and the gradient method have been developed very well, but it is difficult to address the data with missing information. Since the maximum likelihood estimation method and the expectation maximization algorithm can deal with the problem of missing information, these two algorithms have received more and more attention. The maximum likelihood estimation method proposed by Panuskal [53] is the initial probabilistic model identification method, but it did not consider the situation of missing information at that time. To deal with the parameter estimation problem in the absence of data, Dempster [54] proposed the expectation maximization algorithm. This algorithm has been used for parameter estimation of the Gaussian mixture model [55], linear variable parameter model [56], and state space model [57], and a series of expectation maximization variants algorithms have been developed.

Notably, the parametric model identification method can describe the controlled object analytically and achieve better identification results. In the development of recent decades, a fairly complete theoretical system has been formed. However, these methods are mainly for the identification of linear systems. However, most of the controlled objects often contain many complex nonlinear uncertain items in the actual system, and their model structure parameters also show time-varying characteristics, making it impossible to obtain the accurate system dynamic model. Recently, since the non-parametric models can approximate the dynamics of arbitrary complex processes in infinite dimensions, the nonparametric identification methods have begun to become the focus of scholars.

3.2. Non-Parametric Model Identification Method

The model reconstructed by the non-parametric model identification method is called a non-parametric model. It does not mean that there are no parameters in the model but that it does not need to determine the structure and order of the model in advance, which is the advantage of the non-parametric model identification method. Non-parametric model identification methods include some classic identification methods, such as correlation analysis and spectral analysis, etc. It also includes neural network (NN) models which have been developed rapidly in recent years. A neural network has been widely used in nonlinear system control because of its high nonlinearity, approximation ability, and strong self-learning ability. At present, non-parametric model identification methods mainly include: Back-Propagation (BP) neural network non-parametric model identification and Radial Basis Function (RBF) neural network non-parametric model identification. The overview of the non-parametric model identification methods is illustrated in Figure 3.

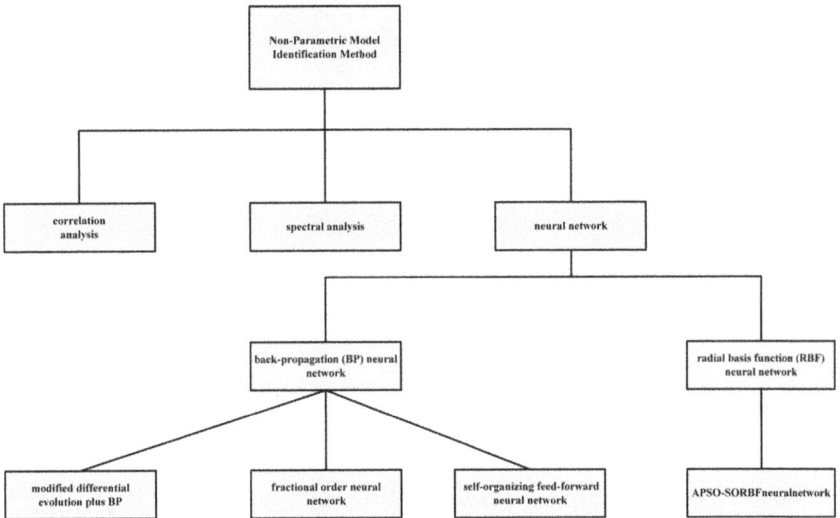

Figure 3. Overview of Non-Parametric Model Identification Methods.

For the non-parametric model identification method using the BP neural network, since the BP neural network can approximate any nonlinear mapping relationship, and the BP algorithm belongs to the global approximation algorithm, it has better generalization ability. Generally speaking, when using a neural network to identify nonlinear systems, it is often combined with classical parameter identification methods to optimize the weights of NN.

Coban [58] proposed a new recurrent neural network, the context layered locally recurrent neural network (CLLRNN), which is effective in the identification of input-output relationships in both linear and nonlinear dynamic systems. To maximize the weights of the neural network model, Nguyen [59] proposed a hybrid modified differential evolution plus back-propagation (MDE-BP) approach. The suggested training method was evaluated in comparison to existing algorithms, including the classic DE and BP algorithms. As a result, the proposed strategy can improve the identification process's accuracy. In [60], Aguilar proposed a fractional order neural network (FONN) for system identification by combining neural network and fractional order calculus methodologies. When compared to existing techniques, the suggested FONN model achieved higher accuracy with fewer parameters. Li [61] developed a new bilevel learning paradigm for self-organizing feedforward neural networks (FFNN). The hybrid binary particle swarm optimization (BPSO) algorithm is used as an upper level optimizer in this interactive learning algorithm to self-organize network architecture, while the Levenberg–Marquardt (LM) algorithm is used as a lower level optimizer to optimize the connection weights of an FFNN. When compared to conventional learning algorithms, experimental results show that the bilevel learning algorithm produces much more compact FFNNs with superior generalization capabilities. Singh [62] developed a gradient evolution-based counter propagation network (GE-CPN) for approximating the noncanonical form of a nonlinear system. Learning from nonlinear systems with parametric uncertainty is a key characteristic of GE-CPN networks. Furthermore, this demonstrated that reparameterization of neural network models is required and beneficial for approximation of noncanonical systems.

As a feedforward network, RBF neural network has attracted extensive attention recently because of its fixed basis function and linear parameter network structure, which can approximate any continuous function with arbitrary precision. For the identification and modeling of nonlinear dynamic systems, Qiao [63] designed a novel self-organizing radial basis function (SORBF) neural network. Based on the neuron activity and mutual information (MI), the SORBF neural network's hidden neurons can be added or removed to reach the desired network complexity while maintaining overall computing efficiency for identification and modeling. Meanwhile, parameter adjustment can considerably increase model performance. Slimani [64] utilized the descent gradient and the genetic algorithm technique to developed an optimization technique of neural networks radial basis function multi-model identification of nonlinear system. Errachdi [65] developed a no-preprocessing online radial basis function (RBF) neural network technique. The suggested online RBF neural network approach is then combined with a kernel principal component analysis (KPCA), which made RBF neural network training efficient and fast by reducing memory requirements of the models. In [66], with the use of adaptive particle swarm optimization, a self-organizing radial basis function (SORBF) neural network was constructed to increase both accuracy and parsimony (APSO). The presented APSO-SORBF neural network is capable of producing a network model with a compact structure and outstanding accuracy. In [67], to self-organize the structure and parameters of the RBFNN, a distance concentration immune algorithm (DCIA) was devised. A sample with the most frequent occurrence of maximum error was constructed to govern the parameters of the new neuron in order to increase forecasting accuracy and reduce computation time.

The above studies have introduced many mature identification algorithms from linear system identification into the RBF network framework. At the same time, many scholars have extended the RBF network framework to solve the problem of parameter model identification, which makes up for the limitations of traditional parametric model identification methods for nonlinear system identification. When the structure and order of the system model are known, even if the controlled object contains many complex nonlinear uncertain terms, or its model parameters show time-varying characteristics, the RBF neural network framework can identify it accurately. This not only makes full use of the available system information but also maximizes the accurate feature description of the original system.

4. Model Identification-Based Optimal Control for SAS

In the optimal control community, researchers are trying to introduce the model identification methods to the classical optimal control for a single agent system (SAS) with an unknown system model.

Bhasin [68] proposed an actor-critic-identifier (ACI) ADP framework, in which an identifier NN is utilized to approximate the unknown system information and then embedded into the actor-critic NN architectures. The ACI ADP framework is shown in Figure 4. However, the input system dynamics are still assumed to be known. To further remove this assumption, Modares [31] designed an new ACI algorithm for the unknown constrained-input nonlinear systems. The proposed ACI algorithm contains an identifier NN with an experience replay technique (ERT) to fully approximate the unknown system information (including system dynamics and input dynamics). Then, a gradient method was used to estimate the weights of critic-actor NNs. Actually, the idea of ERT is very similar to concurrent learning, both of which use recorded historical data and current data to estimate the unknown information of the system. Although this technique can relax the PE condition for parameter convergence during the online learning, it requires more computation time and computer memory to store historical data. The algorithm was then generalized to solve many control problems, such as the IRL algorithm for constrained input systems [69], the H_∞ tracking control problem [70], and so on.

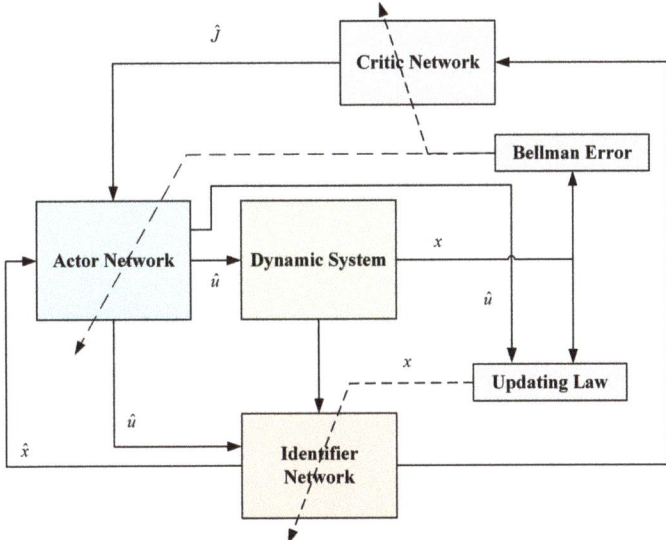

Figure 4. The basic framework of actor-critic-identifier ADP.

To relax the PE condition for parameter convergence during the online learning, Zhao [71] used the ERT to estimate the unknown weights of the identifier NN and critic NN simultaneously, so that the conventional PE condition could be relaxed to a simplified condition on recorded data. However, the proposed algorithm also has the same drawbacks in [31]. Based on the ERT, Yang [72] proposed an event-triggered robust control policy for unknown continuous-time nonlinear systems. To improve the convergence rate of the ERT, a data-based feedback relearning (FR) algorithm for uncertain nonlinear systems with control channel disturbances and actuator faults was developed [73]. Furthermore, a data processing method based on experience replay technology is designed to improve data utilization efficiency and algorithm convergence. To achieve model-free fault compensation, a neural network (NN)-based fault observer is used. To reduce the difficulty of designing NNs for an unknown nonlinear system and improve generalization, the polynomial activation function is redesigned using the sigmoid function/hyperbolic tangent activation function.

To avoid excessive use of NNs and achieve faster convergence, Lv [74] proposed a new identifier-critic (IC) ADP structure with the MRE method. Since the algorithm did not use an actor NN, and it did not need to record historical data, the convergence rate is greatly improved. Later, this IC algorithm was used to solve a series of other control problems [75–77].

5. Model Identification-Based Optimal Control for MASs

More recently, few works on the model identification-based optimal control has been studied for MASs. Based on the work of Modares [31], Tatari [78] proposed an online optimal distributed learning algorithm to find the game theoretic solution of systems on graphs with completely unknown dynamics. In [79], Tatari introduced an online distributed optimal adaptive algorithm for continuous-time nonlinear differential graphical games with unknown systems subject to external disturbances. Shi [80] utilized the MRE filtering technique and designed an adaptive disturbance observer for a class of nonlinear systems with unknown disturbances where the disturbance is assumed to be generated by some unknown dynamics. Tan [81] proposed a novel event-triggered, model-free structure to address the optimal consensus control problem for MASs with unknown dynamics and input constraints.

In the following, as an example, we give the model identification-based optimal control of MASs with unknown dynamics.

Algebraic graph theory: The communication topology between agents in a MAS is described by a directed graph $\mathcal{G} = (\mathcal{V}, \mathcal{E}, \mathcal{A})$ where $\mathcal{V} = \{1, 2, \cdots, N\}$ is a nonempty set of vertices and $\mathcal{E} = \{(i,j) \mid i,j \in \mathcal{V}\} \subseteq \mathcal{V} \times \mathcal{V}$ is the set of edges. Define $\mathcal{A} = \{e_{ij}\} \in \mathbb{R}^{N \times N}$ as a weighted adjacency matrix, where $e_{ij} = 1$ if and only if $(i,j) \in \mathcal{E}$, and $e_{ij} = 0$, otherwise. The neighbor set of the agent i is denoted by $\mathcal{N}_i = \{j \mid (i,j) \in \mathcal{E}\}$. Define a diagonal matrix $\mathcal{D} = diag\{d_i\}$ as the in-degree matrix, where $d_i = \sum_{j \in \mathcal{N}_i} e_{ij}$. The Laplacian matrix $\mathcal{L}$ is defined by $\mathcal{L} = \mathcal{D} - \mathcal{A}$.

In order to take a single leader into account, we introduce an augmented graph $\overline{\mathcal{G}} = (\overline{\mathcal{V}}, \overline{\mathcal{E}})$, where $\overline{\mathcal{V}} = \{0, 1, \cdots, N\}$ and $\overline{\mathcal{E}} \subseteq \overline{\mathcal{V}} \times \overline{\mathcal{V}}$. A nonnegative number e_{i0} is used to describe the interaction relationship between the leader and agent i. Specifically, $e_{i0} > 0$ if agent i can receive information from the leader; otherwise, $e_{i0} = 0$. A leader adjacency matrix $\mathcal{B}$ is defined by $\mathcal{B} = diag(e_{10}, \cdots, e_{N0}) \in \mathbb{R}^{N \times N}$.

Assumption 1. *The communication interaction network $\overline{\mathcal{G}}$ has a spanning tree with the root vertex 0.*

Problem formulation: Consider heterogeneous MASs described by a linear time-invariant system as follows

$$\dot{x}_i(t) = A_i x_i(t) + B_i u_i(t), \ i = 1, 2, \ldots, N, \tag{11}$$

where $x_i(t) \in \mathbb{R}^n$ and $u_i(t) \in \mathbb{R}^m$ are the state vector and the control input vector, respectively. The system matrices $A_i \in \mathbb{R}^{n \times n}$ and input matrices $B_i \in \mathbb{R}^{n \times m}$ are assumed be unknown in this paper. Furthermore, we assume that the pairs (A_i, B_i) $(\forall i = 1, \ldots, N)$ are controllable, and the state and the control input of each agent are available.

The dynamics of the leader agent is described by

$$\dot{x}_0 = A_0 x_0, \tag{12}$$

where $x_0 \in \mathbb{R}^n$.

The local tracking error $\delta_i \in \mathbb{R}^n$, $i = 1, \ldots, N$ can be defined as

$$\delta_i(t) = \sum_{j \in \mathcal{N}_i} e_{ij}(x_i - x_j) + e_{i0}(x_i - x_0), \tag{13}$$

where the pinning gain $e_{i0} \geq 0$. Then, the dynamics of the local tracking error are written by

$$\begin{aligned}\dot{\delta}_i(t) &= \sum_{j \in \mathcal{N}_i} e_{ij}(\dot{x}_i - \dot{x}_j) + e_{i0}(\dot{x}_i - \dot{x}_0) \\ &= \sum_{j \in \mathcal{N}_i} e_{ij}(A_i x_i - A_j x_j) + e_{i0}(A_i x_i - A_0 x_0) + (d_i + e_{i0})B_i u_i - \sum_{j \in \mathcal{N}_i} e_{ij} B_j u_j.\end{aligned} \qquad (14)$$

The overall tracking error vector is given by

$$\begin{aligned}\delta(t) &= ((\mathcal{L} + \mathcal{B}) \otimes I_n)(x - \hat{x}_0) \\ &= ((\mathcal{L} + \mathcal{B}) \otimes I_n)\zeta,\end{aligned} \qquad (15)$$

where $\delta = (\delta_1^T, \delta_2^T, \cdots, \delta_N^T)^T$, $x = (x_1^T, x_2^T, \cdots, x_N^T)^T \in \mathbb{R}^{nN}$, $\hat{x}_0 = (x_0^T, x_0^T, \cdots, x_0^T)^T \in \mathbb{R}^{nN}$, $\zeta = x - \hat{x}_0$ is the global synchronization error.

One of the objectives in this paper is to design a tracking strategy to ensure that all follower agents can follow the leader, that is, $\lim_{t \to \infty} \|x_i(t) - x_0(t)\| = 0$. The second objective is to design a distributed controller that can minimize the performance index function.

In fact, under Assumption 1, $\mathcal{L} + \mathcal{B}$ is invertible. From (15), one can obtain that $\lim_{t \to \infty} \zeta(t) = 0$ if and only if $\lim_{t \to \infty} \|\delta(t)\| = 0$. Thus, once the local neighbor error approaches to zero, we can say that the tracking control problem is solved.

We define the local performance index (value function) for the agent i as follows

$$V_i(\delta_i(t)) = \frac{1}{2}\int_0^\infty (\delta_i^T Q_{ii} \delta_i + U(u_i) + \sum_{j \in \mathcal{N}_i} U(u_j))d\tau, \qquad (16)$$

where $Q_{ii} \geqslant 0$ is a symmetric weight matrix, $U(\cdot) = u_i R_{ii} u_i$ is a positive definite integrand function. We assume that (16) satisfies zero-state observability.

The tracking problem is aimed at finding the Nash equilibrium policies u_i^* for the N player game. That is, for all agent i, there have $V_i^* = V_i(\delta_i(0), u_i^*, u_{\mathcal{N}_i}^*) \leq V_i(\delta_i(0), u_i, u_{\mathcal{N}_i}^*)$, $\forall u_i, (i = 1, \ldots, N)$. Therefore, the tracking problem of MASs with input constraint in this paper can be transformed to solving the N coupled optimization problems, that is

$$V_i^*(\delta_i(t)) = \min_{u_i} \frac{1}{2}\int_0^\infty (\delta_i^T Q_{ii} \delta_i + U(u_i) + \sum_{j \in \mathcal{N}_i} U(u_j))d\tau, \qquad (17)$$

with given (14) while the dynamic informations A_i and B_i, $i = 1, \ldots, N$ are considered completely unknown.

By differentiating each value function V_i, and using (16), the following Lyapunov equation is obtained

$$\begin{aligned}\nabla V_i^T (\sum_{j \in \mathcal{N}_i} e_{ij}(A_i x_i - A_j x_j) + e_{i0}(A_i x_i - A_0 x_0) + (d_i + e_{i0})B_i u_i - \sum_{j \in \mathcal{N}_i} e_{ij} B_j u_j) + \frac{1}{2}\delta_i^T Q_{ii} \delta_i \\ + \frac{1}{2}U(u_i) + \frac{1}{2}\sum_{j \in \mathcal{N}_i} U(u_j) = 0,\end{aligned} \qquad (18)$$

where $\nabla V_i = \partial V_i / \partial \delta_i \in \mathbb{R}^n$ and $V_i(0) = 0$.

Then one can get the Hamiltonian function as follows

$$\begin{aligned}H_i(\delta_i, \nabla V_i, u_i, u_{\mathcal{N}_i}) &= \nabla V_i^T (\sum_{j \in \mathcal{N}_i} e_{ij}(A_i x_i - A_j x_j) + e_{i0}(A_i x_i - A_0 x_0) \\ &+ (d_i + e_{i0})B_i u_i - \sum_{j \in \mathcal{N}_i} e_{ij} B_j u_j) + \frac{1}{2}\delta_i^T Q_{ii} \delta_i + \frac{1}{2}U(u_i) + \frac{1}{2}\sum_{j \in \mathcal{N}_i} U(u_j).\end{aligned} \qquad (19)$$

According to the first-order stationary condition in the Hamiltonians, the optimal control policy for each agent can be obtained as

$$\frac{\partial H_i}{\partial u_i} = 0 \to u_i^* = -(d_i + e_{i0})\lambda R_{ii}^{-T} B_i^T \nabla V_i^*. \tag{20}$$

System identifier using neural networks: Since the system matrices A_i and input matrices B_i are assumed to be unknown, the unknown dynamics of each agent are modeled by using NNs. Then the experience replay technique is used to formulate the identifier weights adaptive update law.

The NN-based identifiers are designed to approximate system dynamic, which is given as follows

$$A_i x_i = A_i^* x_i + \varepsilon_{A_i}, B_i u_i = B_i^* u_i + \varepsilon_{B_i}, i = 1, \ldots, N, \tag{21}$$

where $A_i^* \in \mathbb{R}^{n \times n}, B_i^* \in \mathbb{R}^{n \times m}$ are unknown weights, $x_i \in \mathbb{R}^n, u_i \in \mathbb{R}^m$ are the basis functions, and ε_{A_i} and ε_{B_i} are the reconstruction errors.

Combining (21) and (11), the system can be reformulated as follows

$$\dot{x}_i = \vartheta_{A_i B_i}^* z_i(x_i, u_i) + \varepsilon_{A_i B_i}, i = 1, \ldots, N, \tag{22}$$

where $\vartheta_{A_i B_i}^* = [A_i^* \ B_i^*] \in \mathbb{R}^{n \times d}$, $z_i(x_i, u_i) = [x_i^T \ u_i^T]^T \in \mathbb{R}^d$ is the regressor vector. $\varepsilon_{A_i B_i} = \varepsilon_{A_i} + \varepsilon_{B_i}$ is the model approximation error.

Assumption 2. *On a given compact set $\Omega \subset \mathbb{R}^n$, the approximator reconstruction errors ε_{A_i} and $\varepsilon_{B_i}, i = 1, \ldots, N$ and their gradients are bounded, i.e., $\|\varepsilon_{A_i}\| \leq \bar{\varepsilon}_{A_i}, \|\varepsilon_{B_i}\| \leq \bar{\varepsilon}_{B_i}$, and the approximator basis functions and their gradients are bounded.*

Remark 1. *According to Assumption 2, the model approximation error $\varepsilon_{A_i B_i}$ is bounded, that is, $\|\varepsilon_{A_i B_i}\| \leq \bar{\varepsilon}_{A_i B_i} = \bar{\varepsilon}_{A_i} + \bar{\varepsilon}_{B_i}$.*

A filtered regressor is proposed for (22), which can be expressed as

$$x_i = \vartheta_{A_i B_i}^* h_i(x_i) + c l_i(x_i) + \varepsilon_{x_i}, \tag{23}$$

$$\dot{h}_i(x_i) = -c h_i(x_i) + z(x_i, u_i), h_i(0) = 0,$$
$$\dot{l}_i(x_i) = -C l_i(x_i) + x_i, l_i(0) = 0, \tag{24}$$

where $C = c I_{n \times n}, c > 0$, $h_i(x_i) \in \mathbb{R}^d$ is a filtered regressor version of $z(x_i, u_i)$, $l_i(x_i) \in \mathbb{R}^n$ is a filtered regressor version of the state x_i. $\varepsilon_{x_i} = e^{-Ct} x_i(0) + \int_0^t e^{-C(t-\tau)} \varepsilon_{A_i B_i} d\tau$ is bounded, since $\varepsilon_{A_i B_i}$ is bounded. $x_i(0)$ is the initial state of (22).

To obtain the adaptive tuning law that does not affected by the system instability, both side of the filtered regressor (23) are divided by a normalizing signal $n_{s_i} = 1 + h_i^T h_i + l_i^T l_i$,

$$\bar{x}_i = \vartheta_{A_i B_i}^* \bar{h}_i(x_i) + c \bar{l}_i(x_i) + \bar{\varepsilon}_{x_i}, \tag{25}$$

where $\bar{x}_i = x_i / n_{x_i}, \bar{h}_i = h_i / n_{x_i}, \bar{l}_i = l_i / n_{x_i}, \bar{\varepsilon}_{x_i} = \varepsilon_{x_i} / n_{x_i}$. Obviously, $\bar{\varepsilon}_{x_i}$ is bounded.

Based on (21), (23) and (25), the form of the identifier weights estimator of agent i can be expressed as

$$\hat{\bar{x}}_i = \hat{\vartheta}_{A_i B_i} \bar{h}_i(x_i) + c \bar{l}_i(x_i), i = 1, \ldots, N \tag{26}$$

where $\hat{\vartheta}_{A_i B_i} = [\hat{A}_i \ \hat{B}_i]$ is the estimated value of the identifier weights matrix $\vartheta_{A_i B_i}^*$.

Thus, the state estimation error $e_i \in \mathbb{R}^n, i = 1, \ldots, N$ can be defined as

$$e_i(t) = \hat{\bar{x}}_i - \bar{x}_i = \tilde{\vartheta}_{A_i B_i}(t) \bar{h}_i(x_i) - \bar{\varepsilon}_{x_i}, \tag{27}$$

where $\tilde{\vartheta}_{A_iB_i}(t) = \hat{\vartheta}_{A_iB_i}(t) - \vartheta^*_{A_iB_i}(t), i = 1, \ldots, N$ is the parameter estimation error of agent i at time t.

The experience replay technique is utilized to formulate the identifier weights adaptive tuning law in the following. The idea of this technique is to store or record linearly independent historical data along with current data, so as to improve data utilization.

Then, we set
$$Z_i = [\bar{h}_i(x_i(t_1)), \ldots, \bar{h}_i(x_i(t_{p_i}))] \tag{28}$$
to be the recorded historical data stack of each agent i at the past times $t_1, \ldots, t_{p_i}$.

Remark 2. *It is noted that the number of linearly independent elements in Z_i should be equal to the dimension of the $h_i(x_i)$ in (23), i.e., $rank(Z_i) = d$. This condition aims to satisfying the PE condition and can easily be checked online.*

Then, based on the experience replay technique, a weight tuning law is designed for the identifier of agent i as follows
$$\dot{\hat{\vartheta}}_{A_iB_i}(t) = -\Gamma_i e_i(t)\bar{h}_i^T(x_i(t)) - \Gamma_i \sum_{k=1}^{p_i} e_i(t_k)\bar{h}_i^T(x_i(t_k)), \tag{29}$$

where $\Gamma_i > 0, i = 1, \ldots, N$ is a positive definite learning rate matrix.

It is noted from Remark 2 that, with the aid of experience replay technique, the PE condition can be checked by monitoring the rank of the recorded historical data, but it usually consumes large computing resources, resulting in low learning efficiency. Therefore, how to design an identification method that can take into account the learning efficiency and relaxed the PE condition is an interesting and challenging research direction.

6. Conclusions and Future Work

In this paper, we have reviewed the development of ADP-based learning optimal control, several model identification techniques, and their applications to the learning and control of MASs. Based on these reviews, it is noted that the model identification-based ADP method has made significant progress in both theoretical research and practical applications. However, the model identification-based ADP methods still have many challenges in theory and algorithm design that have not yet been resolved. Through the above summary and analysis of the model identification-based ADP methods, some related issues for future research directions are outlined as follows:

- In fact, the model identification-based ADP method is mainly focused on the design of a single controller currently, but not so much on the design of multiple controllers. It will be a very beneficial work to use the model identification-based ADP method to realize the distributed coordinated control of MASs.
- Most of the existing model identification-based ADP methods need to satisfy the PE condition. However, PE conditions are difficult to verify in practical applications. How to design a novel identification-based ADP method such that the PE condition is easier to be checked and remain low pressure [82].
- For more complex MASs such as power grids and transportation, where their accurate models cannot be obtained, the model identification-based ADP method may be used to solve large-scale practical optimization problems, which have important practical applications.

Author Contributions: Conceptualization, R.L. and Z.P.; methodology, R.L., Z.P. and J.H.; software, R.L. and Z.P.; validation, R.L. and Z.P.; investigation, R.L. and Z.P.; writing—original draft, R.L.; writing—review and editing, R.L., Z.P. and J.H.; visualization, R.L. and Z.P.; supervision, J.H. All authors have read and agreed to the published version of the manuscript.

Funding: This work was supported in part by the National Natural Science Foundation of China under Grant 62203089, Grant 61473061, and Grant 12271083, in part by the Project funded by China Postdoctoral Science Foundation under Grant 2021M700695, and in part by the Sichuan Science and Technology Program under Grant 2022NSFSC0890, Grant 2021YFG0184 and Grant 2018GZDZX0037.

Institutional Review Board Statement: Not applicable.

Informed Consent Statement: Not applicable.

Data Availability Statement: Not applicable.

Conflicts of Interest: The authors declare no conflict of interest.

References

1. Hu, J.; Liu, Z.; Wang, J.; Wang, L.; Hu, X. Estimation, intervention and interaction of multi-agent systems. *Acta Autom. Sin.* **2013**, *39*, 1796–1804. [CrossRef]
2. Ji, Y.; Wang, G.; Li, Q.; Wang, C. Event-triggered optimal consensus of heterogeneous nonlinear multi-agent systems. *Mathematics* **2022**, *10*, 4622. [CrossRef]
3. Hu, J. Second-order event-triggered multi-agent consensus control. In Proceedings of the 31th Chinese Control Conference, Hefei, China, 25–27 July 2012; pp. 6339–6344.
4. Hu, J.; Feng, G. Quantized tracking control for a multi-agent system with high-order leader dynamics. *Asian J. Control* **2011**, *13*, 988–997. [CrossRef]
5. Wang, Q.; Hu, J.; Wu, Y.; Zhao, Y. Output synchronization of wide-area heterogeneous multi-agent systems over intermittent clustered networks. *Inf. Sci.* **2023**, *619*, 263–275. [CrossRef]
6. Chen, B.; Hu, J.; Zhao, Y.; Ghosh, B.K. Finite-time velocity-free rendezvous control of multiple AUV systems with intermittent communication. *IEEE Trans. Syst. Man Cybern. Syst.* **2022**, *52*, 6618–6629. [CrossRef]
7. Peng, Y.; Zhao, Y.; Hu, J. On the role of community structure in evolution of opinion formation: A new bounded confidence opinion dynamics. *Inf. Sci.* **2023**, *621*, 672–690. [CrossRef]
8. Murray, J.J.; Cox, C.J.; Lendaris, G.G.; Saeks, R. Adaptive dynamic programming. *IEEE Trans. Syst. Man Cybern. Syst.* **2002**, *32*, 140–153. [CrossRef]
9. Wang, F.Y.; Zhang, H.; Liu, D. Adaptive dynamic programming: An introduction. *IEEE Comput. Intell. Mag.* **2009**, *4*, 39–47. [CrossRef]
10. Wu, Y.; Liang, Q.; Hu, J. Optimal output regulation for general linear systems via adaptive dynamic programming. *IEEEE Trans. Cybern.* **2022**, *52*, 11916–11926. [CrossRef] [PubMed]
11. Werbos, P. *Approximate Dynamic Programming for Realtime Control and Neural Modelling*; White, D.A., Sofge, D.A., Eds., Van Nostrand: New York, NY, USA, 1992.
12. Bertsekas, D.P. *Dynamic Programming and Optimal Control*; Athena Scientific: Belmont, MA, USA, 1995.
13. Prokhorov, D.V.; Wunsch, D.C. Adaptive critic designs. *IEEE Trans. Neural Netw.* **1997**, *8*, 997–1007. [CrossRef]
14. Bellman, R. Dynamic programming. *Science* **1966**, *153*, 34–37. [CrossRef]
15. Werbos, P. Advanced forecasting methods for global crisis warning and models of intelligence. *Gen. Syst. Yearb.* **1977**, *22*, 25–38.
16. Zhang, H.-G.; Zhang, X.; Luo, Y.-H.; Yang, J. An overview of research on adaptive dynamic programming. *Acta Autom. Sin.* **2013**, *39*, 303–311. [CrossRef]
17. Lewis, F.L.; Vrabie, D. Reinforcement learning and adaptive dynamic programming for feedback control. *IEEE Circuits Syst. Mag.* **2009**, *9*, 32–50. [CrossRef]
18. AbuKhalaf, M.; Lewis, F.L. Nearly optimal control laws for nonlinear systems with saturating actuators using a neural network hjb approach. *Automatica* **2005**, *41*, 779–791. [CrossRef]
19. Vrabie, D.; Lewis, F.L. Neural network approach to continuoustime direct adaptive optimal control for partially unknown nonlinear systems. *Neural Netw.* **2009**, *22*, 237–246. [CrossRef]
20. Modares, H.; Lewis, F.L. Optimal tracking control of nonlinear partiallyunknown constrainedinput systems using integral reinforcement learning. *Automatica* **2014**, *50*, 1780–1792. [CrossRef]
21. Vamvoudakis, K.G.; Lewis, F.L. Online actor-critic algorithm to solve the continuoustime infinite horizon optimal control problem. *Automatica* **2010**, *46*, 878–888. [CrossRef]
22. Zhang, H.; Wei, Q.; Luo, Y. A novel infinitetime optimal tracking control scheme for a class of discretetime nonlinear systems via the greedy hdp iteration algorithm. *IEEE Trans. Syst. Man Cybern. Syst. Part B (Cybernetics)* **2008**, *38*, 937–942. [CrossRef]
23. AlTamimi, A.; Lewis, F.L.; AbuKhalaf, M. Discretetime nonlinear hjb solution using approximate dynamic programming: Convergence proof. *IEEE Trans. Syst. Man Cybern. Syst. Part B (Cybernetics)* **2008**, *38*, 943–949. [CrossRef]
24. Liu, D.; Wang, D.; Zhao, D.; Wei, Q.; Jin, N. Neuralnetworkbased optimal control for a class of unknown discretetime nonlinear systems using globalized dual heuristic programming. *IEEE Trans. Autom. Sci. Eng.* **2012**, *9*, 628–634. [CrossRef]
25. Liu, D.; Wei, Q. Policy iteration adaptive dynamic programming algorithm for discretetime nonlinear systems. *IEEE Trans. Neural Netw. Learn. Syst.* **2013**, *25*, 621–634. [CrossRef] [PubMed]

26. Kiumarsi, B.; Vamvoudakis, K.G.; Modares, H.; Lewis, F.L. Optimal and autonomous control using reinforcement learning: A survey. *IEEE Trans. Neural Netw. Learn. Syst.* **2017**, *29*, 2042–2062. [CrossRef] [PubMed]
27. Hou, Z.S.; Wang, Z. From modelbased control to datadriven control: Survey, classification and perspective. *Inf. Sci.* **2013**, *235*, 3–35. [CrossRef]
28. Peng, Z.; Luo, R.; Hu, J.; Shi, K.; Nguang, S.K.; Ghosh, B.K. Optimal tracking control of nonlinear multiagent systems using internal reinforce Q-learning. *IEEE Trans. Neural Netw. Learn. Syst.* **2022**, *33*, 4043–4055. [CrossRef] [PubMed]
29. Peng, Z.; Zhao, Y.; Hu, J.; Ghosh, B.K. Data-driven optimal tracking control of discrete-time multi-agent systems with two-stage policy iteration algorithm. *Inf. Sci.* **2019**, *481*, 189–202. [CrossRef]
30. Peng, Z.; Zhao, Y.; Hu, J.; Luo, R.; Ghosh, B.K.; Nguang, S.K. Input-output data-based output antisynchronization control of multi-agent systems using reinforcement learning approach. *IEEE Trans. Ind. Inform.* **2021**, *17*, 7359–7367. [CrossRef]
31. Modares, H.; Lewis, F.L.; Naghibi-Sistani, M.B. Adaptive optimal control of unknown constrained-input systems using policy iteration and neural networks. *IEEE Trans. Neural Netw. Learn. Syst.* **2013**, *24*, 1513–1525. [CrossRef]
32. Ding, F.; Wang, F.F. Recursive least squares identification algorithms for linear-in-parameter systems with missing data. *Control Decis.* **2016**, *31*, 2261–2266.
33. Ding, F.; Wang, F.F.; Xu, L.; Wu, M.H. Decomposition based least squares iterative identification algorithm for multivariate pseudo-linear ARMA systems using the data filtering. *J. Franklin Inst.* **2017**, *354*, 1321–1339. [CrossRef]
34. Elisei-Iliescu, C.; Stanciu, C.; Paleologu, C.; Benesty, J.; Anghel, C.; Ciochina, S. Efficient recursive least-squares algorithms for the identification of bilinear forms. *Digit. Signal Process* **2018**, *83*, 280–296. [CrossRef]
35. Huang, W.; Ding, F.; Hayat, T.; Alsaedi, A. Coupled stochastic gradient identification algorithms for multivariate output-error systems using the auxiliary model. *Int. J. Control Autom.* **2017**, *15*, 1622–1631. [CrossRef]
36. Ding, F.; Xu, L.; Meng, D.; Jin, X.-B.; Alsaedi, A.; Hayat, T. Gradient estimation algorithms for the parameter identification of bilinear systems using the auxiliary model. *J. Comput. Appl. Math.* **2020**, *369*, 112575. [CrossRef]
37. Åström, K.J.; Wittenmark, B. *Adaptive Control*; Courier Corporation: Mineola, NY, USA, 2013.
38. Hu, J.; Hu, X. Optimal target trajectory estimation and filtering using networked sensors. In Proceedings of the 27th Chinese Control Conference, Kunming, China, 16–18 July 2008; pp. 540–545.
39. Lion, P.M. Rapid identification of linear and nonlinear systems. *AIAA J.* **1967**, *5*, 1835–1842. [CrossRef]
40. Kreisselmeier, G. Adaptive observers with exponential rate of convergence. *IEEE Trans. Autom. Control* **1977**, *22*, 2–8. [CrossRef]
41. Duarte, M.A.; Narendra, K.S. Combined direct and indirect approach to adaptive control. *IEEE Trans. Autom. Control* **1989**, *34*, 1071–1075. [CrossRef]
42. Slotine, J.E.; Li, W. Composite adaptive control of robot manipulators. *Automatica* **1989**, *25*, 509–519. [CrossRef]
43. Panteley, E.; Ortega, R.; Moya, P. Overcoming the detectability obstacle in certainty equivalence adaptive control. *Automatica* **2002**, *38*, 1125–1132. [CrossRef]
44. Lavretsky, E. Combined composite model reference adaptive control. *IEEE Trans. Autom. Control* **2009**, *54*, 2692–2697. [CrossRef]
45. Chowdhary, G.; Yucelen, T.; Muhlegg, M.; Johnson, E. Concurrent learning adaptive control of linear systems with exponentially convergent bounds. *Int. J. Adapt. Control Signal Process* **2013**, *27*, 280–301. [CrossRef]
46. Cho, N.; Shin, H.; Kim, Y.; Tsourdos, A. Composite MRAC with parameter convergence under finite excitation. *IEEE Trans. Autom. Control* **2018**, *63*, 811–818. [CrossRef]
47. Roy, S.; Bhasin, S.; Kar, I. A UGES switched MRAC architecture using initial excitation. In Proceedings of the 2017 20th IFAC World Congress, Toulouse, France, 9–14 July 2017; pp. 7044–7051.
48. Krause, J.; Khargonekar, P. Parameter information content of measurable signals in direct adaptive control. *IEEE Trans. Autom. Control* **1987**, *32*, 802–810. [CrossRef]
49. Ortega, R. An on-line least-squares parameter estimator with finite convergence time. *IEEE Inst. Electr. Electron. Eng.* **1988**, *76*, 847–848. [CrossRef]
50. Roy, S.; Bhasin, S.; Kar, I. Combined MRAC for unknown MIMO LTI systems with parameter convergence. *IEEE Trans. Autom. Control* **2018**, *63*, 283–290. [CrossRef]
51. Adetola, V.; Guay, M. Finite-time parameter estimation in adaptive control of nonlinear systems. *IEEE Trans. Autom. Control* **2008**, *53*, 807–811. [CrossRef]
52. Aranovskiy, S.; Bobtsov, A.; Ortega, R.; Pyrkin, A. Performance enhancement of parameter estimator via dynamic regressor extension and mixing. *IEEE Trans. Autom. Control* **2017**, *62*, 3546–3550. [CrossRef]
53. Panuska, V.; Rogers, A.E.; Steiglitz, K. On the maximum likelihood estimation of rational pulse transfer-function parameters. *IEEE Trans. Autom. Control* **1968**, *13*, 304–305. [CrossRef]
54. Dempster, A.P.; Laird, N.M.; Rubin, D.B. Maximum likelihood from incomplete data via the EM algorithm. *J. R. Stat. Soc. Series B Stat. Methodol.* **1977**, *39*, 1–22.
55. Sammaknejad, N.; Zhao, Y.; Huang, B. A review of the expectation maximization algorithm in data-driven process identification. *J. Process Control* **2019**, *73*, 123–136. [CrossRef]
56. Yang, X.; Liu, X.; Han, B. LPV model identification with an unknown scheduling variable in the presence of missing observations—A robust global approach. *IET Control Theory Appl.* **2018**, *12*, 1465–1473. [CrossRef]
57. Wang, D.; Zhang, S.; Gan, M.; Qiu J. A novel EM identification method for Hammerstein systems with missing output data. *Trans. Ind. Inform.* **2019**, *16*, 2500–2508. [CrossRef]

58. Coban, R. A context layered locally recurrent neural network for dynamic system identification. *Eng. Appl. Artif. Intell.* **2013**, *26*, 241–250. [CrossRef]
59. Nguyen, S.N.; Ho-Huu, V.A.; Ho, P.H. A neural differential evolution identification approach to nonlinear systems and modelling of shape memory alloy actuator. *Asian J. Control* **2018**, *20*, 57–70. [CrossRef]
60. Aguilar, C.J.Z.; Gómez-Aguilar, J.F.; Alvarado-Martínez, V.M.; Romero-Ugalde, H.M. Fractional order neural networks for system identification. *Chaos Solitons Fractals* **2020**, *130*, 109444. [CrossRef]
61. Li, H.; Zhang, L. A bilevel learning model and algorithm for self-organizing feed-forward neural networks for pattern classification. *IEEE Trans. Neural Netw. Learn. Syst.* **2020**, *32*, 4901–4915. [CrossRef]
62. Singh, U.P.; Jain, S.; Tiwari, A.; Singh, R.K. Gradient evolution-based counter propagation network for approximation of noncanonical system. *Soft Comput.* **2019**, *23*, 4955–4967. [CrossRef]
63. Qiao, J.F.; Han, H.G. Identification and modeling of nonlinear dynamical systems using a novel self-organizing RBF-based approach. *Automatica* **2012**, *48*, 1729–1734. [CrossRef]
64. Slimani, A.; Errachdi, A.; Benrejeb, M. Genetic algorithm for RBF multi-model optimization for nonlinear system identification. In Proceedings of the IEEE International Conference on Control, Automation and Diagnosis, Grenoble, France, 2–4 July 2019; pp. 2–4.
65. Errachdi, A.; Benrejeb, M. Online identification using radial basis function neural network coupled with KPCA. *Int. J. Gen. Syst.* **2017**, *46*, 52–65. [CrossRef]
66. Han, H.G.; Lu, W.; Hou, Y.; Qiao, J.-F. An adaptive-PSO-based self-organizing RBF neural network. *IEEE Trans. Neural Netw. Learn. Syst.* **2016**, *29*, 104–117. [CrossRef] [PubMed]
67. Qiao, J.; Li, F.; Yang, C.; Li, W.; Gu, K. A self-organizing RBF neural network based on distance concentration immune algorithm. *IEEE/CAA J. Autom. Sin.* **2019**, *7*, 276–291. [CrossRef]
68. Bhasina, S.; Kamalapurkar, R.; Johnson, M.; Vamvoudakis, K.G.; Lewis, F.L.; Dixon, W.E. A novel actor-critic-identifier architecture for approximate optimal control of uncertain nonlinear systems. *Automatica* **2013**, *49*, 82–92. [CrossRef]
69. Modares, H.; Lewis, F.L.; Naghibi-Sistani, M.-B. Integral reinforcement learning and experience replay for adaptive optimal control of partially-unknown constrained-input continuous-time systems. *Automatica* **2014**, *50*, 193–202. [CrossRef]
70. Modares, H.; Lewis, F.L.; Jiang, Z.P. H_∞ Tracking control of completely unknown continuous-time systems via off-policy reinforcement learning. *IEEE Trans. Neural Netw. Learn. Syst.* **2015**, *26*, 2550–2562. [CrossRef] [PubMed]
71. Zhao, D.; Zhang, Q.; Wang, D.; Zhu, W. Experience replay for optimal control of nonzero-sum game systems with unknown dynamics. *IEEE Trans. Cybern.* **2015**, *46*, 854–865. [CrossRef] [PubMed]
72. Yang, X.; He, H. Adaptive critic designs for event-triggered robust control of nonlinear systems with unknown dynamics. *IEEE Trans. Cybern.* **2018**, *49*, 2255–2267. [CrossRef]
73. Mu, C.; Zhang, Y.; Sun, C. Data-Based feedback relearning control for uncertain nonlinear systems with actuator faults. *IEEE Trans. Cybern.* **2022**, 1–14. [CrossRef]
74. Lv, Y.; Na, J.; Yang, Q.; Wu, X.; Guo, Y. Online adaptive optimal control for continuous-time nonlinear systems with completely unknown dynamics. *Int. J. Control Autom.* **2016**, *89*, 99–112. [CrossRef]
75. Lv, Y.; Na, J.; Ren, X. Online H_∞ control for completely unknown nonlinear systems via an identifier–critic-based ADP structure. *Int. J. Control Autom.* **2019**, *92*, 100–111. [CrossRef]
76. Lv, Y.; Ren, X.; Na, J. Online Nash-optimization tracking control of multi-motor driven load system with simplified RL scheme. *ISA Trans.* **2020**, *98*, 251–262. [CrossRef]
77. Na, J.; Lv, Y.; Zhang, K.; Zhao, J. Adaptive identifier-critic-based optimal tracking control for nonlinear systems with experimental validation. *IEEE Trans. Syst. Man Cybern. Syst.* **2022**, *52*, 459–472. [CrossRef]
78. Tatari, F.; Naghibi-Sistani, M.B.; Vamvoudakis, K.G. Distributed optimal synchronization control of linear networked systems under unknown dynamics. In Proceedings of the 2017 American Control Conference (ACC), Seattle, WA, USA, 24–26 May 2017; pp. 668–673.
79. Tatari, F.; Vamvoudakis, K.G.; Mazouchi, M. Optimal distributed learning for disturbance rejection in networked non-linear games under unknown dynamics. *IET Control. Theory Appl.* **2018**, *13*, 2838–2848. [CrossRef]
80. Shi, J.; Yue, D.; Xie, X. Optimal leader-follower consensus for constrained-input multiagent systems with completely unknown dynamics. *IEEE Trans. Syst. Man Cybern. Syst.* **2022**, *52*, 1182–1191. [CrossRef]
81. Tan, W.; Peng, Z.; Ji, H.; Luo, R.; Kuang, Y.; Hu, J. Event-triggered model-free optimal consensus for unknown multi-agent systems with input constraints. In Proceedings of the 2022 Chinese Control Conference (CCC), Hefei, China, 25–27 July 2022; pp. 4729–4734.
82. Luo, R.; Peng, Z.; Hu, J.; Ghosh, B.K. Adaptive optimal control of completely unknown systems with relaxed PE conditions. In Proceedings of the IEEE 11th Data Driven Control and Learning Systems Conference, Chengdu, China, 3–5 August 2022; pp. 836–841.

Disclaimer/Publisher's Note: The statements, opinions and data contained in all publications are solely those of the individual author(s) and contributor(s) and not of MDPI and/or the editor(s). MDPI and/or the editor(s) disclaim responsibility for any injury to people or property resulting from any ideas, methods, instructions or products referred to in the content.

Article

Dynamics Analysis for the Random Homogeneous Biased Assimilation Model

Jiangbo Zhang [1,*] and Yiyi Zhao [2]

[1] School of Science, Southwest Petroleum University, Chengdu 610500, China
[2] School of Business Administration, Faculty of Business Administration, Southwestern University of Finance and Economics, Chengdu 611130, China
* Correspondence: jbzhang@amss.ac.cn; Tel.: +86-1518-438-7612

Abstract: This paper studies the evolution of opinions over random social networks subject to individual biases. An agent reviews the opinion of a randomly selected one and then updates its opinion under homogeneous biased assimilation. This study investigates the impact of biased assimilation on random opinion networks, which is different from the previous studies on fixed network structures. If the bias parameters are static, it is proven that the event in which all agents converge to extreme opinions happens almost surely. Next, the opinion polarization event is proved to be a probability one event. While if the bias parameters are dynamic, the opinion evolution is proven to depend on early finite time slots for the dynamical individual bias parameter functions independent of the biased parameter values after the time threshold. Numerical simulations further show that opinion evolution depends on early finite time slots for some nonlinear dynamical individual bias parameter functions.

Keywords: opinion dynamics; bias parameter; homogeneous; polarization; consensus

MSC: 91C99; 91C20

1. Introduction

In our society, opinion formation among individuals and induced dynamics has been extensively studied and debated in the academic literature, including minority opinion dissemination, collective decision making, polarization and fluctuation, fashion emergence, etc. With the extensive development of network communication, such as WeChat groups, Facebook interest groups, Twitter discussion threads, etc., online interactions are becoming increasingly important in many aspects ranging from political decisions to marketing strategies [1,2]. In this setting, it is important to study the way individuals in an online social network update their attitudes.

For the traditional network topologies, the standard DeGroot model employs the discrete-time multi-agent system to simulate how public opinions may influence each other, in which an individual's opinion toward a particular topic is often represented by a real value in the interval $[0, 1]$. It was proven that as long as the underlying graph is connected, all opinions converge to a common value known as the consensus state [3]. Generalizations of this model to continuous-time dynamics and time-varying network structures have been extensively studied in the literature, e.g., [4–7]. Such convergence to consensus still holds for some deterministically switching networks, e.g., [4–6]. However, analyzing the social groups' characteristics is an important way to understand the rule of the social system. Therefore, it is increasingly important to understand how macro-characteristics emerge from the micro-individual psychological and interactive effects. For example, individuals always consider the initial point of view in the process of updating [8]; according to the selective exclusion principle in social psychology, individuals in the group would choose to communicate with people with similar opinions [9]; the research objects of the stochastic

DeGroot model contain stubborn agents who never modify their opinions, such as leaders and rumor disseminators [10]; agents may tend to repulse individual opinions that differ greatly from their own based on self-bias [11]. Beyond consensus, social dynamics can exhibit complex behaviors such as polarization, clustering, and opinion fluctuation [10–12].

Considering the wide existence of biases among individual opinion evolution, especially online comments, WeChat discussions, etc., individual biases were modeled as nonlinear weights on self-opinion and local group opinions, based on which clustering to extreme opinions in the fixed network topologies was revealed recently [13,14]. Studies from social psychology show that people are more likely to accept confirming evidence given by someone similar to themselves [15,16]. A convincing model for this biased opinion assimilation was proposed in [15] as a natural interpretation of confirmation bias. Ref. [14] provided a systemic analysis of a social opinion dynamical model with bias assimilation on fixed network topologies. However, the polarization phenomena were only shown for special fixed networks [13], and rare results on polarization are confirmed for general fixed networks [13,14]. In fact, for social systems designed expressly to facilitate collective decision making regarding complex social issues, the occurrence of polarization would not only depend on the network structures (such as the two-island network shown in [13]) but also on some random accidental factors [17,18].

Previous studies have dealt with convergence and stability analysis of such systems for some fixed network structures, and we focus on how individual opinions evolve for the random network structure. Opinion exchanges among Internet users might promote opinion consensus, polarization, and fluctuations with different psychological effects behind social interactions. This creates some new problems: Assume all internet users have the psychological effect of biased assimilation; then, how do biased individuals' opinions evolve among the random online networks? Do the consensus phenomena always happen similarly to the fixed network topologies [14]? Through this paper, we will answer these questions partly and understand why individual-level polarization would happen, contrary to conventional wisdom, regarding the public opinions of online platforms.

The contribution is that we propose and study opinion dynamics over the random social network with homogeneous bias parameters. Particularly, we focus on how individual biases and randomness affect the opinion limit states. Firstly, we investigate the random bias-induced collective nonlinear network dynamics and provide conditions under which all node states converge to 0 or 1. Next, we prove that opinion polarization happens with a positive probability with homogeneous bias parameters. Finally, we prove that all node states also converge to extremal opinions even if the bias parameters are dynamic, and we show that the opinion evolution only depends on the value ranges of bias parameters in certain early time intervals. Simulations on the time interval thresholds are conducted for some dynamical bias parameters.

The remainder of the paper is organized as follows. In Section 2, we present the social network model for our study and introduce our problems of interest. Section 3 presents our main results on the fixed biased parameters. Then, Section 4 presents our main results on the dynamic biased parameters and provides some numerical simulations of periodical functions of the biased parameters. Finally, Section 5 concludes the paper with a few remarks on potential future directions.

2. Model Formulation

Our opinion formulation process unfolds over the random social network represented by random weighted directed graphs $G(t) = (V, E(t), W)$. Time is slotted at $t = 0, 1, 2, \ldots$. $V = \{1, 2, \ldots, n\}$.

At each time t, each node $i \in V$ randomly selects one node $r_i(t) \in V$ as its neighbor from the network node set V independent of other nodes' selections. This results in a random set of neighbors (clusters), which are denoted by $\{r_i(t)\}$, for $i \in V$ and $t = 0, 1, \ldots$. Note that $(i, j) \in E(t)$ if and only if the agent selected by agent i is $r_i(t) = j$, representing the other node j that influences i.

For $W = [w_{ij}]_{n \times n}$, $0 < w_{ij} < 1$ represents the influence weight between two nodes i and j. Without loss of generality, the node i's self-confidence is represented by $w_{ii} = 1$.

Each node i holds an opinion $x_i(t) \in [0,1]$ at time t. Let b_i be a positive number associated with node i as a bias parameter. The evolution of the $x_i(t), i \in V$ is described by:

$$x_i(t+1) = \frac{x_i(t) + w_{i,r_i(t)}(x_i(t))^{b_i} x_{r_i(t)}(t)}{1 + w_{i,r_i(t)}\left[(x_i(t))^{b_i} x_{r_i(t)}(t) + (1-x_i(t))^{b_i}(1 - x_{r_i(t)}(t))\right]} \quad (1)$$

This model reflects the social psychology phenomenon named biased assimilation. For any node i, $w_{i,r_i(t)}$ is the inspired influence weight that the node i is influenced by the node $r_i(t) \in V$. For the right side of the model (1), the factor $(x_i(t))^{b_i}$ weighting $x_{r_i(t)}(t)$ means the biased manner degree on its previous "relevant disconfirming empirical" opinion $x_i(t)$, while the factor $(1 - x_i(t))^{b_i}$ weighting $1 - x_{r_i(t)}(t)$ means the biased manner degree on its previous "relevant disconfirming empirical" opinion $1 - x_i(t)$.

Note that $(1 - x_i(t))^{b_i}(1 - x_{r_i(t)}(t)) \geq 0$ if $x_i(t) \in [0,1]$; thus, the denominator is not smaller than the numerator. Thus, $x_i(0) \in [0,1]$ for all $i \in V$ guarantees that $x_i(t) \in [0,1]$ for all $t \geq 0$ and $i \in V$. In addition, 0 and 1 represent the extreme opinion of opposing or supporting on the given topic, respectively. Based on the results of [14,15], opinion evolution depends on whether all $b_i > 1$ or all $b_i \in (0,1)$, except the network structure and other parameter constraints. Therefore, it is necessary to induce and classify the bias parameters as follows.

Definition 1. *If all opinion bias parameter $b_i > 1$, $i \in V$ or all opinion bias parameter $b_i \in (0,1)$, $i \in V$, then the biased assimilation model (1) is homogeneous. Correspondingly, if $b_i > 1$ for $i \in V_1$ where V_1 is nonempty, $b_i \in (0,1)$ for $i \in V_2$, $(V_1 \cup V_2) \subset V$, then the biased assimilation model (1) is heterogeneous.*

If there exist at least two agents i,j such that $b_i \in (0,1)$ and $b_j > 1$, then we say the biased assimilation model (1) is *heterogeneous*.

Denote $R(t) = (r_1(t), r_2(t), \ldots, r_n(t))$ as the selection vector for any $t \geq 0$. We impose the following assumptions for the selection rule and bias parameters.

Assumption 1. *$\{r_i(t)\}$ are independent with each other for any $i \in V$ and $t \geq 0$, and $r_i(t)$ is any discrete distribution on $\{1,2,\ldots,n\}$ where $r_i(t) = j$ is a positive probability event for any $i,j \in V$ and $t \in \mathbb{N}$.*

Assumption 2. *The bias assimilation parameter $\{b_i\}$ satisfies $b_i > 0$ for any $i \in V$.*

3. Fixed Bias Parameters

In this section, we investigate the opinion limit analysis for the model (1) where the bias parameters are fixed and homogeneous for $t \geq 0$.

3.1. Probability Space

Let $\{R(t)\}$ be a non-repetitive selection vector set, that is, any $R(t)$ for $t \geq 0$ is a permutations sort of $(1,2,\ldots,n)$. Thus, for a given $t \geq 0$, there is n different vectors for $R(t)$. Denote Ω as the set composed by all n different permutations sort of $(1,2,\ldots,n)$. By Assumption 1, we can construct a probability space $(V, \mathcal{F}, P)$. Furthermore, because the selections are independent among different $t \geq 0$, thus, the probability space reflecting any opinion selection trajectories is independent among $t \in \mathbb{N}$.

3.2. Some Lemmas

In this subsection, we introduce some lemmas for the model (1).

Lemma 1. *For the probability measure P, $\lim_{t\to\infty} x_i(t) = x_i^*$ a.s. is equivalent to $x_i(t)$, which almost uniformly converges to x_i^*. That is to say, $\forall \varepsilon > 0$, $\lim_{t\to\infty} P\left(\cup_{k=t}^{+\infty}(|x_i(t) - x_i^*| \geq \varepsilon)\right) = 0$ for any $i \in V$.*

Lemma 2. *If $x_i(0) \in (0, \frac{1}{2})$ for all $i \in V$, then $x_i(t) \in (0, \frac{1}{2})$ for any $t \in \mathbb{N}$. Similarly, if $x_i(0) \in (\frac{1}{2}, 1)$ for all $i \in V$, then $x_i(t) \in (\frac{1}{2}, 1)$ for any $t \in \mathbb{N}$.*

Proof of Lemma 2 is listed in Appendix A. Lemma 2 explains some realistic social phenomena. If all people own negative opinions (<0.5), then all of them will always keep the negative opinions. A similar phenomenon appears when they all have positive opinions (>0.5). Different from the results in [14], $\max_{i \in V}\{x_i(t)\}$ is not monotonic due to the randomness selection of $\{r_i(t)\}$. Therefore, some of the following main results would be different from the ones on the fixed network topologies. To analyze the nonlinearity of the model (1), we provide the following lemma.

Lemma 3. *For the function $f(x,y) = \frac{x+wx^by}{1+w(x^by+(1-x)^b(1-y))}$ where $w > 0$, $b > 0$ and $x, y \in [0,1]$, we have:*

(i) $f(0,y) \equiv 0$ for any $y \in [0,1]$;
(ii) $f(1,y) \equiv 1$ for any $y \in [0,1]$;
(iii) $f(x,0) = \frac{x}{1+w(1-x)^b}$ and it is a lower convex function;
(iv) $f(x,1) = \frac{x+wx^b}{1+wx^b}$ and it is a upper convex function;
(v) $f(1-x,x) + f(x,1-x) = f(x,y) + f(1-x,1-y) = 1$.

This lemma can be easily obtained and the proof is omitted.

3.3. Results on Homogeneous Bias Parameters

In this subsection, we study the limits of the model (1) where Assumptions 1 and 2 are satisfied. Here, all bias parameters are homogeneous. Specially, if $b_i = 0$ for any $i \in V$, then $x_i(t+1) = \frac{1}{1+w_{ij}}x_i(t) + \frac{w_{ij}}{1+w_{ij}}x_j(t)$. By the standard DeGroot model, all agent opinions reach a consensus. If $b_i = 1$ for any $i \in V$, we obtain the following lemma.

Lemma 4. *For the model (1),*

$$P\{\lim_{t\to\infty} x_i(t) = 1\} = 1$$

for any $i \in V$, if $b_i = 1$ and $x_i(0) \in (\frac{1}{2}, 1]$ for any $i \in V$;

$$P\{\lim_{t\to\infty} x_i(t) = 0\} = 1$$

for any $i \in V$, if $b_i = 1$ and $x_i(0) \in [0, \frac{1}{2})$ for any $i \in V$.

The proof of Lemma 4 is shown in Appendix B. In the following, we generalize the results of Lemma 4 to any homogeneous bias parameters $b_i > 0$.

Theorem 1. *For the model (1), if $\{b_i\}$ are homogeneous, then*

(i) $P\{\lim_{t\to\infty} x_i(t) = 1\} = 1$ for any $i \in V$, if $x_i(0) \in (\frac{1}{2}, 1]$ for any $i \in V$;
(ii) $P\{\lim_{t\to\infty} x_i(t) = 0\} = 1$ for any $i \in V$, if $x_i(0) \in [0, \frac{1}{2})$ for any $i \in V$.

The proof of Theorem 1 is shown in Appendix C. Although the dynamics of model (1) is different from the one in [14], the result of Theorem 1 is similar to Theorem 4 of [14]. Therefore, we can weaken the condition of Theorem 1 and distinguish the results of different network structures, and we provide the following theorem.

Theorem 2. *For the homogeneous parameters $\{b_i\}$, if $x_i(t) \in (\frac{1}{2}, 1)$ infinitely often (i.o.) for certain $i \in V$, then $\lim_{t \to \infty} x_i(t) = 1$ a.s. Similarly, if $x_i(t) \in (0, \frac{1}{2})$ i.o. for certain $i \in V$, then $\lim_{t \to \infty} x_i(t) = 0$ a.s.*

The proof of Theorem 2 is shown in Appendix D. Theorem 2 illustrates that the social group that always owns negative opinions (<0.5) will finally reach the extremely negative attitude (0) after sufficient communication. Similarly, the social group that always owns positive opinions (>0.5) will finally reach an extremely positive attitude (1). These phenomena can usually be found in online interest groups.

Based on Theorem 2, we can show that opinion polarization happens with a positive probability, which is much different from the results on fixed social networks [13,14]. Denote

$$E_{polarization} = \{\exists V_1, V_2 \subset V, V_1 \cup V_2 = V s.t. : \lim_{t \to \infty} x_i(t) = 0, i \in V_1; \lim_{t \to \infty} x_i(t) = 1, i \in V_2\}.$$

Furthermore, we denote

$$E_{con} = \{\lim_{t \to \infty} x_i(t) = 0 \text{ or } 1, \forall i \in V\}.$$

According to Theorem 1, $E_{polarization}$ is a zero probability event if $x_i(0) \in [0, \frac{1}{2})$ or $x_i(0) \in (\frac{1}{2}, 1]$. Generally, we obtain the following theorem.

Theorem 3. *For the model (1), $P\{E_{con}\} = 1$ and $P\{E_{polarization}\} > 0$ if $\{b_i\}$ are homogeneous.*

The proof of Theorem 3 is shown in Appendix E. The result of Theorem 3 is different from anyone in the fixed network topologies [13,14]. The proof of Theorem 3 shows that opinion polarization depends on opinion selection sequence $\{r_i(t), i \in V, t \in \mathbb{N}\}$, not only the initial opinions $\{x_i(0)\}$, which shows that some accidental factors could also affect the opinion evolution for the model (1).

Theorem 3 can explain the social phenomena on online social networks. If there is a group of people who process their information in a biased manner, then opinion polarization happens with a positive probability. For the fixed networks, Ref. [13] shows that opinion polarization happens on the two-island network with strict parameter conditions. Theorem 3 extends it into the case of random selection rules.

According to Theorem 3, we obtain that all opinions will converge to 0 or 1, a.s., for different initial values and biased parameter b_i. Figure 1 shows how opinions $\{x_i(t)\}$ change for different initial values and selection processes when the agent number $n = 20$, the termination time is $T = 200$ and the bias parameters $b_i = 2.2$ for any $i \in V$.

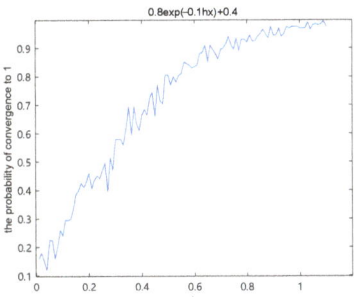
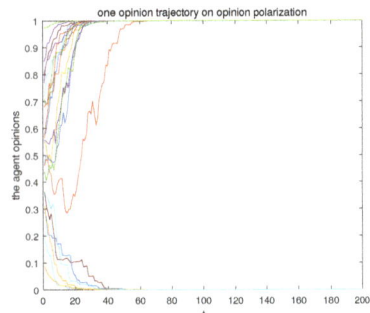

Figure 1. The different opinion evolutions on the model (1), where lines of different colors mean different agent opinion trajectories. (**left**) The opinion consensus phenomenon; (**right**) The opinion polarization phenomenon.

4. Dynamic Biased Parameters

In this section, we will analyze the generalized form of the model (1) where the biased assimilation parameters $\{b_i\}$ are functions of time t, that is,

$$x_i(t+1) = \frac{x_i(t) + w_{i,r_i(t)}(x_i(t))^{b_i(t)} x_{r_i(t)}(t)}{1 + w_{i,r_i(t)}\left[(x_i(t))^{b_i(t)} x_{r_i(t)}(t) + (1-x_i(t))^{b_i(t)}(1-x_{r_i(t)}(t))\right]}. \quad (2)$$

4.1. Results on Dynamic Bias Parameters

For the function $b_i(t)$, we assume that $b_i(t) > 0$ for any $i \in V$ and $t \geq 0$. In this section, we call $\{b_i(t)\}$ homogeneous if $b_i(t) > 1$ for any $i \in V$ and $t \in \mathbb{N}$ simultaneously, or $b_i(t) \in (0,1)$ for any $i \in V$ and $t \in \mathbb{N}$ simultaneously. Obviously, if $b_i(t) \in (0,1)$ for all $i = 1, 2, \ldots, n$ and $t \geq 0$, then opinions will almost surely converge to 1. While if $b_i(t) > 1$ for all $i = 1, 2, \ldots, n$ and $t \geq 0$, then opinions will almost surely converge to 0. Generally, we obtain the following lemma.

Lemma 5. *For the model (2), $P\{E_{con}\} = 1$ and $P\{E_{polarization}\} > 0$ if $\{b_i(t)\}$ are homogeneous.*

Note that in the proof of Theorem 3, the analysis on b_i only depends on the current period t. Thus, the proof of Theorem 3 can be naturally extended to the case of Lemma 5; thus, the proof of Lemma 5 is omitted. Different from the previous results, we propose that agent opinion evolution on the model (2) only depends on the early time slots.

Theorem 4. *For the model (2), there exists a finite time threshold $T_1^* > 0$, if $b_i(t) > 1$, $x_i(0) \in (0, \frac{1}{2})$ for any $i \in V$ and $t \in (0, T_1^*)$, then*

$$\lim_{t \to \infty} x_i(t) = 0 \text{ a.s. for any } i \in V;$$

Similarly, there exists a finite time threshold $T_2^* > 0$, if $b_i(t) \in (0,1)$, $x_i(0) \in (\frac{1}{2}, 1)$ for any $i \in V$ and $t \in (0, T_2^*)$, then

$$\lim_{t \to \infty} x_i(t) = 1 \text{ a.s. for any } i \in V.$$

This theorem shows that there always exists a finite time threshold T^*, such that the opinion evolution only depends on the time interval $[0, T^*]$. However, it is difficult to provide a mathematical expression of T^*. In the following, we demonstrate the threshold T^* for some periodic functions and monotone functions.

4.2. Simulations on Dynamic Bias Parameters

In this subsection, simulations that explore the threshold T^* of the model (2) are presented based on MATLAB software. Set agent number $n = 20$ and the termination time $T = 200$. The initial opinions are equally distributed on the interval $[0,1]$ and the simulation number is 200. We use $\hat{T}_1^* = \min_{t \in \mathbb{N}}\{t : x_i(t) \in (0, \frac{1}{2}) \forall i \in V\}$ (or $\hat{T}_2^* = \min_{t \in \mathbb{N}}\{t : x_i(t) \in (\frac{1}{2}, 1) \forall i \in V\}$) to substitute for T_1^* (or T_2^*) of Theorem 4. Here, the estimated threshold T^* is a weighted combination of $\hat{T}_1^*$ and $\hat{T}_1^*$ where the weight parameters are the frequencies of the events that opinions converge to 0 or 1, respectively. Specially, we set all $b_i(t)$ to be the same for $i \in V$.

(1) Figure 2 shows that the average estimated threshold T^* oscillatory decreases as h changes from 0 to 1.1, where the biased assimilation functions are $b_i(t) = 0.8 mod(x, 10 + 30h)/(10 + 30h) + 0.8$ and $b_i(t) = 0.8 \exp(-0.1hx) + 0.4$, respectively.

In fact, as h increases from 0 to 1, the probabilities of opinions converge to 1 and decrease to 0. By Theorem 4, there exists a time threshold $T_1^* > 0$, such that $\lim_{t \to \infty} x_i(t) = 0$ a.s. for any $i \in V$ if $b_i(t) > 1$, $x_i(0) \in (0, \frac{1}{2})$ for any $i \in V$ and $t \in (0, T_1^*)$.

The following figures (Figure 3) show how the first range where $b_i(t) > 1$ enlarges when h increases from 0 to 1. According to Lemma 5, this is corresponding to Figure 2 where the frequency of opinions converging to 0 increases.

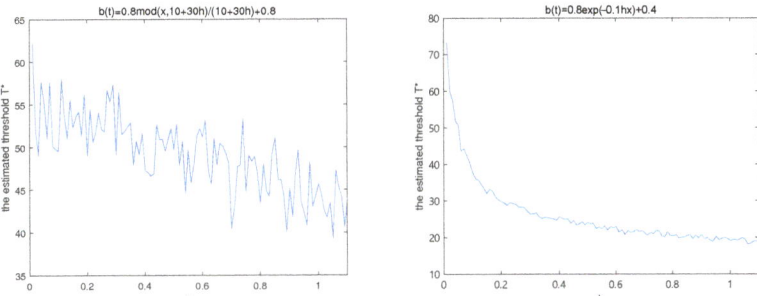

Figure 2. The change of the average estimated threshold T^*. (**left**): $\{b_i(t)\}$ are periodic functions; (**right**): $\{b_i(t)\}$ are monotone decreasing functions.

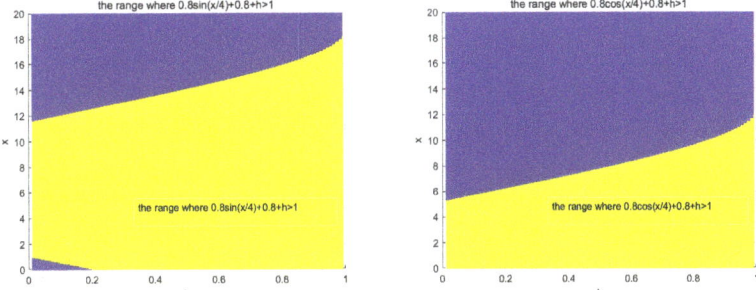

Figure 3. The first ranges where $b_i(t) > 1$ change as h increases from 0 to 1. (**left**): $\{b_i(t)\}$ are sine functions; (**right**): $\{b_i(t)\}$ are cosine functions.

Similarly, Figure 4 shows that the frequency of opinions converging to 1 also oscillatory decreases as h changes from 0 to 1, where the biased assimilation function $b_i(t) = 0.8\sin(\frac{t}{4+10h}) + 0.8$ and $b_i(t) = 0.8\cos(\frac{t}{4+10h}) + 0.8$. The analysis is similar to Figure 2.

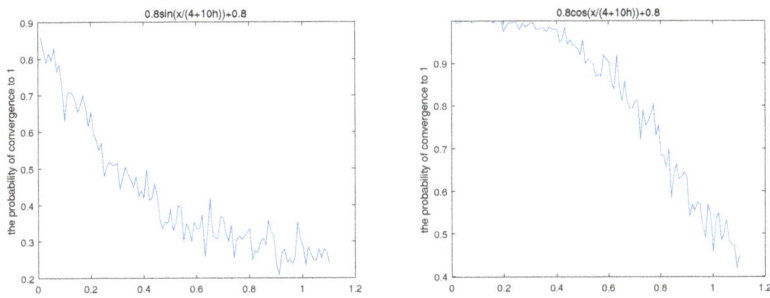

Figure 4. The probability that opinions converge to 1. (**left**): $\{b_i(t)\}$ are sine functions with different periods; (**right**): $\{b_i(t)\}$ are cosine functions with different periods.

(2) Probability of consensus for exponential functions of $b_i(t)$: In this part, the probabilities of opinion consensus to 0 or 1 for the model (1) are demonstrated.

If $b_i(t) = 0.8e^{-0.1t} + 0.4 + h$ where h changes from 0 to 1, then $\{b_i(t) > 1\} = [0, 10\ln\left(\frac{8}{6-10h}\right)]$. When h increases from 0 to 1, $10\ln\frac{8}{6-10h}$ increases. Thus, the range $[0, T_2^*]$ where $b_i(t) > 1$ enlarges, and the frequency of opinions converging to 0 increases. This is corresponding to Figure 5. The similar analysis can be obtained for $b_i(t) = 0.8e^{-0.1ht} + 0.4$.

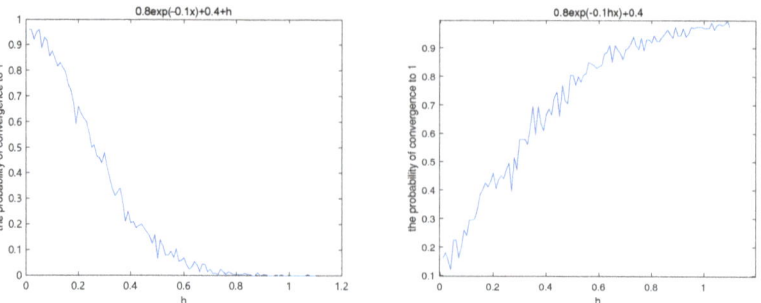

Figure 5. The probability that opinions converge to 1. (**left**): $\{b_i(t)\}$ are exponential functions; (**right**): $\{b_i(t)\}$ are exponential functions.

(3) Probability of consensus for index periodic functions of $b_i(t)$: In this part, the probabilities of opinion consensus to 0 or 1 for the model (1) are demonstrated.

According to Figure 6, when h changes from 0 to 1, the first ranges where $0 < 0.8mod(x, 10+30h)/(10+30h) + 0.8 < 1$ and $0.8mod(5+15h+x, 10+30h)/(10+30h) + 0.8 > 1$ increase. Figure 7 reflects the probability where opinions converging to 1 oscillatory increase and oscillatory decrease, respectively.

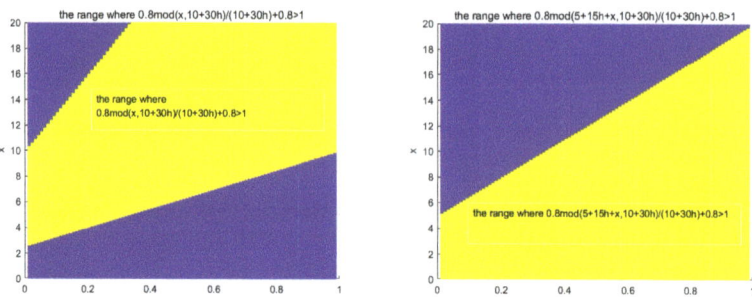

Figure 6. The first ranges where $b_i(t) > 1$ change as h increases from 0 to 1. (**left**): $\{b_i(t)\}$ are index periodic functions; (**right**): $\{b_i(t)\}$ are index periodic functions.

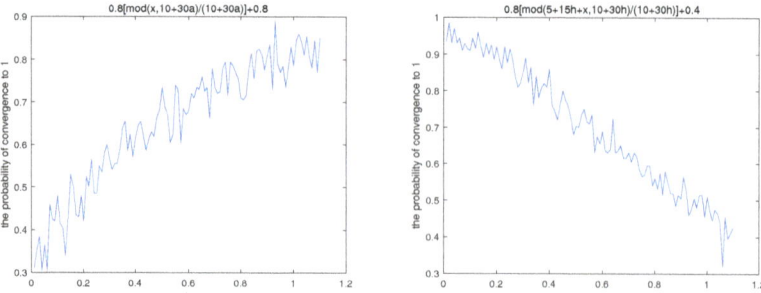

Figure 7. The probability that opinions converge to 1. (**left**): $\{b_i(t)\}$ are index periodic functions; (**right**): $\{b_i(t)\}$ are index periodic functions.

5. Conclusions

We have systematically analyzed online social opinion dynamics subject to individually biased assimilation. With initial opinions being independently and identically distributed, at each time step, peers review the selected opinions of a randomly selected clique with biased assimilation. The contributions are that a series of results on the asymptotic behaviors of the social opinions at a system level were provided, focusing on polarization and consensus. The results show that convergence happens almost surely and polarization happens with a positive probability. Future works include studying the limit states if $\{b_i, i \in V\}$ are heterogeneous, extending the results to general network structures, and validating the established opinion formations with real-world social network data.

Author Contributions: Conceptualization, J.Z. and Y.Z.; methodology, J.Z.; software, Y.Z.; validation, Y.Z. and J.Z.; formal analysis, J.Z.; investigation, Y.Z.; writing—original draft preparation, J.Z.; writing—review and editing, J.Z. and Y.Z.; visualization, Y.Z.; supervision, J.Z. All authors have read and agreed to the published version of the manuscript.

Funding: This research was funded by the MOE Project of Humanities and Social Sciences under grant 21YJA630122, Central Government Funds of Guiding Local Scientific and Technological Development for Sichuan Province of China under grant 2021ZYD0012, the Natural Science Foundation of Sichuan Province (No. 2022NSFSC0529), Sichuan Science and Technology Program (No. 2021YJ0084).

Data Availability Statement: Not applicable.

Conflicts of Interest: The authors declare no conflict of interest.

Appendix A

Fix the time t, for any $i \in V$, denote $r_i(t) = j$. By the model (1),

$$x_i(t+1) = \frac{x_i(t) + w_{ij}(x_i(t))^{b_i} x_j(t)}{1 + w_{ij}[(x_i(t))^{b_i} x_j(t) + (1 - x_i(t))^{b_i}(1 - x_j(t))]} \stackrel{(i)}{\leq} \frac{x_i(t) + w_{ij}(x_i(t))^{b_i} x_j(t)}{1 + w_{ij}(x_i(t))^{b_i}} \stackrel{(ii)}{\leq} \frac{1}{2},$$

where (i) is deduced by $(1 - x_i(t))^{b_i} > (x_i(t))^{b_i}$ for any $x_i(t) \in [0, \frac{1}{2})$ and $b_i > 0$ and (ii) holds because $x_i(t), x_j(t) < \frac{1}{2}$.

Therefore, $x_i(t+1) < \frac{1}{2}$. The non-negativity of $\{x_i(t)\}$ is obviously obtained. In a sum, $x_i(t) \in [0, \frac{1}{2})$ for any $t \in \mathbb{N}$ if $x_i(0) \in (0, \frac{1}{2})$.

For another condition, if $x_i(0) \in (\frac{1}{2}, 1)$, then $1 - x_i(0) \in (0, \frac{1}{2})$. By the model (1),

$$1 - x_i(t+1) = \frac{1 - x_i(t) + w_{ij}(1 - x_i(t))^{b_i}(1 - x_j(t))}{1 + w_{ij}[(x_i(t))^{b_i} x_j(t) + (1 - x_i(t))^{b_i}(1 - x_j(t))]}.$$

This indicates that $\{1 - x_i(t)\}$ has the similar evolution path of $\{x_i(t)\}$. Thus, the conclusion holds.

Appendix B

If $b_i = 1$ for any $i \in V$ and we assume $r_i(t) = j$, then

$$x_i(t+1) = \begin{cases} \frac{1 + w_{ij} x_j(t)}{1 + w_{ij}(1 - x_i(t) - x_j(t) + 2x_i(t) x_j(t))} x_i(t) < x_i(t), & \text{if } x_j(t) \in (0, \frac{1}{2}); \\ \frac{1 + w_{ij} x_j(t)}{1 + w_{ij}(1 - x_i(t) - x_j(t) + 2x_i(t) x_j(t))} x_i(t) > x_i(t) & \text{if } x_j(t) \in (\frac{1}{2}, 1). \end{cases} \quad (A1)$$

Thus, if $x_i(0) \in (0, \frac{1}{2})$, by (A1), then $x_i(1) < x_i(0)$ for any $r_i(0) \in V$ and $i \in V$. Consequently, $x_i(t+1) < x_i(t)$ for any $t \in \mathbb{N}$. Note that $\{x_i(t)\}$ is decreasing and has the lower bound 0 for a given $i \in V$. We assume that $\lim_{t \to \infty} x_i(t) = x_i^*$ for $i \in V$, by (A1), we obtain that

$$x_i^* = \frac{1 + w_{ij} x_j^*}{1 + w_{ij}(1 - x_i^* - x_j^* + 2x_i^* x_j^*)} x_i^*.$$

Then, $x_i^* = 0$ or $x_j^* = \frac{1}{2}$ for any $i, j \in V$. Due to $x_i(0) \in (0, \frac{1}{2})$, $x_i^* = 0$ holds a.s. Similarly, $P\{\lim_{t \to \infty} x_i(t) = 1\} = 1$ for any $i \in V$, if $b_i = 1$ and $x_i(0) \in (\frac{1}{2}, 1)$ for any $i \in V$. The conclusion follows.

Appendix C

Consider the time t, $t \in \mathbb{N}$. Denote $r_i(t) = j$. Then,

$$x_i(t+1) - x_i(t) = \frac{w_{ij} x_i^{b_i+1}(t) x_j(t)}{1 + w_{ij}[x_i^{b_i}(t) x_j(t) + (1-x_i(t))^{b_i}(1-x_j(t))]} \left[\frac{1 - x_i(t)}{x_i(t)} - \frac{1 - x_j(t)}{x_j(t)} \left(\frac{1 - x_i(t)}{x_i(t)} \right)^{b_i} \right]. \quad (A2)$$

Step I: When $0 < b_i < 1$ and $x_k(0) \in (\frac{1}{2}, 1)$, by the model (1) and Lemma 3,

$$1 - x_i(t+1) = \frac{1 - x_i(t) + w_{ij}(1 - x_i(t))^{b_i}(1 - x_j(t))}{1 + w_{ij}\left[(1 - x_i(t))^{b_i}(1 - x_j(t)) + x_i^{b_i}(t) x_j(t)\right]}.$$

Set $z_i(t) = 1 - x_i(t)$, then $z_i(0) \in (0, \frac{1}{2})$. By Lemma 2, $z_i(t) \in (0, \frac{1}{2})$ for any $i \in V$. Obviously, $\{z_i(t), i \in V\}$ satisfies the model (1). In the following, we divided into two cases for analyzing the monotonic decreasing of $\{z_i(t)\}$.

If $z_j(t) < z_i(t)$ and $r_i(t) = j$, then

$$z_i(t+1) - z_i(t) = w_{ij}(z_i(t))^{b_i+1} z_j(t) \left(\frac{1 - z_i(t)}{z_i(t)} - \frac{1 - z_j(t)}{z_j(t)} \left(\frac{1 - z_i(t)}{z_i(t)} \right)^{b_i} \right)$$

$$\stackrel{(a)}{\leq} w_{ij}(z_i(t))^{b_i+1} z_j(t) \frac{z_j(t) - z_i(t)}{z_i(t) z_j(t)} < 0,$$

where (a) is deduced by $1 < \left(\frac{1-z_i(t)}{z_i(t)} \right)^{b_i} < \frac{1-z_i(t)}{z_i(t)}$.

If $z_i(t) < z_j(t)$ and $r_i(t) = j$, then $\frac{1-z_i(t)}{z_i(t)} > \frac{1-z_j(t)}{z_j(t)} > 1$. Hence,

$$\frac{1 - z_i(t)}{z_i(t)} - \frac{1 - z_j(t)}{z_j(t)} \left(\frac{1 - z_i(t)}{z_i(t)} \right)^{b_i} \stackrel{(i)}{<} \frac{1 - z_i(t)}{z_i(t)} - \left(\frac{1 - z_i(t)}{z_i(t)} \right)^{b_i+1} \stackrel{(ii)}{<} 0$$

where (i) comes from $\frac{1-z_i(t)}{z_i(t)} > \frac{1-z_j(t)}{z_j(t)}$ and (ii) is deduced by $f(x) = x - x^{b_i+1} < 0$ for $x > 1$. Consequently,

$$z_i(t+1) - z_i(t) = w_{ij}(z_i(t))^{b_i+1} z_j(t) \left(\frac{1 - z_i(t)}{z_i(t)} - \frac{1 - z_j(t)}{z_j(t)} \left(\frac{1 - z_i(t)}{z_i(t)} \right)^{b_i} \right) < 0.$$

Assume $\lim_{t \to \infty} x_i(t) = x_i^*$; by the model (1), it must satisfy

$$x_i^* = x_j^* \quad (A3)$$

for any $i, j \in V$. In fact, for $n = 2$, if $\lim_{t \to \infty} x_1(t) = L_1$, $\lim_{t \to \infty} x_2(t) = L_2$ and $L_1 \neq L_2$, then

$$L_1 = \frac{L_1 + w_{12}L_1^{b_1}L_2}{1 + w_{12}\left(L_1^{b_1}L_2 + (1-L_1)^{b_1}(1-L_2)\right)}$$

if $r_1(t) = 2$ for sufficiently large t and

$$L_1 = \frac{L_1 + w_{11}L_1^{b_1+1}}{1 + w_{11}\left(L_1^{b_1+1} + (1-L_1)^{b_1+1}\right)}$$

if $r_1(t') = 1$ for another certain sufficiently large t'. This leads to a contradiction. The same result can be similarly deduced for $n = 3, 4, \ldots$. For simplicity, we denote $\lim_{t \to \infty} x_i(t) = x^*$.

Then, we obtain that

$$x^* = \frac{x^* + w_{ij}(x^*)^{b_i+1}}{1 + w_{ij}\left((x^*)^{b_i+1} + (1-x^*)^{b_i+1}\right)}$$

for any $i = 1, 2$ and $j = 1, 2$. Then, $x^* = 0$ or 1. By the monotonous increasing property, $x^* = 1$. Similarly, $\lim_{t \to \infty} x_i(t) = 1$ holds for any $n \geq 2$. Then, $\lim_{t \to \infty} x_i(t) = 1$ for any $i \in V$.

Step II: When $b_i > 1$ and $x_k(0) \in [0, \frac{1}{2})$, by Lemma 2, $x_k(t) \in [0, \frac{1}{2})$ for any $t \in \mathbb{N}$. Then, $\frac{1-x_k(t)}{x_k(t)} > 1$ and $\left(\frac{1-x_k(t)}{x_k(t)}\right)^{b_i} > \frac{1-x_k(t)}{x_k(t)}$ for any $k \in V$. Therefore,

$$\frac{1-x_i(t)}{x_i(t)} - \frac{1-x_j(t)}{x_j(t)}\left(\frac{1-x_i(t)}{x_i(t)}\right)^{b_i} < \frac{1-x_i(t)}{x_i(t)}\left(1 - \frac{1-x_j(t)}{x_j(t)}\right) < 0.$$

By $x_i(t+1) - x_i(t) < 0$ for any $i \in V$ and $t \in \mathbb{N}$, we obtain that $\lim_{t \to \infty} x_i(t)$ exists a.s. for any $i \in V$. With a similar analysis of **Step I**, $\lim_{t \to \infty} x_i(t) = 0$ a.s. for any $i \in V$.

Step III: Similarly, when $b_i > 1$ and $x_k(0) \in (\frac{1}{2}, 1]$, for any $t \in \mathbb{N}$,

$$\frac{1-x_i(t)}{x_i(t)} - \frac{1-x_j(t)}{x_j(t)}\left(\frac{1-x_i(t)}{x_i(t)}\right)^{b_i} \overset{(a)}{>} \frac{1-x_i(t)}{x_i(t)} - \left(\frac{1-x_i(t)}{x_i(t)}\right)^{b_i} \overset{(b)}{>} 0$$

where (a) comes from $\frac{1-x_j(t)}{x_j(t)} < 1$ for any $x_j(t) \in (\frac{1}{2}, 1]$, (b) is deduced by $\frac{1-x_i(t)}{x_i(t)} > \left(\frac{1-x_i(t)}{x_i(t)}\right)^{b_i}$ for $\frac{1-x_i(t)}{x_i(t)} < 1$ and $b_i > 1$. By $x_i(t+1) - x_i(t) > 0$ for any $i \in V$ and $t \in \mathbb{N}$, we obtain that $\lim_{t \to \infty}(t)$ exists a.s. for any $i \in V$. With a similar analysis of **Step I**, $\lim_{t \to \infty} x_i(t) = 1$ a.s. for any $i \in V$.

Step IV: By Lemma 3 and the model (1), $x_i(t+1) = f(x_i(t), x_j(t))$ if $r_i(t) = j$. Obviously, $f(x, y)$ is continuously differentiable on $[0,1] \times [0,1]$, and we obtain

$$\frac{\partial}{\partial x}f(x,y) = \frac{1}{(1+w(x^b y + (1-x)^b(1-y)))^2}\Big[1 + w[b + (1-b)x]x^{b-1}y +$$
$$w(1-x+bx)(1-x)^{b-1}(1-y) + bw^2(1+x)x^{b-1}(1-x)^b(1-y)y\Big] \overset{(i)}{>} 0 \quad (A4)$$

where (i) holds obviously if $b \in (0,1)$, (i) holds because $b - (b-1)x > 0$ if $b \geq 1$ and

$$\frac{\partial}{\partial y}f(x,y) = \frac{1}{(1+w(x^b y + (1-x)^b(1-y)))^2}\Big[wx^b(1-x) + w^2 x^b(1-x)^b + wx(1-x)^b\Big] > 0. \quad (A5)$$

We assume $x_i(0) \in [\eta_1, \eta_2] \in (0, \frac{1}{2})$ for any $i \in V$. Note that $b_i \in (0,1)$ for any $i \in V$. Set $w^* = \max_{i,j \in V} w_{ij}$. By Lemma 3 and the model (1), we have

$$x_i(1) = f(x_i(0), x_j(0)) \stackrel{(i)}{<} f(\eta_2, \eta_2) \stackrel{(ii)}{<} \frac{\eta_2 + w_{ij}\eta_2^{b_i+1}}{1 + w_{ij}(\eta_2^2 + (1-\eta_2)^2)} \leq \frac{\eta_2 + w^*\eta_2^{b_i+1}}{1 + w^*(\eta_2^2 + (1-\eta_2)^2)}$$

where (i) is obtained by the inequalities (A4) and (A5), (ii) comes from $x^{b_i} > x$ for $x \in [\eta_1, \eta_2], b_i \in (0,1)$. Consequently, we prove that

$$Q(x) \triangleq \frac{x + wx^{b+1}}{1 + w(x^2 + (1-x)^2)} < x \quad (A6)$$

for $b \in (0,1), x, w \in (0,1)$. Note that the inequality (A6) is equivalent to $x^b < x^2 + (1-x)^2$. Denote $k(x) = x^b - x^2 - (1-x)^2$. Then, $k(0) = -1 < 0$, $k(1) = 0$ and $k(x)$ has no stationary point by $k'(x) = 0$, $x \in (0,1)$. Thus $k(x) < 0$ for $x \in (0,1)$. Therefore, the inequality (A6) holds.

According to (A6), we obtain that for any $i \in V$,

$$x_i(2) = f(x_i(1), x_j(1)) < f(Q(\eta_2), Q(\eta_2)) \stackrel{(i)}{<} Q(Q(\eta_2))$$
$$x_i(3) = f(x_i(2), x_j(2)) < f(Q(Q(\eta_2)), Q(Q(\eta_2))) < Q(Q(Q(\eta_2)))$$
$$\cdots\cdots\cdots$$
$$x_i(t+1) = f(x_i(t), x_j(t)) < f(Q^{(t)}(\eta_2), Q^{(t)}(\eta_2)) < Q^{(t+1)}(\eta_2)$$

where (i) is deduced by the inequality (A6) and the function $Q^{(t)}(x)$ is defined by $Q^{(t)}(x) = Q(Q\cdots(Q(x))\cdots)$. Obviously, it is not difficult to prove that $\max_{i \in V} x_i(t) < Q^{(t)}(\eta_2) \to 0$ as $t \to \infty$. In fact, by the inequality (A6), $Q^{(t)}(\eta_2)$ is monotonic decreasing and has a lower bound 0. We assume $\lim_{t \to \infty} Q^{(t)}(\eta_2) = Q^*$. According to $\frac{Q^* + w(Q^*)^{b+1}}{1 + w((Q^*)^2 + (1-Q^*)^2)} \leq Q^*$ for $Q^* \in [0, \frac{1}{2}), Q^* = 0$.

Now, we have completed the proof.

Appendix D

Step I: We prove that $P\{\overline{\lim}_{t \to \infty} x_i(t) \in (1 - \frac{1}{1 + (\frac{M}{M-1})^{\frac{1}{b_m}}}, 1) | x_i(t) > \frac{1}{2}, i.o.\} = 0$ for any $\eta \in (0, \frac{1}{2})$, $M \in \{2, 3, 4, \ldots\}$ and $b_m = \max_{i \in V}\{b_i\}$. Denote $\mathcal{A}_M = \{\overline{\lim}_{t \to \infty} x_i(t) \in (1 - \frac{1}{1+(\frac{M}{M-1})^{\frac{1}{b_m}}}, 1)\}$. In fact, if $x_i(t) > \frac{1}{2}, i.o.$, without loss of generality, we set $x_i(t) > \frac{1}{2}$ for $t > T$. By the upper limit definition, in the event $\{\overline{\lim}_{t \to \infty} x_i(t) < 1 - \eta\}$ where $\eta < \frac{1}{1+(\frac{M}{M-1})^{\frac{1}{b_m}}}$, for any

$$\varepsilon \in (0, \min\{\frac{1}{3}\eta, \frac{(1-\eta)\left(\frac{M-1}{M}\right)^{\frac{1}{b_i}} - \eta}{1 + \left(\frac{M-1}{M}\right)^{\frac{1}{b_i}}}, \frac{\eta\left(1 - \frac{4}{3}\eta\right)^{b_i+1}}{3M + 6M(1 - \frac{4}{3}\eta)^{b_i+1}}\}), \quad (A7)$$

there exists $T_1 > 0$, for any $t \geq T_1$, $x_i(t) < 1 - \eta + \varepsilon$ and $x_i(T_1) \in (1 - \eta - \varepsilon, 1 - \eta + \varepsilon) \subset [\frac{1}{2}, 1]$. However, it is obvious that $r_i(T_1) = i$ holds with a positive probability by Assumption 1, then

$$x_i(T_1 + 1) - x_i(T_1) \stackrel{(i)}{=} \frac{x(1-x)(x^{b_i} - (1-x)^{b_i})}{1 + x^{b_i+1} + (1-x)^{b_i+1}} \stackrel{(ii)}{\geq} \frac{(1-x)x^{b_i+1}}{M(1 + 2x^{b_i+1})} \stackrel{(iii)}{>} 2\varepsilon, \quad (A8)$$

where (i) holds by setting $x = x_i(T_1)$ and $r_i(T_1) = i$, (ii) comes from $x^{b_i+1} - (1-x)^{b_i+1} > \frac{x^{b_i+1}}{M}$ because

$$\frac{M-1}{M} > \left(\frac{1-x}{x}\right)^{b_i} \iff \left(\frac{M-1}{M}\right)^{\frac{1}{b_i}} > \frac{1-x}{x} \iff \left(\frac{M-1}{M}\right)^{\frac{1}{b_i}} > \frac{\eta+\varepsilon}{1-\eta-\varepsilon}$$

(Because $\frac{1-x}{x}$ is decreasing on $x \in (1-\eta-\varepsilon, 1-\eta+\varepsilon)) \iff \varepsilon < \frac{(1-\eta)\left(\frac{M-1}{M}\right)^{\frac{1}{b_i}} - \eta}{1 + \left(\frac{M-1}{M}\right)^{\frac{1}{b_i}}}$,

(iii) holds because $1 - x > 1 - (1 - \eta + \varepsilon) > \frac{2}{3}\eta$ and

$$\min_{x \in (1-\eta-\varepsilon, 1-\eta+\varepsilon)} \left\{ \frac{\frac{2}{3}\eta x^{b_i+1}}{1 + 2x^{b_i+1}} \right\} = \frac{\frac{2}{3}\eta(1-\eta-\varepsilon)^{b_i+1}}{1 + 2(1-\eta-\varepsilon)^{b_i+1}} > 2\varepsilon$$

which is equivalent to

$$\eta(1-\eta-\varepsilon)^{b_i+1} \stackrel{(a)}{>} \eta(1-\frac{4}{3}\eta)^{b_i+1} \stackrel{(b)}{>} \frac{3M\varepsilon}{\eta - 6M\varepsilon}$$

where (a) and (b) hold based on the inequality (A11).

According to the inequality (A8), $x_i(T_1+1) > 1 - \eta - \varepsilon + 2\varepsilon = 1 - \eta + \varepsilon$, which contradicts the definition of $\varlimsup_{t \to \infty} x_i(t) < 1 - \eta$.

Consequently, note that $\lim_{M \to \infty} \frac{1}{1+\left(\frac{M}{M-1}\right)^{\frac{1}{b_m}}} = \frac{1}{2}$ and $\frac{1}{1+\left(\frac{M}{M-1}\right)^{\frac{1}{b_m}}}$ is monotonous as M increases. In addition, $P\{\mathcal{A}_M | x_i(t) > \frac{1}{2}, i.o.\} = 0$ for any $M = 2, 3, 4, \ldots$. According to

$$P\{\varlimsup_{t \to \infty} x_i(t) \in (\frac{1}{2}, 1) | x_i(t) > \frac{1}{2}, i.o.\} \le \sum_{M=2}^{\infty} P\{\mathcal{A}_M | x_i(t) > \frac{1}{2}, i.o.\} = 0,$$

we obtain that $P\{\varlimsup_{t \to \infty} x_i(t) \in (\frac{1}{2}, 1) | x_i(t) > \frac{1}{2}, i.o.\} = 0$.

Similarly, $P\{\varliminf_{t \to \infty} x_i(t) \in (0, \frac{1}{2}) | x_i(t) < \frac{1}{2}, i.o.\} = 0$.

Step II: We prove that $P\{\varliminf_{t \to \infty} x_i(t) \in (\frac{1}{2}, 1) | \varlimsup_{t \to \infty} x_i(t) = 1\} = 0$ for any $i \in V$ by contradiction. For any trajectory of the opinion evolution, we extract any subsequence of $\{x_i(t)\}$ which converges to 1, then others are constituted as $\{x_i(t_k)\}$. Set $\{t_k\} \cup \{t'_s\} = \{1, 2, 3, \ldots\}$. Set $\varlimsup_{k \to \infty} x_i(t_k) \le 1 - \eta < 1$, where $\eta \in (0, \frac{1}{2})$. By $\eta < \frac{1}{2}$,

$$\frac{1 + (1-\eta)^{b_i}}{1 + (1-\frac{\eta}{3})^{b_i+1} + \eta^{b_i+1}} > \frac{1}{2}. \tag{A9}$$

Without loss of generality, we only need to analyze the case that $t_{k+1} - t_k > 1$ always holds when k is larger than a certain threshold. In fact, we can always take the subsequence $\{t_{k_s}\}$ of $\{t_k\}$ s.t. $t_{k_{s+1}} - t_{k_s} > 1$ when s is sufficiently large.

Note that $r_i(t_k) = i$ holds with a positive probability by Assumption 1 for any $t_k \in \mathbb{N}$. Based on the definition of $\{t_k\}$, for any

$$\varepsilon \in (0, \min\{\frac{1}{3}\eta, \frac{(1-\frac{\eta}{3})^{b_i+1} + \eta^{b_i+1} - (1-\eta)^{b_i}}{2(1-\eta)^{b_i} - (1-\frac{\eta}{3})^{b_i+1} - \eta^{b_i+1}}(1-\eta)\}) \tag{A10}$$

where $\frac{(1-\frac{\eta}{3})^{b_i+1}+\eta^{b_i+1}-(1-\eta)^{b_i}}{2(1-\eta)^{b_i}-(1-\frac{\eta}{3})^{b_i+1}-\eta^{b_i+1}} > 0$ comes from the inequality (A9), there exists $K > 0$, for any $k \geq K$, $t \geq t_K - 1$ and $t \neq t_k$, $x_i(t_k) < 1 - \eta + \varepsilon$ and $x_i(t) > 1 - \eta + 2\varepsilon$. If $r_i(t_k - 1) = i$, by the inequality (A5), then

$$x_i(t_k) \stackrel{(a)}{=} \frac{x + x^{b_i+1}}{1 + x^{b_i+1} + (1-x)^{b_i+1}} \stackrel{(b)}{>} \frac{(1 - \eta + 2\varepsilon) + (1 - \eta + 2\varepsilon)^{b_i+1}}{1 + (1 - \eta + 2\varepsilon)^{b_i+1} + (\eta - 2\varepsilon)^{b_i+1}}$$

where (a) holds by setting $x = x_i(t_k - 1)$, (b) comes from $\frac{d}{dx}\frac{x+x^{b_i+1}}{1+x^{b_i+1}+(1-x)^{b_i+1}} > 0$ and $x \in (1 - \eta + 2\varepsilon, 1]$.

According to the inequality (A10),

$$\frac{1 + (1-\eta)^{b_i}}{1 + (1-\frac{\eta}{3})^{b_i+1} + \eta^{b_i+1}}(1 - \eta + 2\varepsilon) > 1 - \eta + \varepsilon.$$

Specially,

$$\frac{(1 - \eta + 2\varepsilon) + (1 - \eta + 2\varepsilon)^{b_i+1}}{1 + (1 - \eta + 2\varepsilon)^{b_i+1} + (\eta - 2\varepsilon)^{b_i+1}} > \frac{1 + (1-\eta)^{b_i}}{1 + (1-\frac{\eta}{3})^{b_i+1} + \eta^{b_i+1}}(1 - \eta + 2\varepsilon).$$

Therefore, $x_i(t_k) > 1 - \eta + \varepsilon$ holds with a positive probability. Then,

$$\{\varliminf_{t\to\infty} x_i(t) \in (\frac{1}{2}, 1) | \varlimsup_{t\to\infty} x_i(t) = 1\}$$

is a zero probability event. With a similar method,

$$P\{\varlimsup_{t\to\infty} x_i(t) \in (0, \frac{1}{2}) | \varliminf_{t\to\infty} x_i(t) = 0\} = 0$$

for any $i \in V$.

Step III: We prove that $P\{\varliminf_{t\to\infty} x_i(t) \in (0, \frac{1}{1+(\frac{M}{M-1})^{\frac{1}{b_m}}}) | \varlimsup_{t\to\infty} x_i(t) = 1\} = 0$ for any $M \in \{2, 3, 4, \ldots\}$ and $b_m = \min_{i \in V}\{b_i\}$. Denote $\mathcal{B}_M = \{0 < \varliminf_{t\to\infty} x_i(t) < \frac{1}{1+(\frac{M}{M-1})^{\frac{1}{b_m}}}\}$. In fact, by the lower limit definition, in the event $\{\varliminf_{t\to\infty} x_i(t) < \eta\}$ where $\eta < \frac{1}{1+(\frac{M}{M-1})^{\frac{1}{b_m}}}$, for any

$$\varepsilon \in (0, \min\{\frac{1}{3}\eta, \frac{(1-\eta)\left(\frac{M-1}{M}\right)^{\frac{1}{b_i}} - \eta}{1 + \left(\frac{M-1}{M}\right)^{\frac{1}{b_i}}}, \frac{\eta\left(1+\frac{4}{3}\eta\right)^{b_i+1}}{3M + 6M(1-\frac{4}{3}\eta)^{b_i+1}}\}), \quad (A11)$$

there exists $T_2 > 0$, for any $t \geq T_2$, $x_i(t) > \eta - \varepsilon$ and $x_i(T_2) \in (\eta - \varepsilon, \eta + \varepsilon) \subset [0, \frac{1}{2}]$. However, it is obvious that $r_i(T_1) = i$ holds with a positive probability by Assumption 1. With a similar method of **Step I**, $x_i(T_2 + 1) - x_i(T_2) < -2\varepsilon$, then $x_i(T_2 + 1) < \eta - \varepsilon$. This induces a contradiction. In addition, $P\{\mathcal{B}_M | \varlimsup_{t\to\infty} x_i(t) = 1\} = 0$ for any $M = 2, 3, 4, \ldots$.

According to

$$P\{\varliminf_{t\to\infty} x_i(t) \in (0, \frac{1}{2}) | \varlimsup_{t\to\infty} x_i(t) = 1\} \leq \sum_{M=2}^{\infty} P\{\mathcal{B}_M | \varlimsup_{t\to\infty} x_i(t) = 1\} = 0,$$

we obtain that $P\{\varliminf_{t\to\infty} x_i(t) \in (0, \frac{1}{2}) | \varlimsup_{t\to\infty} x_i(t) = 1\} = 0$.

Similarly, $P\{\varlimsup_{t\to\infty} x_i(t) \in (\frac{1}{2}, 1) | \varliminf_{t\to\infty} x_i(t) = 0\} = 0$.

In a sum, if $x_i(t) \in (\frac{1}{2}, 1)$ i.o. for certain $i \in V$, then $\lim_{t \to \infty} x_i(t) = 1$ a.s. Similarly, if $x_i(t) \in (0, \frac{1}{2})$ i.o. for certain $i \in V$, then $\lim_{t \to \infty} x_i(t) = 0$ a.s.

Appendix E

We prove this theorem by contradiction.

Step I: We first prove that $P\{\cap_{i \in V}\{x_i(t) = \frac{1}{2}, i.o.\}\} = 0$. In fact, if $x_1(t) = \frac{1}{2}$ i.o., then there exists a time subsequence $\{t_k\}$ s.t. $x_1(t_k) = \frac{1}{2}$. There must exist a threshold $T > 0$ s.t. $\{t_k, k \geq T\} = \{t_T, t_T + 1, t_T + 2, \ldots\}$. Otherwise, there exists another time subsequence $\{t_s\}$ s.t. $x_1(t_s) > \frac{1}{2}$ (or $x_1(t_s) > \frac{1}{2}$). According to **Step I** in the proof of Theorem 2, $x_1(t)$ converges to 1 (or 0), which contradicts $x_1(t) = \frac{1}{2}$ i.o. Therefore, $\lim_{t \to \infty} x_1(t) = \frac{1}{2}$. In addition, we assume $\lim_{t \to \infty} x_2(t) > \frac{1}{2}$, or $x_2(t) > \frac{1}{2}$ i.o. By **Step I** of Theorem 2, $\lim_{t \to \infty} x_2(t) = 1$. Note that $r_1(t) = 2$ holds with a positive probability by Assumption 1, with the similar method on the proof of **Step I** of Theorem 2, we obtain that $\lim_{t \to \infty} x_1(t) = 1$ which contradicts $\lim_{t \to \infty} x_1(t) = \frac{1}{2}$. Similarly, $\lim_{t \to \infty} x_2(t) < \frac{1}{2}$ does not hold. Therefore, $\lim_{t \to \infty} x_2(t) = 1$. Recursively, for any $i \in V$, $\lim_{t \to \infty} x_i(t) = 1$.

Note that for any $x, y \in [0, 1]$, the solutions of $f(x, y) = \frac{1}{2}$ satisfies $y = \frac{(1-x)^b - \frac{2x-1}{w}}{(1-x)^b + x^b}$. Therefore, by the model (1), $x_i(t+1) = f(x_i(t), x_j(t))$ if $r_i(t) = j$. Primary images of $f(x_i(t), x_j(t)) = \frac{1}{2}$ on $[0, 1]$ are all scatters; thus, $P\{\cap_{i \in V}\{x_i(t) = \frac{1}{2}, i.o.\}\} = 0$. That is to say, $P\{\mathbf{E}_{con}\} = 1$.

Step 2: We prove that the event that $x_i(t) > \frac{1}{2}$ i.o. for $i \in V_1$ and $x_j(t) < \frac{1}{2}$ i.o. for $j \in V_2$ holds with a positive probability. Based on the above proof, we assume that $x_i(0) \in [0, \eta_1) \cup (1 - \eta_2, 1]$, $\eta_1 + \eta_2 < 1$ and $\eta_1, \eta_2 < \frac{1}{2}$. Set $V_1 = \{i = 1, 2, \ldots, \lfloor \frac{n}{2} \rfloor : x_i(0) \in [0, \eta_1)\}$, $V_2 = \{j = \lfloor \frac{n}{2} \rfloor, \ldots, n : x_j(0) \in (1 - \eta_2, 1]\}$, $V_1 \cup V_2 = V$ and both V_1 and V_2 are nonempty.

In fact, by contradiction, we assume $P\{x_i(t) > \frac{1}{2}, i.o. i \in V_1; x_j(t) < \frac{1}{2}, i.o. j \in V_2\} = 0$. Then,

$$\gamma(x_1(t), x_2(t), \ldots, x_n(t)) = \sum_{1 \leq i < j \leq 2} |x_i(t) - x_j(t)| \to 0$$

a.s. as $t \to \infty$. Therefore, for any $\varepsilon \in (0, \frac{1}{4})$, there exists an almost surely finite r.v. $T > 0$, for any $t > T$, $\gamma(x_1(t), x_2(t), \ldots, x_n(t)) < \varepsilon$. Then,

$$\max_{i,j \in V} |x_i(t) - x_j(t)| < \gamma(x_1(t), x_2(t), \ldots, x_n(t)) < \varepsilon. \tag{A12}$$

According to Theorem 2, there is always an agent i satisfying $x_i(t) > \frac{1}{2}$ i.o. a.s., or $x_i(t) < \frac{1}{2}$ i.o. a.s. Without loss of generality, we assume when $t > T$, $x_i(t) < \varepsilon$ or $1 - x_i(t) < \varepsilon$. Then, for any $j \in V$, by the inequality (A12),

$$x_j(t) < x_i(t) + \max_{i,j \in V} |x_i(t) - x_j(t)| < 2\varepsilon < \frac{1}{2}$$

or $1 - x_j(t) < 1 - x_i(t) + \max_{i,j \in V} |x_i(t) - x_j(t)| < 2\varepsilon < \frac{1}{2}$. \tag{A13}

Note that the event $\{\gamma(x_1(t), x_2(t), \ldots, x_n(t)) \to 0, t \to \infty\} = \mathcal{G}_1 \cup \mathcal{G}_2$ where $\mathcal{G}_1 = \{x_i(t) \to 1, \text{ for any } i \in V\}$ and $\mathcal{G}_2 = \{x_i(t) \to 0, \text{ for any } i \in V\}$. For the event $\mathcal{G}_1$, $x_i(t) \to 1$ for any $i \in V_1$; while for the event $\mathcal{G}_2$, $x_i(t) \to 0$ for any $i \in V_2$. Note that $x_i(0) \in [0, \eta_1)$ for $i \in V_1$ and $x_i(t+1) < x_i(t)$ if $r_i(t) \in V_1$, $i \in V_1$ by Theorem 1. In addition, according to Lemma 2, $x_i(T) > \frac{1}{2}$ for any $i \in V$ or $x_i(T) < \frac{1}{2}$ for any $i \in V$. Denote $p_{V_1} = P\{x_i(T) > \frac{1}{2}, i \in V_1\}$, $p_{V_2} = P\{x_i(T) > \frac{1}{2}, i \in V_2\}$, $q_{V_1} = P\{x_i(T) < \frac{1}{2}, i \in V_1\}$ and $q_{V_2} = P\{x_i(T) < \frac{1}{2}, i \in V_2\}$. By the assumption, $p_{V_1} p_{V_2} + q_{V_1} q_{V_2} = 1$. If

$P\{x_i(t) > \frac{1}{2}, i.o. i \in V_1; x_j(t) < \frac{1}{2}, i.o. j \in V_2\} = 0$, by Theorem 2, $p_{V_1}p_{V_2} + q_{V_1}q_{V_2} = 1$. However,

$$p_{V_1}p_{V_2} + q_{V_1}q_{V_2} \overset{(a)}{<} p_{V_1}p_{V_2} + (1-p_{V_1})(1-p_{V_2}) < 1$$

where (a) holds because $\{x_i(T) < \frac{1}{2}, i \in V_1\} \subset \{x_i(T) > \frac{1}{2}, i \in V_1\}^c$. This induces a contradiction. Thus, the event that $x_i(t) > \frac{1}{2}$ i.o. for $i \in V_1$ and $x_j(t) < \frac{1}{2}$ i.o. for $j \in V_2$ holds with a positive probability.

In a sum, $P\{E_{polarization}\} > 0$ if $\{b_i\}$ are homogeneous.

Appendix F

By Lemma 5, there exist two events $E_{con0} = \{\lim_{t\to\infty} x_i(t) = 0, \forall i \in V\}$ and $E_{con1} = \{\lim_{t\to\infty} x_i(t) = 1, \forall i \in V\}$ such that $P\{E_{con0} \cup E_{con1}\} = 1$.

Without loss of generality, we denote the limit of opinions $\{x_i(t)\}$ as x_i^*. According to Lemma 1, for any $\varepsilon > 0$, given $\varepsilon_0 > 0$, there exists $T_1 > 0$ such that for any $t > T_1$, $P(\cup_{t=T_1}^{+\infty}|x_1(t) - x_1^*| \geq \varepsilon_0) \leq \frac{\varepsilon}{n}$. Similarly, there exists $T_i > 0$, such that for any $t > T_i$, $P(\cup_{t=T_i}^{+\infty}|x_i(t) - x_i^*| \geq \varepsilon_0) \leq \frac{\varepsilon}{n}$ for $i = 2, 3, \ldots, n$. Take $T^* = \max\{T_1, T_2, \ldots, T_n\}$, then

$$P(\bigcup_{i=1}^n (\cup_{t=T_i}^{+\infty}|x_i(t) - x_i^*| \geq \varepsilon_0)) \leq \sum_{i=1}^n P(\cup_{t=T_i}^{+\infty}|x_i(t) - x_i^*| \geq \varepsilon_0) \leq \varepsilon.$$

By DeMorgan formula, for any $t > T^*$,

$$P(\cap\{x_i(t) : x_i(t) \in [x_i^* - \varepsilon_0, x_i^* + \varepsilon_0] \cap [0,1]\}) > 1 - \varepsilon. \tag{A14}$$

Set $\varepsilon_0 < \frac{1}{2}$ and $x_i(0) \in (0, \frac{1}{2})$. According to Theorem 1 and its proof, how opinions of the model (2) evolve only depend on their selected opinions $r_i(t)$ and whether $b_i(t) > 1$ or $b_i(t) \in (0,1)$. Thus, if $b_i(t) > 1$ for any $i \in V$ and $t \in \mathbb{N}$, we have $P(\lim_{t\to\infty} x_i(t) = 0) = 1$. Denote $x_i^* = 0$ and $T_1^* = T^*$. If $b_i(t) > 1$ for $t \in (0, T_1^*)$, by the inequality (A14), we obtain that $P(\cap\{x_i(t) : x_i(t) \in [0, \varepsilon_0]\}) > 1 - \varepsilon$ for any $t > T_1^*$.

We will prove that $\{x_i(t) : x_i(T_1^* + 1) \in [0, \varepsilon_0], \forall i \in V\} = \{\lim_{t\to\infty} x_i(t) = 0\}$.

(I) If $0 < b_i(t) < 1$ for any $t > T_1^*$ and $i \in V$, we can prove that $x_i(t+1) < x_i(t)$. In fact, with a similar method of the inequality (A2), we obtain that

$$x_i(t+1) - x_i(t) = w_{ij}x_i^{b_i+1}(t)x_j(t)\left[\frac{1-x_i(t)}{x_i(t)} - \frac{1-x_j(t)}{x_j(t)}\left(\frac{1-x_i(t)}{x_i(t)}\right)^{b_i(t)}\right].$$

Denote $L(x, y; b) = \frac{1-x}{x} - \frac{1-y}{y}\left(\frac{1-x}{x}\right)^b$ and $R(x) = \frac{1}{1+\left(\frac{1-x}{x}\right)^{1-b}}$. Note that $x_i(t+1) - x_i(t) = w_{ij}x_i^{b_i+1}(t)x_j(t)L(x_i(t), x_j(t); b)$ and $x_i(T_1^* + 1) < \varepsilon_0$ for any $i \in V$. Furthermore, by

$$R(x) = \frac{1}{1+\left(\frac{1-x}{x}\right)^{1-b}} \overset{(a)}{>} x$$

where (a) is deduced by $(1-x)^b > x^b$ for $x < \varepsilon_0 < \frac{1}{2}$, we obtain that

$$L(x, y; b) = \frac{1-x}{x} - \frac{1-y}{y}\left(\frac{1-x}{x}\right)^b \overset{(i)}{<} \frac{1-x}{x} - \frac{1-R(x)}{R(x)}\left(\frac{1-x}{x}\right)^b$$

$$= \frac{1-x}{x} - \left(\frac{1-x}{x}\right)^{1-b}\left(\frac{1-x}{x}\right)^b = 0$$

where (i) comes from $\frac{1-y}{y} > \frac{1-R(x)}{R(x)}$ for $y \leq \varepsilon_0 < R(x)$. Thus, $x_i(T_1^* + 2) < x_i(T_1^* + 1)$ for any $i \in V$. Consequently, it holds that $x_i(t+1) < x_i(t)$ for any $t > T_1^*$.

(II) If $b_i(t) > 1$ for any $t > T_1^*$ and $i \in V$, we can also prove that $x_i(t+1) < x_i(t)$. Similarly, $x_i(t+1) - x_i(t) = w_{ij} x_i^{b_i+1}(t) x_j(t) L(x_i(t), x_j(t); b)$. Note that

$$L(x,y;b) = \frac{1-x}{x} - \frac{1-y}{y}\left(\frac{1-x}{x}\right)^b \overset{(i)}{<} \frac{1-x}{x} - \frac{1-y}{y}\frac{1-x}{x} = -\frac{1}{y}\frac{1-x}{x} < 0$$

where (i) holds because $\left(\frac{1-x}{x}\right)^b > \frac{1-x}{x}$ for $b > 1$ and $0 < x < \frac{1}{2}$. For $x_i(T_1^* + 1) < \varepsilon_0$ for any $i \in V$, we obtain that $x_i(T_1^* + 2) < x_i(T_1^* + 1)$ for any $i \in V$. Consequently, it holds that $x_i(t+1) < x_i(t)$ for any $t > T_1^*$.

By the previous conclusion that $x_i(t)$ decreases for any $t > T_1^*$ and for any $b_i(t) > 0$, $i \in V$. With the similar proof of Theorem 1, all opinions converge to 0 a.s.

With a similar method, there exists a time threshold $T_2^* > 0$, such that $\lim_{t \to \infty} x_i(t) = 1$ a.s. for any $i \in V$ if $b_i(t) \in (0,1)$, $x_i(0) \in (\frac{1}{2}, 1)$ for any $i \in V$ and $t \in (0, T_2^*)$. The conclusion holds.

References

1. Qualman, E. *Socialnomics: How Social Media Transforms the Way We Live and Do Business*; Wiley: New York, NY, USA, 2009.
2. Latane, B. Dynamic social impact: The creation of culture by communication. *J. Commun.* **1996**, *46*, 13–25. [CrossRef]
3. Degroot, M.H. Reaching a Consensus. *J. Am. Stat. Assoc.* **1974**, *69*, 118–121. [CrossRef]
4. Jadbabaie, A.; Lin, J.; Morse, A.S. Coordination of groups of mobile autonomous agents using nearest neighbor rules. *IEEE Trans. Autom. Control* **2003**, *48*, 988–1001. [CrossRef]
5. Moreau, L. Stability of multiagent systems with time-dependent communication links. *IEEE Trans. Autom. Control* **2005**, *50*, 169–182. [CrossRef]
6. Cao, M.; Morse, A.S.; Anderson, B.D.O. Reaching a consensus in a dynamically changing environment: A graphical approach. *SIAM J. Control Optim.* **2008**, *47*, 575–600. [CrossRef]
7. Nedic, A.; Olshevsky, A.; Ozdaglar, A.; Tsitsiklis, J.N. On distributed averaging algorithms and quantization effects. *IEEE Trans. Autom. Control* **2009**, *54*, 2506–2517. [CrossRef]
8. Friedkin, N.E.; Johnsen, E.C. Social influence and opinions. *J. Math. Sociol.* **1990**, *15*, 193–206. [CrossRef]
9. Hegselmann, R.; Krause, U. Opinion dynamics and bounded confidence models, analysis, and simulation. *J. Artif. Soc. Soc. Simul.* **2002**, *5*, 1–33.
10. Acemoglu, D.; Como, G.; Fagnani, F.; Ozdaglar, A. Opinion fluctuations and disagreement in social networks. *Math. Oper. Res.* **2013**, *38*, 1–27. [CrossRef]
11. Shi, G.; Proutiere, A.; Baras, M.; Johansson, K.H. The evolution of beliefs over signed social networks. *Oper. Res.* **2016**, *64*, 585–604. [CrossRef]
12. McCarty, N.; Poole, K.T.; Rosenthal, H. *Polarized America: The Dance of Ideology and Unequal Riches*; MIT Press: Cambridge, MA, USA, 2016.
13. Dandekar, P.; Goel, A.; Lee, D.T. Biased assimilation, homophily, and the dynamics of polarization. *Proc. Natl. Acad. Sci. USA* **2013**, *110*, 5791–5796. [CrossRef] [PubMed]
14. Chen, Z.; Qin, J.; Li, B.; Qi, H.; Buchhorn, P.; Shi, G. Dynamics of opinions with social biases. *Automatica* **2019**, *106*, 374–383. [CrossRef]
15. Lord, C.G.; Ross, L.; Lepper, M.R. Biased assimilation and attitude polarization: The effects of prior theories on subsequently considered evidence. *J. Pers. Soc. Psychol.* **1979**, *37*, 2098–2109. [CrossRef]
16. Munro, G.D.; Ditto, P.H.; Lockhart, L.K.; Fagerlin, A.; Gready, M.; Peterson, E. Biased assimilation of sociopolitical arguments: Evaluating the 1996 U.S. presidential debate. *Basic Appl. Soc. Psych.* **2002**, *24*, 15–26. [CrossRef]
17. Waller, I.; Anderson, A. Quantifying social organization and political polarization in online platforms. *Nature* **2021**, *600*, 264–268. [CrossRef] [PubMed]
18. Kozitsin, I.V. Formal models of opinion formation and their application to real data: Evidence from online social networks. *J. Math. Sociol.* **2020**, *46*, 120–147. [CrossRef]

Disclaimer/Publisher's Note: The statements, opinions and data contained in all publications are solely those of the individual author(s) and contributor(s) and not of MDPI and/or the editor(s). MDPI and/or the editor(s) disclaim responsibility for any injury to people or property resulting from any ideas, methods, instructions or products referred to in the content.

Article

Distributed Disturbance Observer-Based Containment Control of Multi-Agent Systems via an Event-Triggered Approach

Long Jian, Yongfeng Lv, Rong Li, Liwei Kou and Gengwu Zhang *

College of Electrical and Power Engineering, Taiyuan University of Technology, Taiyuan 030024, China; jianlong@tyut.edu.cn (L.J.); lvyilian1989@foxmail.com (Y.L.); lirong@tyut.edu.cn (R.L.); Kouliwei@tyut.edu.cn (L.K.)
* Correspondence: zhanggengwu@tyut.edu.cn

Abstract: This paper studies the containment control problem of linear multi-agent systems (MASs) subject to external disturbances, where the communication graph is a directed graph with the followers being undirected connections. In order to save communication costs and energy consumption, a distributed disturbance observer-based event-triggered controller is employed based on the relative outputs of neighboring followers. Compared with conventional controllers, our observer-based controller utilizes the relative outputs of neighboring followers at the same triggered instant. Furthermore, it is shown that Zeno behavior can be avoided. Finally, the validity of our proposed methodology is demonstrated by a simulation example.

Keywords: multi-agent systems; event-triggered control; disturbance observer; containment control; output feedback

MSC: 93A16

Citation: Jian, L.; Lv, Y.; Li, R.; Kou, L.; Zhang, G. Distributed Disturbance Observer-Based Containment Control of Multi-Agent Systems via an Event-Triggered Approach. *Mathematics* **2023**, *11*, 2363. https://doi.org/10.3390/math11102363

Academic Editors: Jiangping Hu and Zhinan Peng

Received: 25 April 2023
Revised: 15 May 2023
Accepted: 17 May 2023
Published: 19 May 2023

Copyright: © 2023 by the authors. Licensee MDPI, Basel, Switzerland. This article is an open access article distributed under the terms and conditions of the Creative Commons Attribution (CC BY) license (https://creativecommons.org/licenses/by/4.0/).

1. Introduction

Distributed cooperative control of multi-agent systems (MASs) has drawn a great deal of attention, mainly due to its wide applications in engineering systems, such as robotic systems, power sharing in DC microgrids and so forth. A rich body of results about the cooperative control of MASs has been reported, such as consensus control, leader-following tracking control and containment control [1–8]. Although there are many studies on leaderless consensus control and one-leader tracking control, in some practical applications, multiple leaders can complete certain tasks that are difficult for a single agent to complete. In the presence of multiple leaders, the containment control problem has been investigated, that is, all followers tend to the convex hull spanned by all the leaders. There is increasing research on the containment control of different MASs, including simple MASs of double-integrator MASs [9]; homogeneous linear MASs [6]; homogeneous discrete MASs [7]; and heterogeneous high-order MASs [10].

Note that disturbance widely exists in engineering applications and is usually unavoidable. In engineering, a system often works in an environment with various disturbances, which have a certain impact on the control accuracy, while the cooperative control of MASs has strict requirements on the control accuracy. Therefore, how to deal with the interference problem has always been the key to the control design of MASs. Some methods of disturbance rejection have been proposed, including anti-interference methods, disturbance observers, output regulation, and so on [11–16]. In [11], distributed event-based consensus protocols based on the disturbance observer are proposed for MASs with matched disturbances. In [13], a disturbance observer is designed for MASs under deterministic disturbances. Under the state or relative state measurements, disturbance rejection is used to estimate the disturbances [17–20]. However, when the state information is not available, it is necessary to design the output feedback control protocol [21,22]. Therefore, it is of great significance to use the output feedback method to study the containment control problem with external disturbances.

Nowadays, most communication networks between MASs are wireless communication. However, continuous communications among neighboring agents may be equipped with simple embedded microprocessors. High-frequency continuous sampling not only causes high system energy consumption but also leads to bandwidth constraints. Event-triggered control provides an effective strategy to solve this problem [23–31]. In this control strategy, by designing a reasonable trigger strategy, the amount of communication and data updates is reduced, but satisfactory performance is still maintained. Among them, the event-triggered strategy was first applied to MASs in the literature [23]. The consensus problem was addressed in [24,25,27,28,32,33] by using the event-triggered control strategy, and some papers considered leader-following consensus and other issues [34,35], while this paper focuses on its application to containment control problems (see [21,34,36,37]).

Enlightened by the above observations, we integrate a disturbance observer and distributed event-triggered output feedback controller for the containment control problem of linear MASs subject to external disturbances. The main contributions of this paper are at least threefold:

(1) Compared with the works on the consensus [13], this work considers the containment control problem of linear MASs subject to external disturbances;
(2) Compared with most existing strategies [13,38], and based on the event-triggered strategy, the containment control problem can be solved for linear MASs without the need for continuous communications;
(3) The proposed disturbance observer-based event-triggered control uses the relative output information of each agent.

2. Preliminaries and Problem Formulation

2.1. Notations

Let $\mathbf{0}_m$ and $\mathbf{0}_{M \times M}$ be the $m \times 1$ column vector of all zeros and the $M \times M$ matrix of all zeros, respectively. For a matrix X, X^T stands for its transpose, and $\|X\|$ denotes its Euclidean norm. For a square real matrix, $Z > 0 (Z \geq 0)$ means that Z is a positive definite (semi-definite), and $\lambda(Z)$ represents its eigenvalues. $\otimes$ stands for the matrix Kronecker product.

2.2. Graph Theory

A directed graph $\mathcal{G} = (\mathcal{V}, \mathcal{E})$, where $\mathcal{V} = \{1, 2, \cdots, N\}$, $\mathcal{E} \subset \mathcal{V} \times \mathcal{V}$ are the node set and the edge set, respectively. For an edge, $(i, j) \in \mathcal{E}$ means i is a neighbor of j. The self-loop is not considered in this paper, that is, $(i, i) \notin \mathcal{E}$ for any $i \in \mathcal{V}$. For an undirected graph, $(i, j) \in \mathcal{E}$ implies $(j, i) \in \mathcal{E}$. A directed path from node i to node j is a sequence of nodes of the form $i, ..., j$.

A weighted adjacency matrix $\mathcal{A} = [a_{ij}] \in \mathbb{R}^{N \times N}$ is given by $a_{ii} = 0$, $a_{ij} > 0$ if $(i, j) \in \mathcal{E}$. The Laplacian matrix of $\mathcal{G}$ is defined as $L = [l_{ij}] \in \mathbb{R}^{N \times N}$, where $l_{ii} = \sum_{j \neq i} a_{ij}$ and $l_{ij} = -a_{ij}$, where $i \neq j$.

In this paper, suppose that there are $M(M < N)$ followers and $N - M$ leaders. Let $\mathfrak{L} \triangleq \{M+1, ..., N\}$ and $\mathfrak{F} \triangleq \{1, ..., M\}$ denote the leader set and the follower set, respectively. The communication topology among the N agents is represented by a directed graph $\mathcal{G}_{\mathfrak{F} \cup \mathfrak{L}}$. Note that, here, the leaders do not receive any information. Thus, the Laplacian matrix of $\mathcal{G}_{\mathfrak{F} \cup \mathfrak{L}}$ can be partitioned as $L \triangleq \begin{bmatrix} L_{\mathfrak{F}} & L_{\mathfrak{L}} \\ \mathbf{0}_{(N-M) \times M} & \mathbf{0}_{(N-M) \times (N-M)} \end{bmatrix}$, where $L_{\mathfrak{F}} \in \mathbb{R}^{M \times M}$ and $L_{\mathfrak{L}} \in \mathbb{R}^{M \times (N-M)}$.

2.3. Problem Statement

Consider N agents of a linear MAS with a directed graph $\mathcal{G}_{\mathfrak{F}\cup\mathcal{L}}$. The dynamics of the ith agent are described as follows:

$$\dot{x}_i = Ax_i + Bu_i + Dd_i, \quad i \in \mathfrak{F}, \tag{1a}$$
$$\dot{x}_i = Ax_i, \quad i \in \mathcal{L}, \tag{1b}$$
$$y_i = Cx_i, \quad i \in \mathfrak{F} \cup \mathcal{L}. \tag{1c}$$

where $x_i \in \mathbb{R}^n$, $u_i \in \mathbb{R}^m$ and $y_i \in \mathbb{R}^q$ are the ith agent's state, control input and output state, respectively. A, B, C and D are known constant matrices of appropriate dimensions. $d_i \in \mathbb{R}^n$ is a disturbance whose dynamics are given as

$$\dot{d}_i = Sd_i, \quad i \in \mathfrak{F}, \tag{2}$$

with S being a known constant matrix.

To proceed, we also need the assumption and Lemma as follows.

Assumption 1 ([11]). *(A, B) is stabilizable, and (A, C) is detectable.*

Assumption 2 ([11]). *The disturbance is matched, i.e., there exists a matrix F, such that $D = BF$.*

Assumption 3 ([11]). *The eigenvalues of the matrix S are on the imaginary axis, and the pair (S, D) is observable.*

Remark 1. *In some cases, Assumption 2 regarding matched disturbances can be relaxed, as based on output regulation theory [13], mismatched disturbances under uncertain conditions can be transformed into matched disturbances. Assumption 3 is typically used for disturbance rejection. Assume that (S, D) is observable, as any unobservable component will not affect the system state.*

Definition 1 ((Containment control problem) [6]). *Given the MASs (1) and a directed graph $\mathcal{G}_{\mathfrak{F}\cup\mathcal{L}}$, find a certain distributed controller so that the followers asymptotically converge to the convex hull spanned by the states of the leaders, that is, $\lim_{t\to\infty} \|x_{\mathfrak{F}}(t) + (L_{\mathfrak{F}}^{-1} L_{\mathcal{L}} \otimes I_n) x_{\mathcal{L}}(t)\| = 0$.*

Assumption 4 ([6]). *Under the digraph $\mathcal{G}_{\mathfrak{F}\cup\mathcal{L}}$, for each follower $i \in \mathfrak{F}$, there exists at least one leader $k \in \mathcal{L}$ that has a directed path to the follower.*

Lemma 1 ([6]). *Under Assumption 4, all the eigenvalues of $L_{\mathfrak{F}}$ have positive real parts, $-L_{\mathfrak{F}}^{-1} L_{\mathcal{L}}$ is non-negative and $-L_{\mathfrak{F}}^{-1} L_{\mathcal{L}} \mathbf{1}_{N-M} = \mathbf{1}_M$.*

3. Main Results

Assume that the states and relative input measurements are not available for all the followers; then, each follower can only obtain the relative output measurements. Let φ_i be the relative output measurements of ith follower as follows:

$$\varphi_i(t) = \sum_{j \in \mathfrak{F} \cup \mathcal{L}} a_{ij}(y_i(t) - y_j(t)), \tag{3}$$

Similarly, the relative input measurements of the ith follower are as follows:

$$\chi_i(t) = \sum_{j \in \mathfrak{F} \cup \mathcal{L}} a_{ij}(x_i(t) - x_j(t)). \tag{4}$$

By (3) and (4), we have $\varphi_i(t) = C\chi_i(t)$.

Let $x_{\mathfrak{F}} \triangleq [x_1^T, x_2^T, ..., x_M^T]^T \in \mathbb{R}^{nM}$, $x_{\mathfrak{L}} \triangleq [x_{M+1}^T, x_{M+2}^T, ..., x_N^T]^T \in \mathbb{R}^{n(N-M)}$ and $\chi \triangleq [\chi_1^T, \chi_2^T, ..., \chi_M^T]^T \in \mathbb{R}^{nM}$. Then, it follows that the definition of the relative input measurements vector can be written as

$$\chi(t) = (L_{\mathfrak{F}} \otimes I_n) x_{\mathfrak{F}} + (L_{\mathfrak{L}} \otimes I_n) x_{\mathfrak{L}}. \tag{5}$$

Note that the followers can only obtain the relative output measurements. Based on the relative output information, we propose a distributed disturbance observer-based event-triggered containment controller for agent $i \in \mathfrak{F}$ with form

$$\begin{aligned} \dot{\hat{d}}_i &= S\hat{d}_i + G\varphi_i(t_k^i), \\ w_i &= F\hat{d}_i + E\varphi_i(t_k^i), \\ u_i &= -w_i, \, i \in \mathfrak{F}, \, t \in [t_k^i, t_{k+1}^i), \end{aligned} \tag{6}$$

where $\hat{d}_i \in \mathbb{R}^s$ and $w_i \in \mathbb{R}^m$ are the estimates of the disturbance and the output variable, respectively. S, G, F and E are gain matrices to be determined, and t_k^i is the kth event-triggered instant of agent $i \in \mathfrak{F}$. The next event-triggered instant $\{t_k^i, k = 0, 1, ...\}$ is defined by $t_{k+1}^i \triangleq \min\{t > t_k^i \mid f_i(e_i, \chi_i) > 0\}$, where the triggering function $f_i(\cdot)$ is to be designed later, and the measurement error $e_i(t)$ for agent $i \in \mathfrak{F}$ is defined as

$$e_i(t) = \chi_i(t_k^i) - \chi_i(t), \, t \in [t_k^i, t_{k+1}^i).$$

When the triggering condition is satisfied, an event at $t = t_k^i$ is triggered for agent $i \in \mathfrak{F}$, and $e_i(t)$ is reset to zero.

Remark 2. *Compared with the general MASs studied in the literature [33], this paper studies the MASs under the condition of disturbance and adopts the distributed event-triggered controller based on disturbance observers to solve the containment control problem. Many works in the literature do not consider the situation of systems with unknown disturbance, which occurs in most practical engineering applications, making the problem more complex. This article is closer to the complexity of the actual situation and more challenging.*

Remark 3. *With the event-triggered strategy introduced in controller (6), this paper shows that the containment control problem can be solved. For agent i, the event-triggered instants are $\{t_k^i, k = 0, 1, ...\}$. At each event-triggered instant, $\varphi_i(t)$ is sampled by agent i, and its controller is updated accordingly. Noted that in (6), for agent i, all of the outputs required from its neighbors' output are included in $\varphi_i(t)$, which is only updated at its event-triggered instants.*

Define $\varepsilon_i = \hat{d}_i - d_i, i \in \mathfrak{F}$. It follows from (1)–(6) that

$$\begin{aligned} \dot{x}_i &= Ax_i + Bu_i + Dd_i = Ax_i - Bw_i + Dd_i \\ &= Ax_i - BF\hat{d}_i - BEC(\chi_i + e_i) + Dd_i \\ &= Ax_i - BF\varepsilon_i - BEC\chi_i - BECe_i. \end{aligned}$$

For $i \in \mathfrak{F} \cup \mathfrak{L}$,

$$\begin{aligned} \dot{x}_{\mathfrak{F}} &= (I_{\mathfrak{F}} \otimes A) x_{\mathfrak{F}} - (I_{\mathfrak{F}} \otimes BF)\varepsilon \\ &\quad - (I_{\mathfrak{F}} \otimes BEC)\chi - (I_{\mathfrak{F}} \otimes BEC)e \\ \dot{x}_{\mathfrak{L}} &= (I_{\mathfrak{L}} \otimes A) x_{\mathfrak{L}}, \end{aligned} \tag{7}$$

where $e \triangleq [e_1^T, e_2^T, ..., e_N^T]^T \in \mathbb{R}^{nN}$, and $\varepsilon \triangleq [\varepsilon_1^T, \varepsilon_2^T, ..., \varepsilon_N^T]^T \in \mathbb{R}^{sN}$.

Using (7) for (5), it follows that

$$\begin{aligned}
\dot\chi &= (L_{\mathfrak{F}} \otimes I_N)\dot x_{\mathfrak{F}} + (L_{\mathfrak{L}} \otimes I_N)\dot x_{\mathfrak{L}} \\
&= (L_{\mathfrak{F}} \otimes I_N)\big[(I_{\mathfrak{F}} \otimes A)x_{\mathfrak{F}} - (I_{\mathfrak{F}} \otimes BF)\varepsilon \\
&\quad - (I_{\mathfrak{F}} \otimes BEC)\chi - (I_{\mathfrak{F}} \otimes BEC)e\big] + (L_{\mathfrak{L}} \otimes I_N)(I_{\mathfrak{L}} \otimes A)x_{\mathfrak{L}} \\
&= (I_{\mathfrak{F}} \otimes A - L_{\mathfrak{F}} \otimes BEC)\chi - (L_{\mathfrak{F}} \otimes BF)\varepsilon - (L_{\mathfrak{F}} \otimes BEC)e.
\end{aligned} \quad (8)$$

Using (1) and (6), one can obtain that

$$\begin{aligned}
\dot\varepsilon &= \dot{\hat d}_i - \dot d_i \\
&= (I_{\mathfrak{F}} \otimes S)\varepsilon - (L_{\mathfrak{F}} \otimes GC)(\chi + e).
\end{aligned} \quad (9)$$

Next, Algorithm 1 is presented with procedure of controller implementation.

Algorithm 1 Distributed Disturbance Observer-based Event-triggered Control Algorithm

---

Under Assumptions 1–4, for disturbance signals in (2), the distributed disturbance observer-based event-triggered controller (6) can be constructed using the following form:

(i) Solve the following Linear matrix inequality (LMI):

$$A^T P + PA - \theta PBB^T P + \kappa I < 0. \quad (10)$$

to obtain one solution $P > 0$. Then, choose the matrix $EC = B^T P$.

(ii) Take a symmetric matrix $\hat P \in \mathbb{R}^{s \times s} > 0$, $S^T \hat P + \hat P S = -I$.

(iii) Select positive constants κ, θ as the gains to be designed in the proof of Theorem 1.

Theorem 1. *Under Assumptions 1–4, consider the MAS (1) and disturbance signals (2) with the distributed disturbance observer-based event-triggered controller (6) using Algorithm 1, where the triggered times t_k^i is determined:*

$$t_{k+1}^i \triangleq \min\{t > t_k^i \mid \|e_i\| = \gamma_i \|\chi_i\|\}, \quad (11)$$

where $\gamma_i = \frac{\sigma_i}{\rho_3 \bar\lambda^2}$ and the gains ρ_3, σ_i will be defined in the proof. Then, protocol (6) solves the containment control problem.

Proof of Theorem 1. Let $\eta = [\chi^T, \varepsilon^T]^T$. Construct the following Lyapunov function candidate:

$$V = \eta^T \bar P \eta, \quad (12)$$

where $\bar P \triangleq \begin{bmatrix} I_{\mathfrak{F}} \otimes P & 0 \\ 0 & \omega I_{\mathfrak{F}} \otimes \hat P \end{bmatrix} > 0$, $\omega > 0$ will be determined later. Evidently, $\bar P$ is definite-positive, so V is also definite-positive.

The time derivative of $V(t)$ along the trajectory of (8) and (9) is given by

$$\begin{aligned}
\dot V(t) &= \chi^T[I_{\mathfrak{F}} \otimes (A^T P + PA) - 2(L_{\mathfrak{F}} \otimes PBB^T P)]\chi \\
&\quad - e^T(L_{\mathfrak{F}} \otimes PBB^T P)\chi - \chi^T(L_{\mathfrak{F}} \otimes PBB^T P)e \\
&\quad - \varepsilon^T(L_{\mathfrak{F}} \otimes D^T P)\chi - \chi^T(L_{\mathfrak{F}} \otimes PD)\varepsilon \\
&\quad - \omega\varepsilon^T(I_{\mathfrak{F}} \otimes (S^T \hat P + \hat P S))\varepsilon \\
&\quad - \omega e^T(L_{\mathfrak{F}} \otimes C^T G^T \hat P)\varepsilon - \omega\chi^T(L_{\mathfrak{F}} \otimes C^T G^T \hat P)\varepsilon \\
&\quad - \omega\varepsilon^T(L_{\mathfrak{F}} \otimes \hat P GC)\chi - \omega\varepsilon^T(L_{\mathfrak{F}} \otimes \hat P GC)e.
\end{aligned} \quad (13)$$

Under Assumption 4 and Lemma 1, choose a unitary matrix $U \in \mathbb{C}^{M \times M}$, $U^H L_{\mathfrak{F}} U = \Lambda$, where Λ is an upper-triangular matrix with λ_i, $i = 1, ..., M$, as its diagonal entries.

Let $\zeta \triangleq (U^H \otimes I_n)\chi = [\zeta_1^T, \zeta_2^T, ..., \zeta_M^T]^T \in \mathbb{R}^{nM}$, $\bar{\varepsilon} = (U^H \otimes I_s)\varepsilon = [\bar{\varepsilon}_1^T, \bar{\varepsilon}_2^T, ..., \bar{\varepsilon}_M^T]^T \in \mathbb{R}^{sM}$ and $\bar{e} = (U^T \otimes I_n)e = [\bar{e}_1^T, \bar{e}_2^T, ..., \bar{e}_M^T]^T \in \mathbb{R}^{nM}$.

Then, it follows from (13) that

$$\begin{aligned}
\dot{V}(t) &= \zeta^T [I_{\mathfrak{F}} \otimes (A^T P + PA) - 2(\Lambda \otimes PBB^T P)]\zeta \\
&\quad - \bar{e}^T (\Lambda \otimes PBB^T P)\zeta - \zeta^T (\Lambda \otimes PBB^T P)\bar{e} \\
&\quad - \bar{\varepsilon}^T (\Lambda \otimes D^T P)\zeta - \zeta^T (\Lambda \otimes PD)\bar{\varepsilon} - \omega \varepsilon^T \varepsilon \\
&\quad - \omega \bar{e}^T (\Lambda \otimes C^T G^T \hat{P})\bar{\varepsilon} - \omega \zeta^T (\Lambda \otimes C^T G^T \hat{P})\bar{\varepsilon} \\
&\quad - \omega \bar{\varepsilon}^T (\Lambda \otimes \hat{P}GC)\zeta - \omega \bar{\varepsilon}^T (\Lambda \otimes \hat{P}GC)\bar{e} \\
&= \sum_{i=1}^M \zeta_i^T (A^T P + PA - 2\lambda_i PBB^T P)\zeta_i \\
&\quad - \sum_{i=1}^M \bar{e}_i^T (\lambda_i PBB^T P)\zeta_i - \sum_{i=1}^M \zeta_i^T (\lambda_i PBB^T P)\bar{e}_i \\
&\quad - \sum_{i=1}^M \lambda_i \bar{\varepsilon}_i^T (D^T P + \omega \hat{P}GC)\zeta_i - \sum_{i=1}^M \lambda_i \zeta_i^T (PD + \omega C^T G^T \hat{P})\bar{\varepsilon}_i \\
&\quad - \omega \sum_{i=1}^M \bar{e}_i^T (\lambda_i C^T G^T \hat{P})\bar{\varepsilon}_i - \omega \sum_{i=1}^M \bar{\varepsilon}_i^T (\lambda_i \hat{P}GC)\bar{e}_i - \omega \varepsilon^T \varepsilon.
\end{aligned} \quad (14)$$

For any $x, y \in \mathbb{R}^n$ and $\beta > 0$, we use Young's inequalities $x^T y \leq \frac{\beta}{2} \|x\|^2 + \frac{1}{2\beta} \|y\|^2$ ([24]), yields,

$$\begin{aligned}
&-\bar{e}_i^T (\lambda_i PBB^T P)\zeta_i \\
&\leq \frac{\lambda_i \|PBB^T P\|}{2\beta_1} \|\zeta_i\|^2 + \frac{\beta_1 \lambda_i \|PBB^T P\|}{2} \|\bar{e}_i\|^2.
\end{aligned} \quad (15)$$

$$\begin{aligned}
&-\zeta_i^T (\lambda_i PBB^T P)\bar{e}_i \\
&\leq \frac{\lambda_i \|PBB^T P\|}{2\beta_1} \|\bar{e}_i\|^2 + \frac{\beta_1 \lambda_i \|PBB^T P\|}{2} \|\zeta_i\|^2.
\end{aligned} \quad (16)$$

$$\begin{aligned}
&-\lambda_i \bar{\varepsilon}_i^T (D^T P + \omega \hat{P}GC)\zeta_i \\
&\leq \frac{\lambda_i \|D^T P + \omega \hat{P}GC\|}{2\beta_2} \|\zeta_i\|^2 + \frac{\beta_2 \lambda_i \|D^T P + \omega \hat{P}GC\|}{2} \|\bar{\varepsilon}_i\|^2,
\end{aligned} \quad (17)$$

$$\begin{aligned}
&-\lambda_i \zeta_i^T (PD + \omega C^T G^T \hat{P})\bar{\varepsilon}_i \\
&\leq \frac{\lambda_i \|PD + \omega C^T G^T \hat{P}\|}{2\beta_2} \|\bar{\varepsilon}_i\|^2 + \frac{\beta_2 \lambda_i \|PD + \omega C^T G^T \hat{P}\|}{2} \|\zeta_i\|^2,
\end{aligned} \quad (18)$$

$$\begin{aligned}
&-\bar{e}_i^T (\lambda_i C^T G^T \hat{P})\bar{\varepsilon}_i \\
&\leq \frac{\lambda_i \|C^T G^T \hat{P}\|}{2\beta_3} \|\bar{\varepsilon}_i\|^2 + \frac{\lambda_i \beta_3 \|C^T G^T \hat{P}\|}{2} \|\bar{e}_i\|^2,
\end{aligned} \quad (19)$$

$$\begin{aligned}
&-\bar{\varepsilon}_i^T (\lambda_i \hat{P}GC)\bar{e}_i \\
&\leq \frac{\lambda_i \|\hat{P}GC\|}{2\beta_3} \|\bar{e}_i\|^2 + \frac{\lambda_i \beta_3 \|\hat{P}GC\|}{2} \|\bar{\varepsilon}_i\|^2,
\end{aligned} \quad (20)$$

where β_1, β_2 and β_3 are positive constants.

Let $\underline{\lambda} = \min_{i=1,...,M}\{Re(\lambda_i)\}$ and $\overline{\lambda} = \max_{i=1,...,M}\{Re(\lambda_i)\}$, where λ_i, $i = \{1,...,M\}$ are the eigenvalues of $L_{\mathfrak{F}}$. When $0 < \theta \leq 2\underline{\lambda}$ and under Algorithm 1, it follows from (7), (15)–(20) that

$$\dot{V}(t)$$
$$\leq -\kappa \sum_{i=1}^{M} \|\chi_i\|^2 - \omega \sum_{i=1}^{M} \|\varepsilon_i\|^2 + \sum_{i=1}^{M} \rho_1 \lambda_i^2 \|\chi_i\|^2$$
$$+ \rho_2 \overline{\lambda}^2 \sum_{i=1}^{M} \|\varepsilon_i\|^2 + \sum_{i=1}^{M} \rho_3 \overline{\lambda}^2 \|e_i\|^2 \qquad (21)$$
$$= -\sum_{i=1}^{M} (\kappa - \rho_1)\|\chi_i\|^2 - \sum_{i=1}^{M} (\omega - \rho_2 \overline{\lambda}^2)\|\varepsilon_i\|^2 + \sum_{i=1}^{M} \rho_3 \overline{\lambda}^2 \|e_i\|^2,$$

where $\rho_1 = \frac{\|PBB^T P\|}{2\beta_1} + \frac{\beta_1 \|PBB^T P\|}{2} + \frac{\|D^T P + \omega \hat{P} GC\|}{2\beta_2} + \frac{\beta_2 \|PD + \omega C^T G^T \hat{P}\|}{2}$, $\rho_2 = \frac{\beta_2 \|D^T P + \omega \hat{P} GC\|}{2} + \frac{\lambda_i \|PD + \omega C^T G^T \hat{P}\|}{2\beta_2} + \frac{\|C^T G^T \hat{P}\|}{2\beta_3} + \frac{\beta_3 \|\hat{P} GC\|}{2}$ and $\rho_3 = \frac{\beta_1 \|PBB^T P\|}{2} + \frac{\|PBB^T P\|}{2\beta_1} + \frac{\beta_3 \|C^T G^T \hat{P}\|}{2} + \frac{\|\hat{P} GC\|}{2\beta_3}$.

Then, by choosing σ_i and κ, the following condition is enforced:

$$\|e_i\|^2 \leq \frac{\sigma_i}{\rho_3 \overline{\lambda}^2} \|\chi_i\|^2,$$

where choosing $0 < \sigma_i < \kappa - \rho_1$. It is noted that $\gamma_i = \sqrt{\frac{\sigma_i}{\rho_3 \overline{\lambda}^2}}$, and choosing $\sigma_i < \rho_3 \overline{\lambda}^2$, so $\gamma_i < 1$ can be guaranteed.

From (21) and choosing $\omega \gg 0$ such that $\omega \geq \rho_2 \overline{\lambda}^2$, one can obtain that

$$\dot{V}(t) \leq -\sum_{i=1}^{N} (\kappa - \rho_1 - \sigma_i)\|\chi_i\|^2$$
$$- (\omega - \rho_2 \overline{\lambda}^2) \sum_{i=1}^{N} \|\varepsilon_i\|^2 \leq 0.$$

Thus, by the definition of $V(t)$, $\dot{V}(t) = 0$ implies that $\chi_i(t) = 0$. According to [39], it implies that $\lim_{t\to\infty} \|x_{\mathfrak{F}}(t) + (L_{\mathfrak{F}}^{-1} L_{\mathfrak{L}} \otimes I_n) x_{\mathfrak{L}}(t)\| = 0$. Therefore, the containment control problem stated in Definition 1 is solved. □

Feasibility Analysis

In this section, the development analyzes the feasibility of the proposed controller (6) by excluding Zeno behavior (i.e., in the event time defined in (11) within a finite time interval, an infinite number of triggers occur). The result is summarized in the following theorem.

Theorem 2. *Consider the linear MAS (1), controller (6) and triggering condition (11). No agent will exhibit Zeno behavior.*

Proof of Theorem 2. Without loss of generality, to prove that the Zeno behavior does not exist, it is only necessary to prove that $\tau \triangleq t_{k+1}^i - t_k^i > 0$ has a positive lower bound.

According to the definition of $e_i(t)$, there exists $\mid \|\chi_i(t_k^i)\| - \|\chi_i(t)\| \mid \leq \|e_i(t)\|$. Using (11), we have

$$\frac{\|\chi_i(t_k^i)\|}{1+\gamma_i} \leq \|\chi_i(t)\| \leq \frac{\|\chi_i(t_k^i)\|}{1-\gamma_i}. \qquad (22)$$

By substituting (8) with the time derivative of $\|e_i(t)\|$ over the interval $[t_k^i, t_{k+1}^i)$, we can obtain that

$$\begin{aligned}&\frac{d}{dt}\|e_i(t)\|\\ &\leq \|\dot{e}_i(t)\| = \|-\dot{\chi}_i(t)\|\\ &= \| - A\chi_i + BEC\sum_{j\in\mathcal{N}_i}a_{ij}(\chi_i - \chi_j) + BF\sum_{j\in\mathcal{N}_i}a_{ij}(\varepsilon_i - \varepsilon_j)\\ &\quad + BEC\sum_{j\in\mathcal{N}_i}a_{ij}(e_i - e_j)\|\\ &\leq \|A + BEC(|\mathcal{N}_i|+1)\|\|e_i(t)\| + \|BF(|\mathcal{N}_i|+1)\|\|\varepsilon_i(t)\|\\ &\quad + \|A\chi_i(t_k^i) + BEC\sum_{j\in\mathcal{N}_i}a_{ij}(\chi_i(t_k^i) - \chi_j(t_k^i))\|. \end{aligned} \quad (23)$$

From (23), we can obtain that $\|e_i(t)\|$ will not approach zero unless $\|\varepsilon_i(t)\|$ approaches zero, which implies the existence of $0 < R < \infty$, such that $\frac{\|\varepsilon_i(t)\|}{\|e_i(t)\|} < R$. Substituting (5) and (9) into (23), one has

$$\frac{d}{dt}\|e_i(t)\| \leq \zeta_i\|e_i(t)\| + \phi_k^i, \quad (24)$$

where $\zeta_i = \|A + BEC(|\mathcal{N}_i|+1)\|$ and $\phi_k^i = \|BF(|\mathcal{N}_i|+1)\|R + \max_{t\in[t_k^i, t_{k+1}^i]}\|A\chi_i(t_k^i) + BEC\sum_{j\in\mathcal{N}_i}a_{ij}(\chi_i(t_k^i) - \chi_j(t_k^i))\|$

Then, it follows that

$$\|e_i(t)\| \leq \frac{\phi_k^i}{\zeta_i}\left[\exp(\zeta_i(t - t_k^i)) - 1\right]. \quad (25)$$

At this point, we need to present a sufficient condition $\|e_i(t)\| \leq \frac{\gamma_i}{\sqrt{2+2\gamma_i^2}}\|\chi_i(t_k^i)\|$ that ensures that the triggering condition (11) holds.

Let $s_k^i = \frac{\gamma_i}{\sqrt{2+2\gamma_i^2}}\|\chi_i(t_k^i)\|$. Using (24) gives

$$\|e_i(t_{k+1}^i)\| = s_k^i \leq \frac{\phi_k^i}{\zeta_i}\left[\exp(\zeta_i(t_{k+1}^i - t_k^i)) - 1\right],$$

which yields $t_{k+1}^i - t_k^i \geq (1/\zeta_i)\ln(\zeta_i s_k^i/\phi_k^i + 1)$.

Next, we will discuss two cases.

The first case is when $\chi_i(t_k^i) \neq 0$. Since $\chi_i(t_k^i) \neq 0$, it can be seen that $s_k^i > 0$. Thus, $t_{k+1}^i - t_k^i = (1/\zeta_i)\ln(\zeta_i s_k^i/\phi_k^i + 1) > 0$.

The second case is when $\chi_i(t_k^i) = 0$ as $k \to \infty$. Then, from (22), one has $\chi_i(t) = 0$, and thus,

$$\begin{aligned}\dot{\chi}_i =& A\chi_i + BEC\sum_{j\in\mathcal{N}_i}a_{ij}(\chi_i - \chi_j) - BF\sum_{j\in\mathcal{N}_i}a_{ij}(\varepsilon_i - \varepsilon_j)\\ &+ BEC\sum_{j\in\mathcal{N}_i}a_{ij}(\chi_i(t_k^i) - \chi_j(t_{k(t)}^i))\\ =& 0. \end{aligned} \quad (26)$$

By simple transposition (22), we obtain

$$\lim_{k\to\infty}\frac{\|\chi_i(t)\|}{\|\chi_i(t_k^i)\|} \leq \frac{1}{1-\gamma_i}. \quad (27)$$

In light of (26), we obtain

$$\phi_k^i \leq \zeta_i\|\chi_i(t)\| + \frac{2-\gamma_i}{1-\gamma_i}\zeta_i\|\chi_i(t)\|. \quad (28)$$

According to (27) and (28), the same as those in [24,40], we have

$$\lim_{k \to \infty}(t^i_{k+1} - t^i_k) \geq \frac{1}{\zeta_i}\ln(\frac{\gamma_i(2-\gamma_i)}{(1-\gamma_i)\sqrt{2+2\gamma_i^2}} + 1).$$

Consequently, Zeno behavior is excluded for all the agents. □

4. Simulation

For illustration, consider an MAS with the communication graph $\mathcal{G}_{\mathfrak{F} \cup \mathfrak{L}}$, where there are six followers $\{1-6\} \in \mathfrak{F}$ and three leaders $\{7-9\} \in \mathfrak{L}$. Assume the dynamics matrices of (1) are:

$$A = \begin{bmatrix} 0 & 1 \\ -0.5 & 0 \end{bmatrix}, \quad B = \begin{bmatrix} 0 \\ 1 \end{bmatrix},$$

$$C = \begin{bmatrix} 1 & 0 \end{bmatrix}, \quad D = \begin{bmatrix} 0 & 0 \\ 0 & 1 \end{bmatrix}.$$

By solving the LMI (10) and the equation in Algorithm 1, the feedback gain matrices S, F, G and E satisfy the condition (6)

$$S = \begin{bmatrix} 0 & 1 \\ 0 & -2 \end{bmatrix}, \quad F = \begin{bmatrix} 0 & 1 \end{bmatrix},$$

$$G = \begin{bmatrix} -2 \\ -3.5 \end{bmatrix}, \quad E = \begin{bmatrix} 1 \end{bmatrix}.$$

The initial conditions of the closed-loop system are randomly chosen. The other parameters are set as follows, $\kappa = 4.6$, $\sigma_i = 0.999$ and $\gamma_i = 0.08$, for all $i = 1, ..., 6$.

The communication graph $\mathcal{G}_{\mathfrak{F} \cup \mathfrak{L}}$ can be given by Figure 1, where nodes 7, 8 and 9 are the three leaders and the others are followers. The red dotted line represents the directed communication connection from the leader to the corresponding follower, and the black solid line represents the communication connection between the followers. Then, matrices $L_{\mathfrak{F}}$ and $L_{\mathfrak{L}}$ are as follows:

$$L_{\mathfrak{F}} = \begin{bmatrix} 3 & 0 & 0 & -1 & -1 & -1 \\ -1 & 1 & 0 & 0 & 0 & 0 \\ -1 & -1 & 2 & 0 & 0 & 0 \\ -1 & 0 & 0 & 2 & 0 & 0 \\ 0 & 0 & 0 & -1 & 2 & 0 \\ 0 & 0 & 0 & 0 & -1 & 2 \end{bmatrix}, L_{\mathfrak{L}} = \begin{bmatrix} 0 & 0 & 0 \\ 0 & 0 & 0 \\ 0 & 0 & 0 \\ 0 & 0 & 1 \\ 0 & 1 & 0 \\ 1 & 0 & 0 \end{bmatrix}.$$

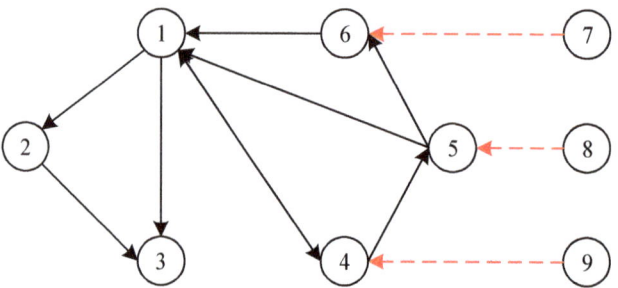

Figure 1. Communication graph $\mathcal{G}_{\mathfrak{F} \cup \mathfrak{L}}$.

The trajectory of the follower is represented by the solid line and that of the leader is represented by the dashed line in Figure 2, which can be clearly obtained in Definition 1, i.e., the containment control problem is indeed solved.

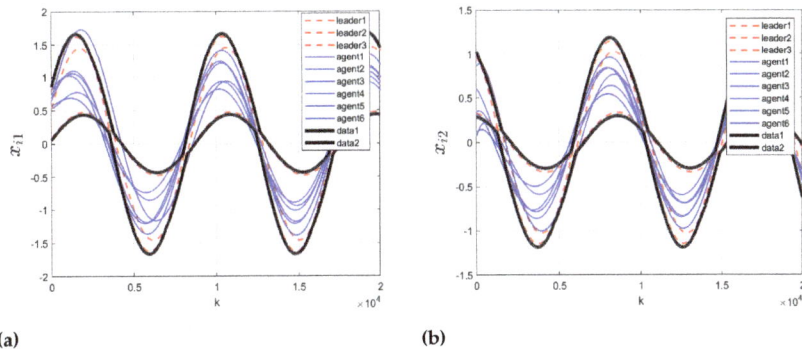

(a)   (b)

Figure 2. The state trajectories of nine agents under controller (6).

Through the three-dimensional effect diagram in Figure 3, the movement trajectories of six agents and three leaders over time can be more clearly seen.

Moreover, the triggering times of six followers are presented in Figure 4. As can be seen, it can effectively reduce the communication among agents.

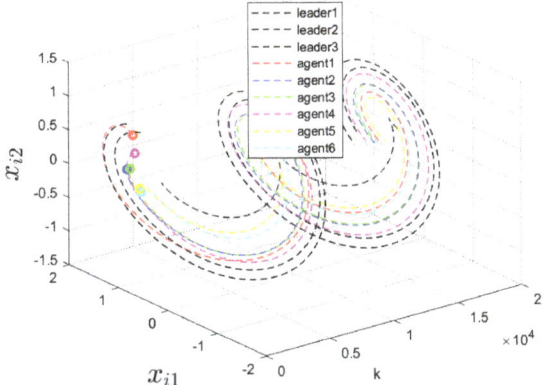

Figure 3. Three-dimensional trajectories of all agents.

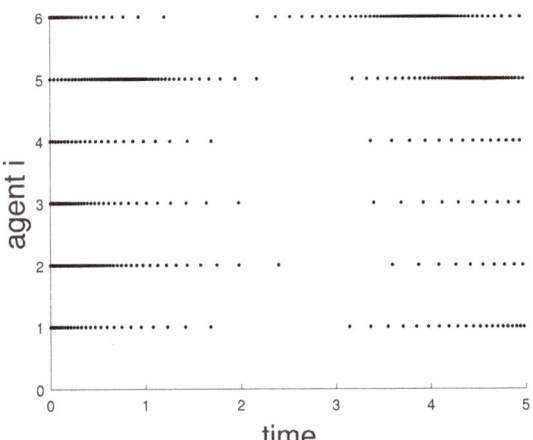

Figure 4. Triggering time of each followers.

5. Conclusions

In this paper, we have considered the containment control of MASs with external disturbances. First, a novel disturbance observer-based control has been developed by the output feedback control. Then, in order to save communication costs and energy consumption, our controller is combined with the event-triggered control. It has been shown that Zeno behavior can be excluded for the proposed controller. Here, we have only considered matched disturbances. Future work will be devoted to investigating the containment control problem with mismatched disturbances. In the meantime, this paper does not consider MASs in the presence of deception attack effects, but attacks often happen [41–43]. In the future, we will consider the containment control problem under deception attacks.

Author Contributions: Conceptualization, L.J.; methodology, L.J.; software, Y.L. and R.L.; validation, L.K. and G.Z.; writing—original draft preparation, L.J.; writing—review and editing, G.Z. All authors have read and agreed to the published version of the manuscript.

Funding: This work is supported in part by the Fundamental Research Program of Shanxi Province 20210302124552, 20210302124030, 201901D211083 and 202103021223048 and in part by the National Natural Science Foundation of China (NSFC) under Grants 62103296 and 62003233.

Institutional Review Board Statement: Not applicable.

Informed Consent Statement: Not applicable.

Data Availability Statement: Not applicable.

Conflicts of Interest: The authors declare no conflict of interest.

References

1. OlfatiSaber, R.; Murray, R. Consensus problems in networks of agents with switching topology and time-delays. *IEEE Trans. Autom. Control* **2004**, *9*, 1520–1533. [CrossRef]
2. Li, Z.; Duan, Z.; Chen, G.; Lin, H. Consensus of multiagent systems and synchronization of complex networks: A unified viewpoint. *IEEE Trans. Circuits Syst. I Regul. Pap.* **2010**, *1*, 213–224.
3. Hu, J.; Wu, Y.; Li, T., Ghosh, B.K. Consensus control of general linear multi-agent systems with antagonistic interactions and communication noises. *IEEE Trans. Autom. Control* **2019**, *5*, 2122–2127. [CrossRef]
4. Hu, J.; Hong, Y. Leader-following coordination of multi-agent systems with coupling time delays. *Physica A* **2007**, *2*, 853–863. [CrossRef]
5. Wu, Y.; Hu, J. Observer-based output regulation of cooperative-competitive high-order multi-agent systems. *J. Frankl. Inst.* **2018**, *10*, 4111–4130. [CrossRef]
6. Li, Z.; Ren, W.; Liu, X.; Fu, M. Distributed containment control of multi-agent systems with general linear dynamics in the presence of multiple leaders. *Int. J. Robust. Nonlinear. Control* **2013**, *5*, 534–547. [CrossRef]
7. Peng, Z.; Luo, R.; Hu, J.; Shi, K.; Nguang, S.K.; Ghosh, B.K. Optimal tracking control of nonlinear multi-agent systems using internal reinforce Q-learning. *IEEE Trans. Neural Netw. Learn. Syst.* **2022**, *8*, 4043–4055. [CrossRef]
8. Kou, L.; Chen, Z.; Xiang, J. Cooperative Fencing Control of Multiple Vehicles for a Moving Target With an Unknown Velocity. *IEEE Trans. Autom. Control* **2021**, *2*, 1008–1015. [CrossRef]
9. Cao, Y.; Stuart, D.; Ren, W.; Meng, Z. Distributed containment control for multiple autonomous vehicles with double-integrator dynamics: Algorithms and experiments. *IEEE Trans. Control Syst. Technol.* **2011**, *4*, 929–938. [CrossRef]
10. Haghshenas, H.; Badamchizadeh, M.A.; Baradarannia, M. Containment control of heterogeneous linear multi-agent systems. *Automatica* **2015**, *54*, 210–216. [CrossRef]
11. Cheng, B.; Li, Z. Consensus disturbance rejection with event-triggered communications. *J. Frankl. Inst.* **2018**, *2*, 956–974. [CrossRef]
12. Rong, L.; Liu, X.; Jiang, G.; Xu, S. Event-Driven Multiagent Consensus Disturbance Rejection With Input Uncertainties via Adaptive Protocols. *IEEE Trans. Syst. Man. Cybern. Syst.* **2022**, *5*, 2911–2919. [CrossRef]
13. Ding, Z. Consensus Disturbance Rejection with Disturbance Observers. *IEEE Trans. Autom. Control* **2015**, *9*, 5829–5837. [CrossRef]
14. Han, T.; Li, J.; Guan, Z.; Cai, C.; Zhang, D.; He, D. Containment control of multi-agent systems via a disturbance observer-based approach. *J. Frankl. Inst.* **2019**, *5*, 2919–2933. [CrossRef]
15. Wu, Y.; Hu, J.; Xiang, L.; Liang, Q.; Shi, K. Finite-Time Output Regulation of Linear Heterogeneous Multi-Agent Systems. *IEEE Trans. Circuits Syst. II Express. Brief.* **2022**, *3*, 1248–1252. [CrossRef]
16. Hu, W.; Liu, L. Cooperative output regulation of heterogeneous linear multi-agent systems by event-triggered control. *IEEE Trans. Cybern* **2016**, *5*, 105–116. [CrossRef]

17. Wang, H.; Yu, W.; Ren, W.; Lu, J. Distributed adaptive finite-time consensus for second-order multiagent systems with mismatched disturbances under directed networks. *IEEE Trans. Cybern* **2019**, *3*, 1347–1358. [CrossRef]
18. Wang, H.; Ren, W.; Yu, W.; Zhang, D. Fully distributed consensus control for a class of disturbed second-order multi-agent systems with directed networks. *Automatica* **2021**, *132*, 109816. [CrossRef]
19. Wei, X.; Yu, W.; Wang, H.; Yao, Y.; Mei, F. An observer-based fixed-time consensus control for second-order multi-agent systems with disturbances. *IEEE Trans. Circuits Syst. II Express. Brief.* **2019**, *2*, 247–251. [CrossRef]
20. Wang, Z.; Wang, D.; Wang, W. Distributed dynamic average consensus for nonlinear multi-agent systems in the presence of external disturbances over a directed graph. *Inform. Sci.* **2019**, *479*, 40–54. [CrossRef]
21. Zhang, J.; Zhang, H.; Zhang, K.; Cai, Y. Observer-Based Output Feedback Event-Triggered Adaptive Control for Linear Multiagent Systems Under Switching Topologies. *IEEE Trans. Neural Netw. Learn. Syst.* **2021**, *12*, 7161–7171. [CrossRef] [PubMed]
22. Zhang, J.; Feng, G. Event-driven observer-based output feedback control for linear systems. *Automatica* **2014**, *7*, 1852–1859. [CrossRef]
23. Dimarogonas, D.V.; Frazzoli, E.; Johansson, K.H. Distributed event-triggered control for multi-agent systems. *IEEE Trans. Autom. Control* **2012**, *5*, 1291–1297. [CrossRef]
24. Hu, W.; Liu, L.; Feng, G. Consensus of linear multi-agent systems by distributed event-triggered strategy. *IEEE Trans. Cybern* **2016**, *1*, 148–157.
25. Cheng, T.H.; Kan, Z.; Klotz, J.R.; Shea, J.M.; Dixon, W.E. Event-triggered control of multi-agent systems for fixed and time-varying network topologies. *IEEE Trans. Autom. Control* **2017**, *10*, 5365–5371. [CrossRef]
26. Fan, Y.; Feng, G.; Wang, Y.; Song, C. Distributed event-triggered control of multi-agent systems with combinational measurements. *Automatica* **2013**, *2*, 671–675. [CrossRef]
27. Seyboth, G.S.; Dimarogonas, D.V.; Johansson, K.H. Event-based broadcasting for multi-agent average consensus. *Automatica* **2013**, *1*, 245–252. [CrossRef]
28. Fan, Y.; Liu, L.; Feng, G.; Wang, Y. Self-triggered consensus for multi-agent systems with zeno-free triggers. *IEEE Trans. Autom. Control* **2015**, *5*, 2779–2784. [CrossRef]
29. Deng, C.; Wen, C.; Huang, J.; Zhang, X.; Zou, Y. Distributed observer-based cooperative control approach for uncertain nonlinear MASs under event-triggered communication. *IEEE Trans. Autom. Control* **2021**, *5*, 2669–2676. [CrossRef]
30. Zhang, X.; Han, Q.; Ge, X.; Ning, B.; Zhang, B. Sampled-data control systems with non-uniform sampling: A survey of methods and trends. *Annu. Rev. Control* **2023**. . . [CrossRef]
31. Qian, Y.; Liu, L.; Feng, G. Output Consensus of Heterogeneous Linear Multi-Agent Systems with Adaptive Event-Triggered Control. *IEEE Trans. Autom. Control* **2018**, *6*, 2606–2613. [CrossRef]
32. Peng, Z.; Luo, R.; Hu, J.; Shi, K.; Ghosh, B.K. Distibuted optimal tracking control of discrete-time multi-agent systems via event-triggered reinforcement learning. *IEEE Trans. Circuits Syst. I Regul. Pap.* **2022**, *9*, 3689–3700. [CrossRef]
33. Jian, L.; Hu, J.; Wang, J.; Shi, K.; Peng, Z.; Yang, Y.; Huang, J. Distributed functional observer-based event-triggered containment control of multi-agent systems. *Int. J. Control. Autom. Syst.* **2020**, *5*, 1094–1102. [CrossRef]
34. Yang, J.; Xiao, F.; Ma, J. Model-based edge-event-triggered containment control under directed topologies. *IEEE Trans. Cybern* **2018**, *7*, 2556–2567. [CrossRef]
35. Hu, J.; Zheng, W.X. Adaptive tracking control of leader-follower systems with unknown dynamics and partial measurements. *Automatica* **2014**, *5*, 1416–1423. [CrossRef]
36. Zhang, J.; Zhang, H.; Cai, Y.; Li, W. Containment control of general linear multi-agent systems by event-triggered control mechanisms. *Neurocomputing* **2020**, *7*, 263–274. [CrossRef]
37. Qian, Y.; Liu, L.; Feng, G. Distributed event-triggered adaptive control for consensus of linear multi-agent systems with external disturbances. *IEEE Trans. Cybern* **2018**, *5*, 2197–2208. [CrossRef]
38. Li, Z.; Duan, Z.; Ren, W.; Feng, G. Containment control of linear multi-agent systems with multiple leaders of bounded inputs using distributed continuous controllers. *Int. J. Robust. Nonlinear. Control* **2019**, *5*, 2101–2121. [CrossRef]
39. Miao, G.; Cao, J.; Alsaedi, A.; Alsaadi, F.E. Event-triggered containment control for multi-agent systems with constant time delays. *J. Frankl. Inst.* **2017**, *15*, 6956–6977. [CrossRef]
40. Hu, W.; Liu, L.; Feng, G. Output consensus of heterogeneous linear multi-agent systems by distributed event-triggered/self-triggered strategy. *IEEE Trans. Cybern* **2017**, *8*, 1914–1924. [CrossRef]
41. Jeong, J.; Lim, Y.; Parivallal, A. An asymmetric Lyapunov-Krasovskii functional approach for event-triggered consensus of multi-agent systems with deception attacks. *Appl. Math. Comput.* **2023**, *439*, 127584. [CrossRef]
42. Kazemy, A.; Lam, J.; Zhang, X. Event-Triggered Output Feedback Synchronization of Master–Slave Neural Networks Under Deception Attacks. *IEEE Trans. Neural Netw. Learn Syst.* **2022**, *33*, 952–961. [CrossRef]
43. Kavikumar, R.; Kwon, O.M.; Lee, S.H.; Lee, S.; Sakthivel, R. Event-Triggered Input–Output Finite-Time Stabilization for IT2 Fuzzy Systems Under Deception Attacks. *IEEE Trans. Fuzzy Syst.* **2023**, *4*, 1139–1151. [CrossRef]

Disclaimer/Publisher's Note: The statements, opinions and data contained in all publications are solely those of the individual author(s) and contributor(s) and not of MDPI and/or the editor(s). MDPI and/or the editor(s) disclaim responsibility for any injury to people or property resulting from any ideas, methods, instructions or products referred to in the content.

Article

Adaptive Consensus of the Stochastic Leader-Following Multi-Agent System with Time Delay

Shoubo Jin * and Guanghui Zhang

School of Mathematics and Statistics, Suzhou University, Suzhou 234000, China; ghzhang@ahszu.edu.cn
* Correspondence: jinshoubo@ahszu.edu.cn; Tel.: +86-186-5570-0163

Abstract: For the multi-agent system with time delay and noise, the adaptive consensus of tracking control problems is discussed by the Lyapunov function. The main purpose of this study is to design an adaptive control protocol for the system, such that even if there exists time delay among agents, the protocol can still ensure the consensus of the stochastic system. The main contribution is to revise the protocols that were previously only applicable to system without time delay. Because the system is inevitably disrupted by time delay and noise during the interactive process, achieving coordination and consensus is difficult. To enable the followers to track the leader, a novel adaptive law depending on the Riccati equation is firstly proposed, and the adaptive law is different from previous mandatory control law completely depending on a known function. The ability to be altered online based on the state of system is a major feature of the adaptive law. When there are interactive noise and time delay between the followers and leader of the system, a special Lyapunov function is constructed to prove the adaptive consensus. And the upper bound of time delay is obtained by using the Itô integral theory. Finally, if the time delay of the system approaches zero, it is shown that the adaptive law still ensures that each follower tracks the leader under simpler conditions.

Keywords: time delay; multi-agent system; adaptive law; white noise

MSC: 93A16; 68T42

1. Introduction

The multi-agent system can complete a complex task through mutual coordination among agents, which has become a research hotspot in current academic research. Centralized control and distributed control are two main aspects of current research on multi-agent applications. The focus of current research is distributed control, since it is more fault-tolerant to the environment and has lower cost requirements than centralized control. The application scope of distributed control in multi-agent systems includes unmanned aerial vehicles, smart grid, target tracking, traffic control and other fields [1–3]. The core of many distributed control systems is to seek a suitable control protocol that makes it possible for all agents to reach the same state, which is called the consensus of system. Currently, the research topics of the consensus focus on random disturbance control, finite time control, event-triggered control, distributed optimal control and so on.

Since Visek et al. [4] proposed a special mathematical model and discovered that all agents ultimately reach the same state under specific conditions, the multi-agent system has quickly attracted the attention of a large number of scholars. Recently, Qin et al. [5] and Amirkhani et al. [6] reviewed the theoretical progress of the consensus and introduced some difficulties in the system. In order to achieve the consensus, it is often necessary to constrain the topology of system and construct an appropriate control protocol. For an undirected graph, connectivity is usually required, while it is balanced for a directed graph. This paper mainly studies the adaptive consensus on a directed graph. Our goal is to build an adaptive protocol that enables the followers to track a certain objective. Moreover, the problem is disturbed by noise and has a hysteresis effect. Up to now, numerous academics have investigated the leader-following consensus from various angles. Jiang et al. [7] discussed the tracking issue when the equations of state contain time-varying matrices. A similar consensus was analyzed in the event-triggered mechanism [8–10]. Zhang et al. [11] extended the tracking problem to stochastic system and utilized mathematical expectation to analyze the problem. The multi-agent system mentioned in these references all have definite models. However, the internal structure of the system is often uncertain in complex environments, so the adaptive control methods are proposed to continuously update the structure of the system.

Adaptive control technology is a method that automatically adjusts its own control parameters with the change of the environment to achieve the best performance. Adaptive consensus can be defined as that the state of all agents is finally consistent due to the adaptive control technology. Adaptive law can be seen as the changing law of the control parameters, and it is usually represented by a differential equation. Adaptive algorithms are usually characterized by information and intelligence; the information of this paper mainly comes from the state of system, and the intelligence is determined by the adaptive laws. The algorithm is often combined with machine learning theory and applied to some game scenarios. Adaptive control was initially applied in the aerospace field, and Whitaker is crucial to the advancement of the method. Currently, this special technology has found extensive use in fields such as aerospace, power, transportation, robotics, etc. The creation of a suitable adaptive law is the crux of the challenge for this technology. For a multi-agent system, when the mandatory gain is independent of the states of output and input, Li et al. [12] and Cheng et al. [13] respectively analyzed the average consensus. Zong et al. [14] investigated the random weak consensus under mandatory gain. For the adaptive gain that can be dynamically modified according to the current state, Knotek et al. [15] established an adaptive control law with decay gain, and the edge-based adaptive techniques for a nonlinear multi-agent system were taken into consideration by Yu et al. [16]. Luo et al. [17] analyzed a gradient-descent-based adaptive law and gave a scheme for the optimal control problem of uncertain multi-agent system. Li et al. [18] proposed a value iteration strategy and used the gradient descent method to update the weights. For a self-organized system, some important self-organized models were discussed in [19], and a self-organized interlimb coordination control was analyzed in [20]. For the optimal control problem of discrete systems, Peng et al. [2] designed a strategy for the adaptive adjustment of weight vectors based on neural network approximation. Nevertheless, these studies did not consider the effects of noise and time delay. Since the system is inevitably disturbed by time delay and noise at the same time, it is necessary to study the adaptive consensus under noise and time delay.

Currently, there have been many research conclusions about the consensus under noisy environments, but less research has been conducted on the topic of adaptive consensus. In fact, the interactive network among agents is subject to noise, so the stochastic multi-agent system should be considered. Itô integral theory provides an important tool for the adaptive problems of stochastic system. When agents have noise perturbations during communication, Duan et al. [21] designed an adaptive control protocol and proved that the tracking error of the problem is bounded. Huang [22] discussed the adaptive consensus of uncertain system, and proved agents can obtain average consensus in the almost sure sense. Xiao et al. [23] proposed the adaptive finite-time control protocols for a leaderless system, and proved similar properties hold for systems with a leader. The bipartite adaptive consensus of the stochastic system were taken into account in [24,25]. However, these references did not consider the interference of time delay. Time delay often degrades the performance of the control system and disrupts the stability of the system. Furthermore, the presence of time delay causes the great difficulties in the analysis and synthesis of the control system.

When the system is jointly disturbed by time delay and noise during the interactive process, the dynamical model of the system has a more complex form. There are currently only a few papers that consider the adaptive consensus in this situation. When the adaptive gain is mandatory, Zong et al. [14] analyzed the tracking problem in the case of the joint disturbance of noise and time delay. Also, a neural network approach was employed to analyze the topic for mandatory gain [26]. In practical applications, the mandatory gain has to be accurately selected based on the actual situation, which is often quite difficult. This paper will consider an adaptive control law that can dynamically adjust on the basis of state. For the tracing problem of multi-agent system, we first propose an adaptive control protocol and design a novel adaptive law, then the Lyapunov function is used to prove the adaptive consensus of the system. Finally, when the time delay trends to zero, we simplify the conditions for the system to attain the adaptive consensus. The significance of this paper is to revise the control protocol that were previously only applicable to a system without time delay. Our proposed adaptive control protocol can ensure the consensus of a system under the interference of noise and time delay. The contributions are as follows:

(1) For a stochastic multi-agent system, a novel adaptive control law is firstly proposed when there is a lag phenomenon in the interactive process. The control laws in [12–14] were all mandatory and often required precise selection to determine the specific form. The adaptive control proposed in this paper can be dynamically adjusted based on the current state of the system, thus avoiding the difficulty of precise selection.

(2) No matter whether the stochastic multi-agent system has time delay or not, the adaptive control law can ensure the consensus. However, the adaptive laws in [15,16] were only applied to multi-agent systems without delay and noise. Additionally, the sufficient conditions of consensus in this paper are simpler for the case without delay.

(3) Compared with some early references in [21,24], the final tracking error in this paper has a smaller value under the adaptive law. Furthermore, when the intensity of noise approaches zero, the final dynamic error will trend to zero. However, many previous conclusions can only converge to a non-zero constant.

2. Theoretical Basis

The system in this work includes one leader and N followers, denoted as $v_0, v_1, \cdots,$ and v_N, respectively. $\mathcal{G} = (\mathcal{V}, \mathcal{N}, \mathcal{A})$ represents a digraph among the followers. $\mathcal{V} = \{v_1, v_2, \cdots, v_N\}$ and $\mathcal{N} \subseteq \mathcal{V} \times \mathcal{V}$ is the set of the followers and edges, respectively. $\mathcal{A} = [e_{ij}] \in \mathcal{R}^{N \times N}$ is called adjacency matrix, its elements satisfy $e_{ij} = 1$ if and only if $(v_i, v_j) \in \mathcal{N}$, or else, $e_{ij} = 0$. $\mathcal{N}_i = \{v_j \in V : (v_j, v_i) \in \mathcal{N}\}$ is the neighbor set, and $L_\mathcal{G} = [l_{ij}]$ is the Laplace matrix, where $l_{ii} = \sum_{j \neq i} e_{ij}$ and $l_{ij} = -e_{ij}, i \neq j$. In addition, assuming $\widetilde{\mathcal{G}}$ is a digraph composed of all agents, and the matrix $L_{\widetilde{\mathcal{G}}}$ is defined by $\begin{bmatrix} 0 & 0_{1 \times N} \\ -E_0 \cdot 1_N & L_\mathcal{G} + E_0 \end{bmatrix}$, where $E_0 = \text{diag}\{e_{10}, e_{20}, \cdots, e_{N0}\}$ and $1_N = [1, 1, \cdots, 1]^T$. The difference between the two digraphs is that $\widetilde{\mathcal{G}}$ contains the node of leader, while $\mathcal{G}$ does not.

Supposing the leader v_0 is globally reachable in this paper, which means a directed path from each follower v_i to the leader v_0 can be found. When all elements of the adjacency matrix $\mathcal{A}$ satisfy $\sum_{j=1}^{N} e_{ij} = \sum_{j=1}^{N} e_{ji}$, the digraph is a balanced graph. The following lemmas are introduced.

Lemma 1 ([27]). *Assuming $\widetilde{\mathcal{G}}$ is a digraph, the three properties are equivalent:*
(1) *The node v_0 is globally reachable.*
(2) *For the matrix $H = L_\mathcal{G} + E_0$, the real parts of all eigenvalues are positive.*
(3) *Further suppose the digraph $\mathcal{G}$ is balanced, then $H + H^T$ is positive definite.*

Lemma 2 ([28]). *For the matrices M_1, M_2, M_3 and M_4, the Kronecker product of two matrices is represented by the symbol $\otimes$. Assuming the four matrices have appropriate dimensions, then the following properties hold:*

(1) $M_1 \otimes (M_2 + M_3) = M_1 \otimes M_2 + M_1 \otimes M_3$.
(2) $(M_1 \otimes M_2) \otimes M_3 = M_1 \otimes (M_2 \otimes M_3)$.
(3) $(M_1 \otimes M_2)(M_3 \otimes M_4) = M_1 M_3 \otimes M_2 M_4$.
(4) $(M_1 \otimes M_2)^T = M_1^T \otimes M_2^T$.
(5) $tr(M_1 \otimes M_2) = tr(M_1) tr(M_2)$.

3. The Adaptive Consensus

Considering a multi-agent system, its dynamic behavior can be expressed as

$$\dot{x}_i(t) = A x_i(t) + B u_i(t), \quad i = 1, 2, \cdots, N. \tag{1}$$

In the equation, $u_i(t) \in \mathbf{R}^p$ denotes the input and needs to be devised, $x_i(t) \in \mathbf{R}^n$ represents the state of the position. A is a $n \times n$ order constant matrix, B is a $n \times p$ order constant matrix, and the two matrices are known. The model of leader is represented as

$$\dot{x}_0(t) = A x_0(t). \tag{2}$$

In order to obtain the adaptive consensus of system (1), the key issue is to construct a control protocol $u_i(t)$ containing adaptive gain based on the communication graph among agents, and then use the state $x_i(t)$ to design the adjustment method of the adaptive gain. The adaptive method can rely on relatively little prior knowledge about the model. If the system (1) is not disturbed by time delay and noise, a general control protocol can be represented as $u_i(t) = cK \sum_{j \in N_i} e_{ij}(x_j(t) - x_i(t))$, where c is a coupling weight and K is a feedback gain matrix. The protocol was investigated in [29,30], who pointed out that the constant c is related to the global information of the system. When there exists noise interference and time delay in the process of communication, this paper proposes an adaptive control protocol, designs an novel adaptive control law, and analyzes the impact of time delay on the system.

For n dimensional probability space $(\Omega, \mathcal{F}, P)$, the standard Brownian motions in the space are denoted by $W_i(t) \in \mathbf{R}^n$, the standard white noise is written as $\eta_i(t) \in \mathbf{R}^n$ and satisfies $\int_0^t \eta_i(s) ds = W_i(t)$. For the system (1), the control protocol perturbed by noise and time delay is designed as

$$u_i(t) = s_i(t) K \left[\sum_{j \in N_i} e_{ij}(x_j(t-\tau) - x_i(t-\tau)) + e_{i0}(x_0(t-\tau) - x_i(t-\tau)) + e_{i0} \sigma_{0i} \eta_i(t) \right]. \tag{3}$$

In the protocol, $\tau > 0$ is time delay, σ_{0i} is noise intensity, the constants e_{ij} and e_{i0} indicate the weights of digraphs in the multi-agent system, the matrix $K \in \mathbf{R}^{p \times n}$ is called a feedback gain matrix. The adaptive gain $s_i(t)$ satisfies $\underline{\theta} \leq s_i(t) \leq \overline{\theta}$, where $\underline{\theta}$ and $\overline{\theta}$ are two positive constants. The difficulty of solving adaptive control problems lies in designing an appropriate adaptive control law. For this control protocol (3), in order to obtain the adaptive consensus of the system, the main difficulty is to construct a differential equation that the gain $s_i(t)$ satisfies.

When the control protocol (3) does not contain time delay and noise, many scholars have already studied the adaptive consensus. Li et al. [31] considered the adaptive tracking problem of system with a leader. The adaptive event-triggering theory was discussed for a linear time-varying system in [32]. Deng et al. [33] analyzed the adaptive tracking problem of high-order system. However, time delay and noise are inevitable in the process of agent interaction. For leaderless multi-agent system, Wu et al. [34] designed an adaptive control protocol in noisy environments. The adaptive consensus with multiplicative noise was analyzed in [35]. Duan et al. [21] discussed one order leader-following system with noise in the absence of time delay. In this section, the adaptive problem of system (1) and (2) will be studied under the control protocol (3), which not only considers the impact of noise, but also considers the effect of time delay, so it is more in line with real scenarios.

If the adaptive gain is mandatory, such as $s_i(t) = s(t) = \frac{1}{1+t}$ or $\frac{\log(1+t)}{1+t}$, there have been many results. The mean square consensus was achieved in [12,13]. Zong et al. [14] investigated the adaptive protocol of the system under time delay and noise. Nevertheless, the mandatory gain has to be accurately selected in order to satisfy the limiting conditions, which is often quite difficult. Therefore, the adaptive gain that can be dynamically adjusted according to the state has obvious advantages in practical applications. In order to solve the consensus of the system (1)–(3), we construct a novel adaptive law as

$$\dot{s}_i(t) = \varepsilon_i(t)^T \sum_{j=1}^{N} h_{ij} \Gamma \varepsilon_j(t) - (s_i(t) - \delta). \tag{4}$$

where the constant $\delta > \frac{1}{\lambda_{\min}(H^T + H)}$, the dynamic error $\varepsilon_i(t) = x_i(t) - x_0(t)$, and the symbol h_{ij} is the element of H. The adaptive law (4) can continuously improve the structure of the model by extracting model's information, thereby enabling the model to more and more accurate. It is worth noting that the adaptive laws proposed in most of the literature are different, such as the mandatory adaptive law [13,14], the decaying adaptive law [15], the edge-based adaptive law [16], etc. The advantages of the adaptive law (4) is that it can be applied to multi-agent systems with noise and time delay. In order to prove the consensus of system, the solution of the algebraic Riccati equation is used to build the matrix Γ. Let $K = B^T P$, the matrix $\Gamma = PBK$ is called adaptive gain matrix in (4), and P is a positive matrix and satisfies the algebraic Riccati equation

$$A^T P + PA - PBB^T P + kI = 0, (k > 0). \tag{5}$$

The above equation has been widely applied to prove the stability of the system since it was proposed. Generally, the matrix P can be used to construct Lyapunov functions, combined with the special form of the Riccati equation, it is easy to verify the conditions of the stability theorem.

Remark 1. *The adaptive law (4) has a simpler structure and can be rewritten as*

$$\begin{pmatrix} \dot{s}_1(t) \\ \dot{s}_2(t) \\ \vdots \\ \dot{s}_N(t) \end{pmatrix} = \begin{pmatrix} \varepsilon_1(t)^T & 0 & \cdots & 0 \\ 0 & \varepsilon_2(t)^T & \cdots & 0 \\ \vdots & \vdots & & \vdots \\ 0 & 0 & \cdots & \varepsilon_N(t)^T \end{pmatrix} (H \otimes PBK) \begin{pmatrix} \varepsilon_1(t) \\ \varepsilon_2(t) \\ \vdots \\ \varepsilon_N(t) \end{pmatrix} - \begin{pmatrix} s_1(t) - \delta \\ s_2(t) - \delta \\ \vdots \\ s_N(t) - \delta \end{pmatrix}.$$

Although many different forms of adaptive laws have been proposed, most cannot be represented by the Kronecker products, which will make previous adaptive laws appear more complex. In addition, for the mandatory gain $s(t)$ proposed in many literature, the two constraints $\int_0^\infty s(t)dt = \infty$ and $\int_0^\infty s^2(t)dt < \infty$ need to be used, such as the continuous mandatory gain in references [13,14] and the discrete mandatory gain in reference [34]. The adaptive gain proposed in this article will automatically adjust according to the current state.

Let $\varepsilon(t) = [(x_1(t) - x_0(t))^T, (x_2(t) - x_0(t))^T, \cdots, (x_N(t) - x_0(t))^T]^T$, the dynamic error equation can be abbreviated as

$$d\varepsilon(t) = [(I_N \otimes A)\varepsilon(t) - (S(t)H \otimes BK)\varepsilon(t - \tau)]dt - (S(t)E_0C_0 \otimes BK)dW. \tag{6}$$

where $S(t) = \text{diag}\{s_1(t), s_2(t), \cdots, s_N(t)\}$ is a diagonal matrix, I_N is an identity matrix, $C_0 = \text{diag}\{\sigma_{01}, \sigma_{02}, \cdots, \sigma_{0N}\}$ is the matrix corresponding to noise intensity, dW is nN dimensional standard Brownian motion, and $E_0 = \text{diag}\{e_{10}, e_{20}, \cdots, e_{N0}\}$ reflects the interaction of the system. Equation (5) is known as a stochastic differential equation, which includes a differential part and random part. The random part can reflect the changes of disturbance. The following theorem demonstrates the adaptive consensus of the system (1)–(3) when the adaptive law adopts the Equation (4).

Theorem 1. *Assuming that the digraph $\widetilde{\mathcal{G}} = (\widetilde{\mathcal{V}}, \widetilde{\mathcal{N}}, \widetilde{\mathcal{A}})$ for a system of $N+1$ agents is made up of N followers and one leader, and that its subgraph $\mathcal{G}$ for all followers is a balanced graph. For the multi-agent system determined by the Equations (1)–(3), if there exists a positive constant ξ satisfying*

$$k > \xi \bar{\theta}^2 \lambda_{\max}(HH^T) \lambda_{\max}^4(P) \lambda_{\max}^2(BB^T) + 4\xi^{-1}\tau^2 \bar{\theta}^2 \lambda_{\max}(HH^T) \lambda_{\max}^2(P) \lambda_{\max}^2(BB^T)$$
$$+ 4\xi^{-1}\tau^2 \lambda_{\max}(AA^T) \tag{7}$$

then the mean square bounded consensus can be gained under the adaptive law (4), i.e.,

$$\lim_{t \to +\infty} E|x_i(t) - x_0(t)|^2 = \epsilon_1 \tag{8}$$

where E represents the expectation, and ϵ_1 is a small constant independent of time t.

Proof. The Lyapunov function is chosen as follows,

$$\begin{aligned} V_1(t) &= V_{11}(t) + V_{12}(t) \\ &= \varepsilon(t)^T (I_N \otimes P) \varepsilon(t) + w_1 \int_{t-\tau}^{t} |\varepsilon(s)|^2 ds + w_2 \int_{-\tau}^{0} \int_{t+\theta}^{t} |\varepsilon(s)|^2 ds d\theta \\ &\quad + w_3 \int_{-\tau}^{0} \int_{t+\theta}^{t} |\varepsilon(s-\tau)|^2 ds d\theta + V_{12}(t), \end{aligned}$$

where the function $V_{12}(t) = \sum_{i=1}^{N} (s_i(t) - \delta)^2$. The Lyapunov function is mainly divided into three parts, the first part $\varepsilon(t)^T (I_N \otimes P) \varepsilon(t)$ is similar to the construction of Lyapunov functions in most references, the special integral part was referred to as a degenerate functional and was used by Kolmanovskii et al. [36]. The last part $V_{12}(t)$ is a commonly used form in most of the literature when discussing adaptive consensus, and after taking the derivative of this function, the adaptive law can be used to eliminate some unnecessary terms in the following calculations. If the time t is less than $-\tau$ in the double integral $\int_{-\tau}^{0} \int_{t+\theta}^{t} |\varepsilon(s-\tau)|^2 ds d\theta$, we assume $\varepsilon_i(t)$ equals to the initial value $\varepsilon_i(0)$.

Applying the Itô formula and the error closed-loop systems (6), the random differentiation is expressed as

$$dV_1(t) = \mathcal{L}_1 V_1(t) dt + 2\varepsilon(t)^T [S(t) E_0 C_0 \otimes PBK] dW, \tag{9}$$

where the first term is defined as

$$\begin{aligned} \mathcal{L}_1 V_1(t) &= \varepsilon(t)^T [I_N \otimes (A^T P + PA)] \varepsilon(t) - 2\varepsilon(t)^T [S(t) H \otimes PBK] \varepsilon(t-\tau) \\ &\quad + \mathrm{tr}\{S^2(t) E_0^2 C_0^2 \otimes K^T B^T PBK\} + w_1 |\varepsilon(t)|^2 - w_1 |\varepsilon(t-\tau)|^2 \\ &\quad + w_2 \tau |\varepsilon(t)|^2 - w_2 \int_{t-\tau}^{t} |\varepsilon(s)|^2 ds + w_3 \tau |\varepsilon(t-\tau)|^2 \\ &\quad - w_3 \int_{t-\tau}^{t} |\varepsilon(s-\tau)|^2 ds + \dot{V}_{12}(t). \end{aligned}$$

Using the adaptive laws, we can obtain the following equation by combining the derivative rule of the composite function,

$$
\begin{aligned}
&\dot{V}_{12}(t) \\
&= 2\sum_{i=1}^{N}(s_i(t)-\delta)\dot{s}_i(t) \\
&= 2\sum_{i=1}^{N}\left[(s_i(t)-\delta)\varepsilon_i(t)^{\mathrm{T}}\sum_{j=1}^{N}h_{ij}PBK\varepsilon_j(t)\right] - 2\sum_{i=1}^{N}(s_i(t)-\delta)^2 \\
&= 2\sum_{i=1}^{N}\left[s_i(t)\varepsilon_i(t)^{\mathrm{T}}\sum_{j=1}^{N}h_{ij}PBK\varepsilon_j(t)\right] - 2\delta\sum_{i=1}^{N}\left[\varepsilon_i(t)^{\mathrm{T}}\sum_{j=1}^{N}h_{ij}PBK\varepsilon_j(t)\right] - 2\sum_{i=1}^{N}(s_i(t)-\delta)^2 \\
&= 2\varepsilon(t)^{\mathrm{T}}[S(t)H\otimes PBK]\varepsilon(t) - \delta\varepsilon(t)^{\mathrm{T}}[(H^{\mathrm{T}}+H)\otimes PBK]\varepsilon(t) - 2\sum_{i=1}^{N}(s_i(t)-\delta)^2.
\end{aligned}
$$

By Lemma 1, we can obtain the matrix $H+H^{\mathrm{T}}$ is positive definite, which means all eigenvalues are greater than zero. Thus, we can obtain $\delta\lambda_{\min}(H^{\mathrm{T}}+H) > 1$ by the known condition $\delta > \frac{1}{\lambda_{\min}(H^{\mathrm{T}}+H)}$. From the Ricatti equation, we have,

$$
\begin{aligned}
&\mathcal{L}_1 V_1(t) \\
&= \varepsilon(t)^{\mathrm{T}}[I_N\otimes(A^{\mathrm{T}}P+PA)]\varepsilon(t) - \delta\varepsilon(t)^{\mathrm{T}}[(H^{\mathrm{T}}+H)\otimes PBK]\varepsilon(t) \\
&\quad + 2\varepsilon(t)^{\mathrm{T}}[S(t)H\otimes PBK]\big[\varepsilon(t)-\varepsilon(t-\tau)\big] + \mathrm{tr}\big\{S^2(t)E_0^2 C_0^2\otimes K^{\mathrm{T}}B^{\mathrm{T}}PBK\big\} \\
&\quad + w_1|\varepsilon(t)|^2 - w_1|\varepsilon(t-\tau)|^2 + w_2\tau|\varepsilon(t)|^2 - w_2\int_{t-\tau}^{t}|\varepsilon(s)|^2 ds \\
&\quad + w_3\tau|\varepsilon(t-\tau)|^2 - w_3\int_{t-\tau}^{t}|\varepsilon(s-\tau)|^2 ds - 2\sum_{i=1}^{N}(s_i(t)-\delta)^2 \\
&\leq \varepsilon(t)^{\mathrm{T}}[I_N\otimes(A^{\mathrm{T}}P+PA-PBK)]\varepsilon(t) \\
&\quad + 2\varepsilon(t)^{\mathrm{T}}[S(t)H\otimes PBK]\big[\varepsilon(t)-\varepsilon(t-\tau)\big] + \mathrm{tr}\big\{S^2(t)E_0^2 C_0^2\otimes K^{\mathrm{T}}B^{\mathrm{T}}PBK\big\} \\
&\quad + w_1|\varepsilon(t)|^2 - w_1|\varepsilon(t-\tau)|^2 + w_2\tau|\varepsilon(t)|^2 - w_2\int_{t-\tau}^{t}|\varepsilon(s)|^2 ds \\
&\quad + w_3\tau|\varepsilon(t-\tau)|^2 - w_3\int_{t-\tau}^{t}|\varepsilon(s-\tau)|^2 ds - 2\sum_{i=1}^{N}(s_i(t)-\delta)^2 \\
&= -k|\varepsilon(t)|^2 + 2\varepsilon(t)^{\mathrm{T}}[S(t)H\otimes PBK]\big[\varepsilon(t)-\varepsilon(t-\tau)\big] + \mathrm{tr}\big\{S^2(t)E_0^2 C_0^2\otimes K^{\mathrm{T}}B^{\mathrm{T}}PBK\big\} \\
&\quad + w_1|\varepsilon(t)|^2 - w_1|\varepsilon(t-\tau)|^2 + w_2\tau|\varepsilon(t)|^2 - w_2\int_{t-\tau}^{t}|\varepsilon(s)|^2 ds + w_3\tau|\varepsilon(t-\tau)|^2 \\
&\quad - w_3\int_{t-\tau}^{t}|\varepsilon(s-\tau)|^2 ds - 2\sum_{i=1}^{N}(s_i(t)-\delta)^2.
\end{aligned}
\tag{10}
$$

Note the inequality $2ab \leq \xi a^2 + \frac{1}{\xi}b^2$ for any positive constant ξ, we have

$$
\begin{aligned}
&2\varepsilon(t)^{\mathrm{T}}[S(t)H\otimes PBK]\big[\varepsilon(t)-\varepsilon(t-\tau)\big] \\
&\leq \xi\varepsilon(t)^{\mathrm{T}}[S(t)H\otimes PBK][S(t)H\otimes PBK]^{\mathrm{T}}\varepsilon(t) + \xi^{-1}|\varepsilon(t)-\varepsilon(t-\tau)|^2 \\
&\leq \xi\bar{\theta}^2\lambda_{\max}(HH^{\mathrm{T}})\lambda_{\max}^4(P)\lambda_{\max}^2(BB^{\mathrm{T}})|\varepsilon(t)|^2 + \xi^{-1}|\varepsilon(t)-\varepsilon(t-\tau)|^2.
\end{aligned}
\tag{11}
$$

Now, we can obtain from the above inequality,

$$\begin{aligned}
&\mathcal{L}_1 V_1(t) \\
&\leq -k|\varepsilon(t)|^2 + \xi\bar{\theta}^2\lambda_{\max}(HH^T)\lambda_{\max}^4(P)\lambda_{\max}^2(BB^T)|\varepsilon(t)|^2 + \xi^{-1}|\varepsilon(t) - \varepsilon(t-\tau)|^2 \\
&\quad + \bar{\theta}^2\lambda_{\max}^3(P)\lambda_{\max}^2(BB^T)\max\{(e_{i0}\sigma_{0i})^2\} + w_1|\varepsilon(t)|^2 - w_1|\varepsilon(t-\tau)|^2 \\
&\quad + w_2\tau|\varepsilon(t)|^2 - w_2\int_{t-\tau}^t |\varepsilon(s)|^2 ds + w_3\tau|\varepsilon(t-\tau)|^2 \\
&\quad - w_3\int_{t-\tau}^t |\varepsilon(s-\tau)|^2 ds - 2\sum_{i=1}^N (s_i(t) - \delta)^2 \\
&= -\Lambda_1|\varepsilon(t)|^2 - \Lambda_2|\varepsilon(t-\tau)|^2 + \xi^{-1}|\varepsilon(t) - \varepsilon(t-\tau)|^2 \\
&\quad + \bar{\theta}^2\lambda_{\max}^3(P)\lambda_{\max}^2(BB^T)\max\{(e_{i0}\sigma_{0i})^2\} \\
&\quad - w_2\int_{t-\tau}^t |\varepsilon(s)|^2 ds - w_3\int_{t-\tau}^t |\varepsilon(s-\tau)|^2 ds - 2\sum_{i=1}^N (s_i(t) - \delta)^2,
\end{aligned} \tag{12}$$

where the two constants in the above inequality are denoted as

$$\Lambda_1 = k - \xi\bar{\theta}^2\lambda_{\max}(HH^T)\lambda_{\max}^4(P)\lambda_{\max}^2(BB^T) - w_1 - w_2\tau$$

and $\Lambda_2 = w_1 - w_3\tau$.

According to the error closed-loop equation, it obtains from the Hölder inequality,

$$\begin{aligned}
&|\varepsilon(t) - \varepsilon(t-\tau)|^2 \\
&= \left|\int_{t-\tau}^t d\varepsilon(s)\right|^2 \\
&= \left|\int_{t-\tau}^t [(I_N \otimes A)\varepsilon(s) - (S(s)H) \otimes (BK)\varepsilon(s-\tau)]ds - \int_{t-\tau}^t (S(s)E_0C_0) \otimes (BK)dW\right|^2 \\
&\leq 4\left|\int_{t-\tau}^t (I_N \otimes A)\varepsilon(s)ds\right|^2 + 4\left|\int_{t-\tau}^t (S(s)H) \otimes (BK)\varepsilon(s-\tau)ds\right|^2 \\
&\quad + 4\left|\int_{t-\tau}^t (S(s)E_0C_0) \otimes (BK)dW\right|^2 \\
&\leq 4\tau\lambda_{\max}(AA^T)\int_{t-\tau}^t |\varepsilon(s)|^2 ds + 4\tau\bar{\theta}^2\lambda_{\max}(HH^T)\lambda_{\max}^2(P)\lambda_{\max}^2(BB^T)\int_{t-\tau}^t |\varepsilon(s-\tau)|^2 ds \\
&\quad + 4\bar{\theta}^2\lambda_{\max}^2(P)\lambda_{\max}^2(BB^T)\max\{(e_{i0}\sigma_{0i})^2\}\left|\int_{t-\tau}^t dW\right|^2.
\end{aligned}$$

So, we obtain

$$\begin{aligned}
&\mathcal{L}_1 V_1(t) \\
&\leq -\frac{\Lambda_1}{\lambda_{\max}(P)}\varepsilon(t)^T(I_N \otimes P)\varepsilon(t) - \Lambda_2|\varepsilon(t-\tau)|^2 - [w_2 - 4\xi^{-1}\tau\lambda_{\max}(AA^T)]\int_{t-\tau}^t |\varepsilon(s)|^2 ds \\
&\quad - [w_3 - 4\xi^{-1}\tau\bar{\theta}^2\lambda_{\max}^2(BB^T)\lambda_{\max}^2(P)\lambda_{\max}(HH^T)]\int_{t-\tau}^t |\varepsilon(s-\tau)|^2 ds \\
&\quad - \alpha_1\left[w_1\int_{t-\tau}^t |\varepsilon(s)|^2 ds + w_2\int_{-\tau}^0\int_{t+\theta}^t |\varepsilon(s)|^2 ds d\theta + w_3\int_{-\tau}^0\int_{t+\theta}^t |\varepsilon(s-\tau)|^2 ds d\theta + V_{12}(t)\right] \\
&\quad + \alpha_1\left[w_1\int_{t-\tau}^t |\varepsilon(s)|^2 ds + w_2\int_{-\tau}^0\int_{t+\theta}^t |\varepsilon(s)|^2 ds d\theta + w_3\int_{-\tau}^0\int_{t+\theta}^t |\varepsilon(s-\tau)|^2 ds d\theta + V_{12}(t)\right] \\
&\quad + \bar{\theta}^2\lambda_{\max}^3(P)\lambda_{\max}^2(BB^T)\max\{(e_{i0}\sigma_{0i})^2\} - 2\sum_{i=1}^N (s_i(t) - \delta)^2 \\
&\quad + 4\xi^{-1}\bar{\theta}^2\lambda_{\max}^2(P)\lambda_{\max}^2(BB^T)\max\{(e_{i0}\sigma_{0i})^2\}\left|\int_{t-\tau}^t dW\right|^2
\end{aligned} \tag{13}$$

where α_1 is a positive constant that will be determined later.

From the known condition (7), we have

$$\frac{k - \xi\bar{\theta}^2\lambda_{\max}(HH^T)\lambda_{\max}^4(P)\lambda_{\max}^2(BB^T)}{\tau} > 4\xi^{-1}\tau\bar{\theta}^2\lambda_{\max}(HH^T)\lambda_{\max}^2(P)\lambda_{\max}^2(BB^T) + 4\xi^{-1}\tau\lambda_{\max}(AA^T)$$

We can select w_2 and w_3 to satisfy

$$w_2 > 4\xi^{-1}\tau\lambda_{\max}(AA^T), \quad w_3 > 4\xi^{-1}\tau\bar{\theta}^2\lambda_{\max}(HH^T)\lambda_{\max}^2(P)\lambda_{\max}^2(BB^T)$$

and

$$w_2 + w_3 < \frac{k - \xi\bar{\theta}^2\lambda_{\max}(HH^T)\lambda_{\max}^4(P)\lambda_{\max}^2(BB^T)}{\tau}$$

From the above equation, we have

$$k - \xi\bar{\theta}^2\lambda_{\max}(HH^T)\lambda_{\max}^4(P)\lambda_{\max}^2(BB^T) - w_2\tau > w_3\tau$$

Now, we can select w_1 to satisfy

$$k - \xi\bar{\theta}^2\lambda_{\max}(HH^T)\lambda_{\max}^4(P)\lambda_{\max}^2(BB^T) - w_2\tau > w_1 > w_3\tau$$

which implies $\Lambda_1 = k - \xi\bar{\theta}^2\lambda_{\max}(HH^T)\lambda_{\max}^4(P)\lambda_{\max}^2(BB^T) - w_1 - w_2\tau > 0$ and $\Lambda_2 = w_1 - w_3\tau > 0$.

On the other hand, the positive constant α_3 is selected to satisfy

$$\alpha_1 \leq 2, \alpha_1 \leq \frac{\Lambda_1}{\lambda_{\max}(P)},$$

$$\alpha_1 \leq \frac{w_2 - 4\xi^{-1}\tau\lambda_{\max}(AA^T)}{w_1 + w_2\tau},$$

$$\alpha_1 \leq \frac{w_3 - 4\xi^{-1}\tau\bar{\theta}^2\lambda_{\max}^2(BB^T)\lambda_{\max}^2(P)\lambda_{\max}(HH^T)}{w_3\tau}. \tag{14}$$

Note $\int_{-\tau}^{0}\int_{t+\theta}^{t}|\varepsilon(s)|^2 ds d\theta \leq \tau\int_{t-\tau}^{t}|\varepsilon(s)|^2 ds$ and $\int_{-\tau}^{0}\int_{t+\theta}^{t}|\varepsilon(s-\tau)|^2 ds d\theta \leq \tau\int_{t-\tau}^{t}|\varepsilon(s-\tau)|^2 ds$, the following inequality can be given from the Equation (13),

$$\begin{aligned}\mathcal{L}_1V_1(t) &\leq -\alpha_1V_1(t) - \Lambda_2|\varepsilon(t-\tau)|^2 - [w_2 - 4\xi^{-1}\tau\lambda_{\max}(AA^T) - \alpha_1(w_1+w_2\tau)]\int_{t-\tau}^{t}|\varepsilon(s)|^2 ds \\ &\quad - [w_3 - 4\xi^{-1}\tau\bar{\theta}^2\lambda_{\max}^2(BB^T)\lambda_{\max}^2(P)\lambda_{\max}(HH^T) - \alpha_1 w_3\tau]\int_{t-\tau}^{t}|\varepsilon(s-\tau)|^2 ds \\ &\quad - (2-\alpha_1)\sum_{i=1}^{N}(s_i(t)-\delta)^2 + \bar{\theta}^2\lambda_{\max}^3(P)\lambda_{\max}^2(BB^T)\max\{(e_{i0}\sigma_{0i})^2\} \\ &\quad + 4\xi^{-1}\bar{\theta}^2\lambda_{\max}^2(P)\lambda_{\max}^2(BB^T)\max\{(e_{i0}\sigma_{0i})^2\}\left|\int_{t-\tau}^{t}dW\right|^2 \\ &\leq -\alpha_1V_1(t) + \bar{\theta}^2\lambda_{\max}^3(P)\lambda_{\max}^2(BB^T)\max\{(e_{i0}\sigma_{0i})^2\} \\ &\quad + 4\xi^{-1}\bar{\theta}^2\lambda_{\max}^2(P)\lambda_{\max}^2(BB^T)\max\{(e_{i0}\sigma_{0i})^2\}\left|\int_{t-\tau}^{t}dW\right|^2\end{aligned}$$

By using $d(e^{\gamma t}V_1(t)) = \gamma e^{\gamma t}V_1(t)dt + e^{\gamma t}dV_1(t)$ and integrating on both sides of the formula, it follows from the Equation (9),

$$\begin{aligned}
e^{\gamma t}\text{EV}_1(t) &= \text{EV}_1(0) + \gamma\text{E}\int_0^t e^{\gamma s}V_1(s)\text{d}s + \text{E}\int_0^t e^{\gamma s}\text{d}V_1(s)\\
&\leq \text{EV}_1(0) - (\alpha_1-\gamma)\text{E}\int_0^t e^{\gamma s}V_1(s)\text{d}s + \left[\bar{\theta}^2\lambda_{\max}^3(P)\lambda_{\max}^2(BB^\text{T})\max\left\{(e_{i0}\sigma_{0i})^2\right\}\right.\\
&\quad \left.+ 4\xi^{-1}\tau n N\bar{\theta}^2\lambda_{\max}^2(P)\lambda_{\max}^2(BB^\text{T})\max\left\{(e_{i0}\sigma_{0i})^2\right\}\right]\frac{e^{\gamma t}-1}{\gamma}\\
&\leq \text{EV}_1(0) + \frac{\mu_1}{\gamma}e^{\gamma t}
\end{aligned}$$

where γ is chosen to satisfy $\gamma < \alpha_1$, the symbol E represents the expectation of the random variable, and the positive constant μ_1 is defined as follows

$$\mu_1 = \bar{\theta}^2\lambda_{\max}^3(P)\lambda_{\max}^2(BB^\text{T})\max\left\{(e_{i0}\sigma_{0i})^2\right\} + 4\xi^{-1}\tau n N\bar{\theta}^2\lambda_{\max}^2(P)\lambda_{\max}^2(BB^\text{T})\max\left\{(e_{i0}\sigma_{0i})^2\right\}.$$

So, we have

$$\text{EV}_1(t) \leq \text{EV}_1(0)e^{-\gamma t} + \frac{\mu_1}{\gamma} \tag{15}$$

Note $|\varepsilon(t)|^2 \leq \frac{\varepsilon(t)^\text{T}(I_N\otimes P)\varepsilon(t)}{\lambda_{\min}(P)} \leq \frac{V_1(t)}{\lambda_{\min}(P)}$, we can obtain

$$\lim_{t\to+\infty}\text{E}|x_i(t) - x_0(t)|^2 = \epsilon_1$$

and ϵ_1 is a small constant independent of time t. □

Remark 2. *Under the random noise disturbance, little papers discuss the adaptive consensus of multi-agent systems in the presence of time delays. Theorem 1 indicates that the adaptive control law (4) can ensure that the dynamic error between the followers and the leader can converge to a small number ϵ_1 in the mean square sense. Looking back at the above proof, it can be found that $\epsilon_1 = \frac{\mu_1}{\gamma\lambda_{\min}(P)} = \Xi\max\left\{(e_{i0}\sigma_{0i})^2\right\}$, where Ξ is a constant. So this boundary ϵ_1 tends to zero when the noise intensity of the system approaches zero.*

Remark 3. *Formula (7) can be transformed into*

$$\tau < \sqrt{\frac{k\xi - \xi^2\bar{\theta}^2\lambda_{\max}(HH^\text{T})\lambda_{\max}^4(P)\lambda_{\max}^2(BB^\text{T})}{4\bar{\theta}^2\lambda_{\max}(HH^\text{T})\lambda_{\max}^2(P)\lambda_{\max}^2(BB^\text{T}) + 4\lambda_{\max}(AA^\text{T})}}.$$

So the upper limit of time delay can be obtained as

$$\frac{k}{4\bar{\theta}\lambda_{\max}^2(P)\lambda_{\max}(BB^\text{T})\sqrt{\lambda_{\max}(HH^\text{T})\left[\bar{\theta}^2\lambda_{\max}(HH^\text{T})\lambda_{\max}^2(P)\lambda_{\max}^2(BB^\text{T}) + \lambda_{\max}(AA^\text{T})\right]}}.$$

Although the constant time delay in this paper cannot be directly extended to time-varying delay, the above formula gives the range of time delay, which can provide some reference for future work.

Now, we further analyze the adaptive control law (4), and investigate whether multi-agent system can still achieve the consensus under $\tau = 0$. For this case, the adaptive law is kept unchanged, and the control protocol is constructed as

$$u_i(t) = s_i(t)K\left[\sum_{j\in N_i}e_{ij}(x_j(t) - x_i(t)) + e_{i0}(x_0(t) - x_i(t)) + e_{i0}\sigma_{0i}\eta_i(t)\right], \tag{16}$$

where $\eta_i(t) \in \mathbf{R}^n$ is n dimensional standard white noise. The abbreviated form of the error dynamic equation is represented by

$$d\varepsilon(t) = [I_N \otimes A - S(t)H \otimes BK]\varepsilon(t)dt - (S(t)E_0C_0 \otimes BK)dW. \tag{17}$$

Theorem 2. *Assuming the digraph $\widetilde{\mathcal{G}} = (\widetilde{\mathcal{V}}, \widetilde{\mathcal{N}}, \widetilde{\mathcal{A}})$ has the same properties as Theorem 1. If the control protocol of the multi-agent system (1) and (2) satisfies (16) and the adaptive law is shown in (4), then the system can achieve the mean square bounded consensus, i.e.,*

$$\lim_{t \to +\infty} E|x_i(t) - x_0(t)|^2 = \epsilon_2 \tag{18}$$

where ϵ_2 is a small constant independent of time t.

Proof. The Laypunov function is denoted as

$$V_2(t) = V_{21}(t) + V_{22}(t) = \varepsilon(t)^\mathrm{T}(I_N \otimes P)\varepsilon(t) + \sum_{i=1}^{N}(s_i(t) - \delta)^2,$$

where P satisfies the Equation (5). We can obtain from the Itô formula

$$dV_2(t) = \mathcal{L}_2 V_2(t)dt - 2\varepsilon(t)^\mathrm{T}[S(t)E_0C_0 \otimes PBK]dW, \tag{19}$$

and the operator $\mathcal{L}_2$ satisfies

$$\mathcal{L}_2 V_2(t) = \varepsilon(t)^\mathrm{T}[I_N \otimes (PA + A^\mathrm{T}P)]\varepsilon(t) - 2\varepsilon(t)^\mathrm{T}[S(t)H \otimes PBK]\varepsilon(t) + \dot{V}_{12}(t)$$
$$+ \mathrm{tr}\left\{S(t)^2 E_0^2 C_0^2 \otimes K^\mathrm{T} B^\mathrm{T} PBK\right\},$$

Using the similar method, we can obtain the following equality from the adaptive law (4)

$$\dot{V}_{22}(t) = 2\varepsilon(t)^\mathrm{T}[S(t)H \otimes PBK]\varepsilon(t) - \delta\varepsilon(t)^\mathrm{T}[(H^\mathrm{T} + H) \otimes PBK]\varepsilon(t) - 2\sum_{i=1}^{N}(s_i(t) - \delta)^2.$$

Lemma 1 indicates that the minimum eigenvalue of the matrix $H^\mathrm{T} + H$ satisfies $\lambda_{\min}(H^\mathrm{T} + H) > 0$. Using the known condition $\delta > \frac{1}{\lambda_{\min}(H^\mathrm{T}+H)}$, we have

$$\mathcal{L}_2 V_2(t) = \varepsilon(t)^\mathrm{T}[I_N \otimes (PA + A^\mathrm{T}P)]\varepsilon(t) - \delta\varepsilon(t)^\mathrm{T}[(H + H^\mathrm{T}) \otimes (PBK)]\varepsilon(t)$$
$$+ \mathrm{tr}\left\{S(t)^2 E_0^2 C_0^2 \otimes K^\mathrm{T} B^\mathrm{T} PBK\right\} - 2\sum_{i=1}^{N}(s_i(t) - \delta)^2$$
$$\leq \varepsilon(t)^\mathrm{T}[I_N \otimes (PA + A^\mathrm{T}P)]\varepsilon(t) - \delta\lambda_{\min}(H + H^\mathrm{T})\varepsilon(t)^\mathrm{T}[I_N \otimes (PBK)]\varepsilon(t)$$
$$+ \mathrm{tr}\left\{S(t)^2 E_0^2 C_0^2 \otimes K^\mathrm{T} B^\mathrm{T} PBK\right\} - 2\sum_{i=1}^{N}(s_i(t) - \delta)^2$$
$$\leq -k|\varepsilon(t)|^2 - 2\sum_{i=1}^{N}(s_i(t) - \delta)^2 + \bar{\theta}^2 \max\{(e_{i0}\sigma_{0i})^2\}\lambda_{\max}^3(P)\lambda_{\max}^2(BB^\mathrm{T})$$
$$\leq -\frac{k}{\lambda_{\max}(P)}\varepsilon(t)^\mathrm{T}(I_N \otimes P)\varepsilon(t) - 2\sum_{i=1}^{N}(s_i(t) - \delta)^2$$
$$+ \bar{\theta}^2 \max\{(e_{i0}\sigma_{0i})^2\}\lambda_{\max}^3(P)\lambda_{\max}^2(BB^\mathrm{T})$$
$$\leq -\min\left\{\frac{k}{\lambda_{\max}(P)}, 2\right\}V_2(t) + \bar{\theta}^2 \max\{(e_{i0}\sigma_{0i})^2\}\lambda_{\max}^3(P)\lambda_{\max}^2(BB^\mathrm{T})$$

From the formula $d(e^{\gamma t}V_2(t)) = \gamma e^{\gamma t}V_2(t)dt + e^{\gamma t}dV_2(t)$, we can obtain the following inequality from the Equation (19),

$$\begin{aligned}
e^{\gamma t}EV_2(t) &= EV_2(0) + \gamma E\int_0^t e^{\gamma s}V_2(s)ds + E\int_0^t e^{\gamma s}dV_2(s) \\
&\leq EV_2(0) - \left(\min\left\{\frac{k}{\lambda_{\max}(P)}, 2\right\} - \gamma\right)E\int_0^t e^{\gamma s}V_2(s)ds \\
&\quad + \bar{\theta}^2 \max\{(e_{i0}\sigma_{0i})^2\}\lambda_{\max}^3(P)\lambda_{\max}^2(BB^T)\frac{e^{\gamma t}-1}{\gamma} \\
&\leq EV_2(0) + \frac{\mu_2}{\gamma}e^{\gamma t}
\end{aligned}$$

where $\gamma < \min\left\{\frac{k}{\lambda_{\max}(P)}, 2\right\}$ and $\mu_2 = \bar{\theta}^2 \max\{(e_{i0}\sigma_{0i})^2\}\lambda_{\max}^3(P)\lambda_{\max}^2(BB^T)$. Hence, divide the inequality by $e^{\gamma t}$, it obtains

$$EV_2(t) \leq EV_2(0)e^{-\gamma t} + \frac{\mu_2}{\gamma}. \tag{20}$$

Finally, the mean square bounded consensus is obtained as follows

$$\lim_{t\to+\infty} E|x_i(t) - x_0(t)|^2 = \epsilon_2$$

where ϵ_2 is a small constant independent of time t. □

Remark 4. *Under the same adaptive law (4), the conditions of Theorem 2 are much simpler than those of Theorem 1, which can greatly expand the application range of the adaptive law in the problem. Moreover, Theorems 1 and 2 show that the adaptive law (4) can ensure that followers can track leader in the mean square sense, regardless of whether the stochastic multi-agent system has a time delay or not.*

Remark 5. *Hu et al. [37] designed a dynamic output-feedback controller by using the relative state information, and achieved the consensus by adjusting the internal state of the controller. Compared with the literature, the consensus in this paper can be achieved by adjusting the adaptive gain of system. Although both control strategies can achieve the consensus, [37] did not consider the impact of time delay, and the adaptive gain is mandatory.*

4. Simulation

To analyze the validity of main conclusions, assuming that the system covers one leader and three followers, we conduct numerical simulations in one- and two-dimensional space respectively, and verify that the adaptive control law of this paper can make all followers track the target regardless of whether the system has time delay.

Example 1. *For the system in one-dimensional space, let the leader be globally reachable, and the digraph $\mathcal{G}_1$ formed by the followers be balanced, its adjacency matrix is represented by $\mathcal{A} = \begin{pmatrix} 0 & 0 & 1 \\ 1 & 0 & 0 \\ 0 & 1 & 0 \end{pmatrix}$. Using the definition of Laplacian matrix $L_{\mathcal{G}_1}$, we can obtain the matrix $H = L_{\mathcal{G}_1} + E_0 = \begin{pmatrix} 2 & 0 & -1 \\ -1 & 1 & 0 \\ 0 & -1 & 2 \end{pmatrix}$, where $E_0 = \begin{pmatrix} 1 & 0 & 0 \\ 0 & 0 & 0 \\ 0 & 0 & 1 \end{pmatrix}$ is the communication matrix between the leader and the followers. The leader-following multi-agent system is represented by*

$$\dot{x}_i(t) = -0.3x_i(t) + 0.4u(t), \quad \dot{x}_0(t) = -0.3x_0(t).$$

For the above one-dimensional multi-agent system, taking $k = 0.8$, the matrix $P = 1.0432$ can be obtained from the Riccati equation $A^T P + PA - PBB^T P + kI = 0$. After simple calculation, we obtain the adaptive gain matrix $\Gamma = PBB^T P = 0.1741$. Since the minimum eigenvalue of $H^T + H$ is 1, we take the constant $\delta = 1.02$ to ensure $\delta > \frac{1}{\lambda_{\min}(H^T+H)}$, so the adaptive law can be represented by $\dot{s}_i(t) = 0.1741(x_i(t) - x_0(t))^T \sum_{j=1}^{N} h_{ij}(x_j(t) - x_0(t)) - (s_i(t) - 1.02)$. For the system, if τ is 0.13, the noise intensity is 0.23, the constant $\bar{\theta}$ is 1.4, and the constant $\bar{\varsigma}$ is 0.052, then the condition of Theorem 1 holds as $k = 0.8 > \bar{\varsigma}\bar{\theta}^2\lambda_{\max}(HH^T)\lambda_{\max}^4(P)\lambda_{\max}^2(BB^T) + 4\bar{\varsigma}^{-1}\tau^2\bar{\theta}^2\lambda_{\max}(HH^T)\lambda_{\max}^2(P)\lambda_{\max}^2(BB^T) + 4\bar{\varsigma}^{-1}\tau^2\lambda_{\max}(AA^T) = 0.6447$. At this point, Figure 1 shows the trend of tracking error over time, and Figure 2 shows the trajectory of the adaptive gain. Under the combined effects of noise and time delay, it can be seen that the state errors eventually converge to a small range. For the system with $\tau = 0$, we maintain the topological structure and dynamic equations of the problem invariable, which means the conditions of Theorem 2 hold. Under the same adaptive control law, the noise intensity is increased to 0.9, the system can still attain the mean square bounded consensus. Figures 3 and 4 show the trajectories of dynamic error and adaptive gain of each agents.

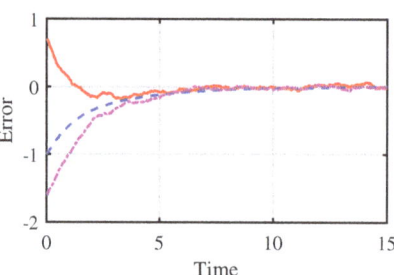

Figure 1. Dynamic error of one–dimensional system with time delay.

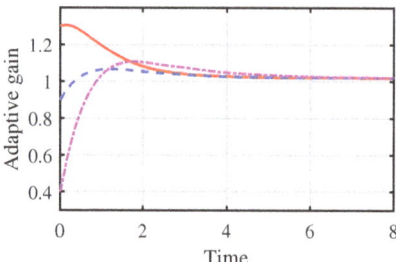

Figure 2. Adaptive gain of one–dimensional system with time delay.

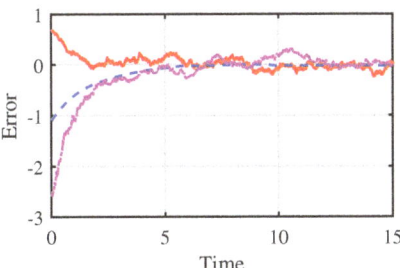

Figure 3. Dynamic error of one–dimensional system without time delay.

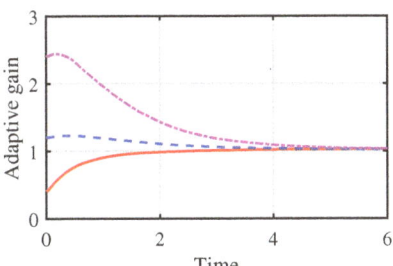

Figure 4. Adaptive trajectory of one–dimensional system without time delay.

Example 2. *In the two dimensional space, assuming the leader is globally reachable, the digraph $\mathcal{G}_2$ composed of followers is balanced, and the matrix $H = L_{\mathcal{G}_2} + E_0 = \begin{pmatrix} 2 & -1 & 0 \\ -1 & 1 & 0 \\ 0 & 0 & 2 \end{pmatrix}$. The system is represented by*

$$\dot{x}_i(t) = \begin{pmatrix} 0 & -0.8 \\ 0.8 & 0 \end{pmatrix} x_i(t) + \begin{pmatrix} 0.4 & 0.1 \\ 0.1 & 0.4 \end{pmatrix} u(t), \quad \dot{x}_0(t) = \begin{pmatrix} 0 & -0.8 \\ 0.8 & 0 \end{pmatrix} x_0(t).$$

For two dimensional system with time delay, the constant k in the Riccati equation is taken as 0.4, we can obtain $P = \begin{pmatrix} 1.4388 & -0.0350 \\ -0.0350 & 1.6538 \end{pmatrix}$ and the adaptive gain matrix $\Gamma = PBB^TP = \begin{pmatrix} 0.3441 & 0.1721 \\ 0.1721 & 0.4559 \end{pmatrix}$. Due to $\lambda_{\min}(H^T + H) = 0.7693$, the constant δ in the adaptive control law is taken as 1.35 in order to satisfy $\delta > \frac{1}{\lambda_{\min}(H^T+H)}$. Let the time delay $\tau = 0.0041$, the noise intensity $\sigma_{0i} = 0.33$, the constant $\bar{\theta} = 3.4$, and $\varsigma = 0.0041$, we can obtain that the condition of Theorem 1 holds as $k = 0.4 > \varsigma\bar{\theta}^2\lambda_{\max}(HH^T)\lambda_{\max}^4(P)\lambda_{\max}^2(BB^T) + 4\varsigma^{-1}\tau^2\bar{\theta}^2\lambda_{\max}(HH^T)\lambda_{\max}^2(P)\lambda_{\max}^2(BB^T) + 4\varsigma^{-1}\tau^2\lambda_{\max}(AA^T) = 0.3881$. Figures 5 and 6 describe the trajectory of dynamic error and adaptive gain of system with time delay in a noisy environment. It can be seen that all components of the three followers can track the target. When the time delay disappears, we maintain the above adaptive law unchanged, and then the conditions of Theorem 2 hold. Let the noise intensity $\sigma_{0i} = 0.24$, and the trends of the dynamic error and the adaptive gain of the system are shown in Figures 7 and 8.

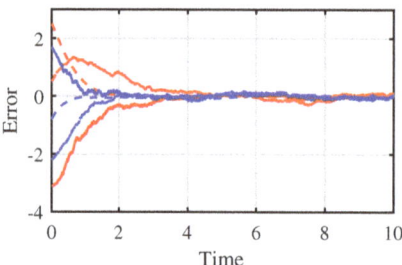

Figure 5. Dynamic error of two–dimensional system with time delay.

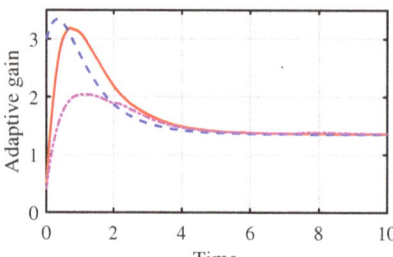

Figure 6. Adaptive gain of two–dimensional system with time delay.

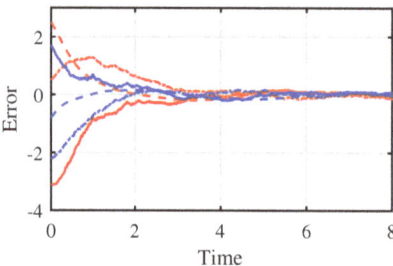

Figure 7. Dynamic error of two–dimensional system without time delay.

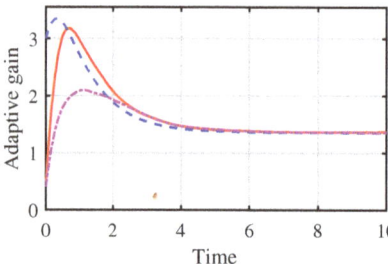

Figure 8. Adaptive gain of two–dimensional system without time delay.

In order to compare the differences between the adaptive control protocol proposed in this paper and some previous papers, we once again simulate the one-dimensional multi-agent system in Example 1, and take the noise intensity as 0.2. Under the mandatory gain $a_i(t) = \frac{\log(1+t)}{1+t}$ and the adaptive law (4), we simulate the dynamic error and the gain of two situations, respectively, as shown in Figures 9 and 10. The black curve represents the situation of mandatory gain, the other colors represent the changes of three agents under the control law (4). From the two figures, it can be seen that the adaptive control protocol proposed in this paper has a faster rate of convergence, so three followers can track the leader in a shorter time. In addition, the mandatory gain will eventually converge to zero, while the adaptive gain (4) will converge to a non-zero constant.

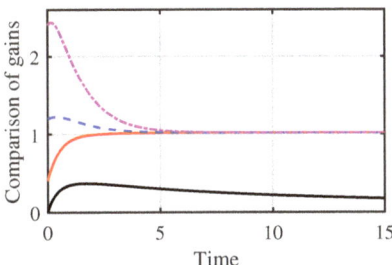

Figure 9. Comparison of two different gains.

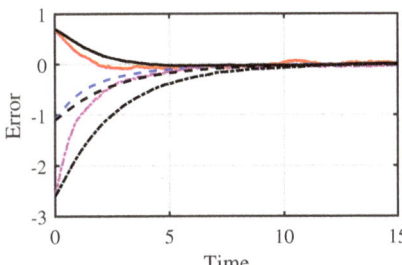

Figure 10. Comparison of dynamic errors under two different gains.

5. Conclusions

For the tracking issues, adaptive control is analyzed in cases both with and without time delay. Firstly, the adaptive control protocol of the stochastic system is given in the presence of time delay, and the adaptive law is designed. The adaptive control law depends on the solution of the Riccati equation and can be abbreviated into matrix form by the Kronecker products. Then, it was proved that the followers can track the target in the mean square sense, and the dynamic error can obtain to a very little constant. Compared with the previous references, the final dynamic error has a smaller value, and when the noise intensity converges to zero, this dynamic error value also trends to zero. It should be noted that the method of proof can not be directly extended to the case of variable delay. In the future, it is meaningful to further explore the adaptive consensus of multi-agent system with variable delay, and the output feedback control with time delay also needs additional investigation.

Author Contributions: Conceptualization, methodology, writing original draft preparation. S.J.; writing—review and editing, G.Z. All authors have read and agreed to the published version of the manuscript.

Funding: This work is Major Projects of Natural Science Research in Anhui Universities (2022AH040207 and KJ2021A1101).

Institutional Review Board Statement: Not applicable.

Informed Consent Statement: Not applicable.

Data Availability Statement: Not applicable.

Conflicts of Interest: The authors declare no conflict of interest.

References

1. Wang, Y.; He, L.; Huang, C.Q. Adaptive time-varying formation tracking control of unmanned aerial vehicles with quantized input. *ISA Trans.* **2019**, *85*, 76–83. [CrossRef]
2. Peng, Z.; Luo, R.; Hu, J.; Shi, K.; Ghosh, B.K. Distributed optimal tracking control of discrete-time multiagent systems via event-triggered reinforcement learning. *IEEE Trans. Circuits Syst. Regul. Pap.* **2022**, *69*, 3689–3700. [CrossRef]
3. Wan, Y.; Qin, J.; Li, F.; Yu, X.; Kang, Y. Game theoretic-based distributed charging strategy for PEVs in a smart charging station. *IEEE Trans. Smart Grid* **2021**, *V12*, 538–547. [CrossRef]
4. Vicsek, T.; Czirók, A.; Ben-Jacob, E.; Cohen, I.; Shochet, O. Novel type of phase transition in a system of self-driven particles. *Phys. Rev. Lett.* **1995**, *75*, 1226–1229. [CrossRef] [PubMed]
5. Qin, J.; Ma, Q.; Shi, Y.; Wang, L. Recent advances in consensus of multi-agent systems: A brief survey. *IEEE Trans. Ind. Electron.* **2017**, *64*, 4972–4983. [CrossRef]
6. Amirkhani, A.; Barshooi, A.H. Consensus in multi-agent systems: A review. *Artif. Intell. Rev.* **2022**, *55*, 3897–3935. [CrossRef]
7. Jiang, J.H.; Jiang, Y.Y. Leader-following consensus of linear time-varying multi-agent systems under fixed and switching topologies. *Automatica* **2020**, *113*, 108804. [CrossRef]
8. Wang, X.X.; Liu, Z.X.; Chen, Z.Q. Event-triggered fault-tolerant consensus control with control allocation in leader-following multi-agent systems. *Sci. China* **2021**, *64*, 879–889. [CrossRef]
9. Jeong, J.; Lim, Y.; Parivallal, A. An asymmetric lyapunov-krasovskii functional approach for event-triggered consensus of multi-agent systems with deception attacks. *Appl. Math. Comput.* **2023**, *439*, 127584. [CrossRef]
10. Murugesan, S.; Liu, Y.C. Resilient finite-time distributed event-triggered consensus of multi-agent systems with multiple cyber-attacks. *Commun. Nonlinear Sci. Numer. Simul.* **2023**, *116*, 106876. [CrossRef]
11. Zhang, R.H.; Zhang, Y.Y.; Zong, X.F. Stochastic leader-following consensus of discrete-time nonlinear multi-agent systems with multiplicative noises. *J. Frankl. Inst.* **2022**, *359*, 7753–7774. [CrossRef]
12. Li, T.; Zhang, J.F. Mean square average-consensus under measurement noises and fixed topologies: Necessary and sufficient conditions. *Automatica* **2009**, *45*, 1929–1936. [CrossRef]
13. Cheng, L.; Hou, Z.G.; Tan, M. Necessary and sufficient conditions for consensus of double-integrator multi-agent systems with measurement noises. *IEEE Trans. Autom. Control* **2011**, *56*, 1958–1963. [CrossRef]
14. Zong, X.F.; Li, T.; Zhang, J.F. Consensus conditions of continuous-time multi-agent systems with time-delays and measurement noises. *Automatica* **2019**, *99*, 412–419. [CrossRef]
15. Knotek, T.; Hengster-Movric, K.; Ebek, M. Distributed adaptive consensus protocol with decaying gains. *Int. J. Robust Nonlinear Control* **2020**, *30*, 6166–6188. [CrossRef]
16. Yu, Z.; Huang, D.; Jiang, H.; Hu, C. Consensus of second-order multi-agent systems with nonlinear dynamics via edge-based distributed adaptive protocols. *J. Frankl. Inst.* **2016**, *353*, 4821–4844. [CrossRef]
17. Luo, R.; Peng, Z.N.; Hu, J.P. On model identification based optimal control and it's applications to multi-agent learning and control. *Mathematics* **2023**, *11*, 906. [CrossRef]
18. Li, M.; Qin, J.; Ma, Q.; Zheng, W.X.; Kang, Y. Hierarchical optimal synchronization for linear systems via reinforcement learning: A stackelberg-nash game perspective. *IEEE Trans. Neural Netw. Learn. Syst.* **2020**, *99*, 1–12. [CrossRef]
19. Duarte, A.; Weissing, F.J.; Pen, I.; Keller, L. An evolutionary perspective on self-organized division of labor in social insects. *Annu. Rev.* **2011**, *42*, 91–110. [CrossRef]
20. Larsen, A.D.; Büscher, T.H.; Chuthong, T.; Pairam, T.; Bethge, H.; Gorb, S.N.; Manoonpong, P. Self-organized stick insect-like locomotion under decentralized adaptive neural control: From biological investigation to robot simulation. *Adv. Theory Simulations* **2023**, *228*, 2300228. [CrossRef]
21. Duan, Y.B.; Yang, Z.W. Design and analysis of the consensus gain for stochastic multi-agent systems. *Control Theory Appl.* **2019**, *36*, 629–635.
22. Huang, Y.X. Adaptive consensus for uncertain multi-agent systems with stochastic measurement noises. *Commun. Nonlinear Sci. Numer. Simul.* **2023**, *120*, 107156. [CrossRef]
23. Xiao, G.L.; Wang, J.R.; Meng, D.Y. Adaptive finite-time consensus for stochastic multi-agent systems with uncertain actuator faults. *IEEE Trans. Control. Netw. Syst.* **2023**, *10*, 1–12. [CrossRef]
24. Wu, Y.Z.; Liang, Q.P.; Zhao, Y.Y. Adaptive bipartite consensus control of general linear multi-agent systems using noisy measurements. *Eur. J. Control* **2021**, *59*, 123–128. [CrossRef]
25. Ma, C.Q.; Qin, Z.Y.; Zhao, Y.B. Bipartite consensus of integrator multi-agent systems with measurement noise. *IET Control Theory Appl.* **2017**, *11*, 3313–3320. [CrossRef]
26. Wen, G.; Chen, C.P.; Liu, Y.J.; Liu, Z. Neural network-based adaptive leader-following consensus control for a class of nonlinear multiagent state-delay systems. *IEEE Trans. Cybern.* **2016**, *47*, 1–10. [CrossRef]
27. Hu, J.P.; Hong, Y.G. Leader-following coordination of multi-agent systems with coupling time delays. *Phys. A* **2007**, *374*, 853–863. [CrossRef]

28. Horn, R.A.; Johnson, C.R. *Matrix Analysis*; Cambridge University Press: Cambridge, UK, 2012.
29. Li, Z.; Duan, Z.; Chen, G.; Huang, L. Consensus of multi-agent systems and synchronization of complex networks: A unified viewpoint. *IEEE Trans. Circuits Syst. I Regul. Pap.* **2010**, *57*, 213–224.
30. Zhang, H.; Lewis, F.; Das, A. Optimal design for synchronization of cooperative systems: State feedback, observer, and output feedback. *IEEE Trans. Autom. Control* **2011**, *56*, 1948–1952. [CrossRef]
31. Li, Z.; Wen, G.; Duan, Z.; Ren, W. Designing fully distributed consensus protocols for linear multi-agent systems with directed graphs. *IEEE Trans. Autom. Control* **2015**, *60*, 1152–1157. [CrossRef]
32. Zhang, W.; Abuzar Hussein Mohammed, A.; Bao, J.; Liu, Y. Adaptive event-triggering consensus for multi-agent systems with linear time-varying dynamics. *J. Syst. Sci. Complex.* **2022**, *35*, 1700–1718. [CrossRef]
33. Deng, C.; Wen C.Y.; Li, X.Y. Distributed adaptive tracking control for high-order nonlinear multiagent systems over event-triggered communication. *IEEE Trans. Autom. Control* **2023**, *68*, 1176–1183. [CrossRef]
34. Wu, Z.H.; Fang, H.J. Delayed-state-derivative feedback for improving consensus performance of second-order delayed multi-agent systems. *Int. J. Syst. Sci.* **2012**, *43*, 140–152. [CrossRef]
35. Jin, S.B.; Yu, Q.J. Construction of adaptive consensus gains for multi-agent systems with multiplicative noise. *Complexity* **2021**, *2021*, 4425511. [CrossRef]
36. Kolmanovskii, V.; Myshkis, A. *Applied Theory of Functional Differential Equations*; Kluwer Academic Publishers: Dordrecht, The Netherlands, 1992.
37. Hu, J.; Wu, Y.; Li, T.; Ghosh, B.K. Consensus control of general linear multi-agent systems with antagonistic interactions and communication noises. *IEEE Trans. Autom. Control* **2019**, *64*, 2122–2127. [CrossRef]

Disclaimer/Publisher's Note: The statements, opinions and data contained in all publications are solely those of the individual author(s) and contributor(s) and not of MDPI and/or the editor(s). MDPI and/or the editor(s) disclaim responsibility for any injury to people or property resulting from any ideas, methods, instructions or products referred to in the content.

Article

Robust Model Predictive Control for Two-DOF Flexible-Joint Manipulator System

Rong Li [1,2,*], Hengli Wang [1], Gaowei Yan [1], Guoqiang Li [2] and Long Jian [1]

1. College of Electrical and Power Engineering, Taiyuan University of Technology, Taiyuan 030024, China
2. Nuclear Emergency and Nuclear Safety Department, China Institute for Radiation Protection, Taiyuan 030006, China
* Correspondence: lirong@tyut.edu.cn

Abstract: This paper presents a practical study on how to improve the $\mathcal{H}_\infty$ performance and meet the input–output constraints of the two-degrees-of-freedom (DOF) flexible-joint manipulator system (FJMS) with parameter uncertainties and external disturbances. For this reason, a robust constrained moving-horizon $\mathcal{H}_\infty$ controller is designed to improve the system $\mathcal{H}_\infty$ performance while still satisfying the input–output constraints of the uncertain system. First, the uncertain controlled system model of the two-DOF FJMS is established via the Lagrange equation method, Spong's assumption, and the linear fractional transformation (LFT) technique. Then, the control requirements and input–output constraints of the uncertain system are transformed into the linear matrix inequality (LMI) via the theory of $\mathcal{H}_\infty$ control and the full-block multiplier technique. Next, the LMI optimization problem refreshed by the current state is addressed at each sample moment with the idea of the moving-horizon control of the model predictive control (MPC), and the calculated gain is implemented to the nonlinear closed-loop system under the state feedback structure. The validity and feasibility of the designed control scheme is finally verified via the results of simulation experiments.

Keywords: two-DOF FJMS; LFT; LMI; moving-horizon control; robust $\mathcal{H}_\infty$ control

MSC: 93-08

1. Introduction

In recent years, due to the advantages of the lighter weight, higher flexibility, lower energy consumption, and higher load ratio of the FJMS compared with the traditional rigid-joint manipulator, the proportion of industrial processing, medical treatment, aerospace engineering, living services, and other application scenarios has increased dramatically. Thus, the control accuracy and robustness of the FJMS have become the key targets of researchers and users [1–5]. For the flexible joints of the manipulator, actuator motors are installed in the individual joints, driving each link to perform the specified actions. However, the rotors inside the motors and links are equipped with harmonic gears for transmission, which subsequently leads to extra errors and vibrations in the angles of the joints, and ultimately greatly influences the control accuracy of the FJMS [6,7]. In addition, because of the objective existence of external disturbances and parameter uncertainties, the conventional dynamic model of the manipulator is frequently impractical, and the above-mentioned factors must be considered when establishing the dynamic model of the FJMS in order to enhance the $\mathcal{H}_\infty$ performance of the controlled system with constraints.

With the improvement in the control accuracy requirements, the existence of the manipulator joint flexibility has already become a non-negligible matter, and a series of methods have been adopted by international scholars to control the FJMS. For example, L. Zouari et al. designed a sliding-mode controller to address the problem of uncertainties in the joint flexibility of the manipulator [8]. The robust controller was designed for the tracking control of the FJMS with the help of the voltage control strategy in [9].

I. Hassanzadeh et al. invoked an approach to the model following adaptive control when controlling a nonlinear FJMS [10]. The reason why the control performance is not good enough is that the system modeling is not modeled and analyzed for flexible joints. In the process of targeted modeling for the flexible joints of the manipulator, the order of the model is greatly increased to double, and thus the sophistication of the model is strengthened, which has attracted many scholars to explore this issue. The method of M. Spong has been the most broadly adopted and convenient in recent years [11–13]. His assumption is that there exists a linear-torsion spring, and the flexible deformation of the joint is equivalently replaced by the torsional deformation of the spring [14].

Up to now, numerous international scholars have conducted in-depth research to explore the control problem of suppressing perturbations for the FJMS, achieving some research results. L. Sun et al. proposed a PD control method with the help of online gravity compensation to achieve the position control of the FJMS [15]. Y. Pan et al. designed a simplified, adaptive command-filtered backstepping controller for the FJMS [16]. K. Rsetam et al. specifically designed the optimal second-order integral sliding-mode controller in order to improve the robustness of a single-link FJMS [17]. In addition, optimal controllers were designed for the discrete controlled systems in [18–20], and optimization control algorithms were designed to improve the control performance of the controlled systems while taking into account the existence of external disturbances and uncertainties in [21–23]. Z.H. Jiang et al. designed a linear-feedback- and neural-network-based controller to handle the control problem caused by the nonlinearity and dynamic instability of the FJMS [24]. Although several of the above approaches have suppressed the perturbations to some extent, they do not take into account the modeling errors due to the uncertainties of the model parameters. Z. Yan et al. proposed a robust control method based on the equivalent-input-perturbation method to achieve the high-precision motion control of an uncertain FJMS with a single link [25]. Although this method considers the errors caused by parameter inaccuracies, obtaining the real values of the parameters is required in the design of this controller, without realizing the true sense of considering the uncertainties of the system. For the problem of parameter uncertainties, the most realistic case is to be aware of the nominal values of the parameters and the possible variation ranges.

To this day, several control methods have been proposed to overcome the modeling uncertainties of the FJMS. K. Rsetam et al. designed a sliding-mode controller based on a cascaded extended state observer for the under-driven FJMS, where the sliding-mode control method was mainly used to diminish the error caused by uncertainties [26]. W. He et al. introduced the full-state feedback strategy in the neural network, which subsequently was used to respond to the uncertainties of the FJMS for guaranteeing the robustness of the system [27]. H. Ma et al. designed an adaptive fuzzy controller to improve the performance of the single-link FJMS via the performance functions, in which the dynamic signals were applied to replace the uncertainties of the system modeling [28]. F. Dong et al. designed a robust controller based only on the possible bounds of the system uncertainties and a consistent positive characterization of the inertia matrix to guarantee the robustness of the FJMS with uncertainties [29]. J.G. Yim et al. proposed a robust nonlinear recursive-control approach to design a virtually robust control for the FJMS, utilizing nonlinear $\mathcal{H}_\infty$ control with energy dissipation to attenuate the $\mathcal{L}_2$ gain from the performance impact of uncertainties [30]. In addition, designed control optimization algorithms were implemented on the real two-DOF manipulator to verify the controllers' effectiveness in [31,32]. However, several of the above methods are not effective at achieving the control performance enhancement of the FJMS with constraints while dealing with the parameter uncertainties of the manipulator.

Most process models are nonlinear, but they are often linearized to perform the simulation and stability analysis. Linearization is the procedure of approximating and eliminating the higher-order nonlinear terms existing in the mathematical equations. Linear models are easier to understand than nonlinear models and are necessary to design the controllers for the controlled systems. T.T Do et al. established the dynamic equations of a general flexible-joint robot using the Lagrange formulation and linearized them on the basis of the Taylor series [33]. X.Z. Lai et al. divided the motion space of an underactuated two-link manipulator into two areas: the swing area and attractive area, and designed control laws for each system, where the controlled model in the attractive area was approximately linearized, while its controller was designed based on optimal control [34]. E. Spyrakos-Papastavridis et al. linearized an n-DOF flexible-joint robot at a desired operating point, and then utilized the LQR controller to obtain the full-state feedback gain of this system [35]. D. Richiedei et al. rationally performed the model linearization in the case of a two-DOF, two-link planar manipulator, producing small displacements to the configuration [36]. X.Z. Lai et al. approximately linearized the dynamic equations of an underactuated three-link gymnast robot in the attractive area and stabilized it at the straight-up equilibrium position using the balancing-control law [37]. A.G. Lynch et al. linearized nonlinear equations of the multibody dynamic systems around the equilibrium point [38]. A. Ghoreishi et al. linearized a single-link flexible robot around the origin (equilibrium point) [39]. One tends to linearize the nonlinear dynamic model around the equilibrium point, illustrating the fact that the nonlinear systems are locally linear at the equilibrium point. Motivated by the literature mentioned above, we selected the controlled system to be the two-DOF FJMS in the vicinity of the vertical equilibrium position.

As a matter of fact, there are inevitably constraints on the manipulator during the movement process, such as the control input constraints, joint angle constraints [40], etc. Accordingly, a controller that could improve the system $\mathcal{H}_\infty$ performance while still satisfying the input–output constraints of the system is necessary for this paper. For this paper, a robust constrained moving-horizon $\mathcal{H}_\infty$ controller is designed to enable the two-DOF FJMS to achieve the above control objectives under the consideration of external disturbances and parameter uncertainties.

The main contributions of this paper are as follows:

1. By means of the LFT technique, the LFT uncertain system of the two-DOF FJMS is constructed, which takes into account the parameter uncertainties of the spring-stiffness coefficients;
2. The $\mathcal{H}_\infty$ norm of system disturbances to the performance output and the input–output constraints of the two-DOF FJMS are transformed into the LMIs via the theory of $\mathcal{H}_\infty$ control and the full-block multiplier technique;
3. The robust constrained moving-horizon $\mathcal{H}_\infty$ controller is designed for this LFT uncertain system, which can improve the $\mathcal{H}_\infty$ performance of the controlled system while ensuring that the input–output constraints of this system are satisfied.

The remainder of this paper is organized as follows. In Section 2, the dynamics of the two-DOF FJMS is modeled and converted to the state-space expression after linearization. In Section 3, the uncertainties of the spring coefficients in the two-DOF FJMS are investigated by means of the LFT technique, and the LFT uncertain model of this system is constructed. In Section 4, the robust constrained moving-horizon $\mathcal{H}_\infty$ controller is designed for the LFT uncertain system. In Section 5, the properties of the closed-loop system under the action of the optimization algorithm are given. In Section 6, the above controller implemented on the two-DOF FJMS for the simulation is described, and the experimental results are compared and analyzed. The conclusions are presented in Section 7.

2. Problem Statement

2.1. Dynamic Modeling of the Two-DOF FJMS

In this section, the dynamic characteristics of the studied two-DOF FJMS are discussed in detail. The simplified physical model of the two-DOF FJMS studied in this paper is established as shown in Figure 1. The manipulator system has two rotatable homogeneous links driven by motors at the shoulder joint and elbow joint, which can be moved in the vertical plane around their respective joints. The two-DOF FJMS moves around the vertical equilibrium position. The shoulder joint of the two-DOF FJMS is fixed, and the origin of the coordinate axis (O) is the point where the shoulder joint is located. After establishing the coordinate frame for the system, the horizontal plane where the O axis is located is taken to be the surface of zero gravitational potential energy. In addition, due to the fact that both joints of the two-DOF FJMS considered in this paper are flexible joints, which are conceived as the linear springs between the motors and the links based on Spong's assumption [14], the internal parts of the flexible joints are especially expanded, as shown in Figure 1, so that this relationship can be visualized.

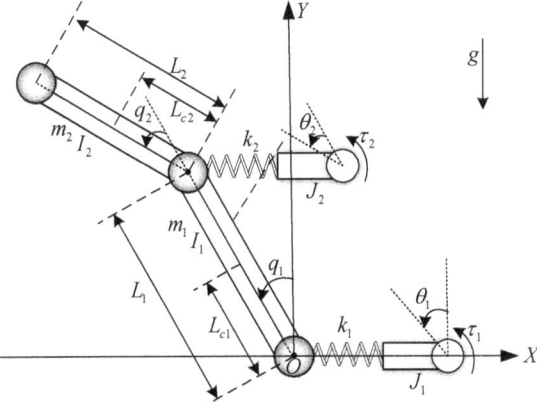

Figure 1. The simplified model of the two-DOF FJMS.

The physical significances of the model parameters in Figure 1 are shown in Table 1. The q_1 and q_2 are the rotation angles of the first and second links, respectively, with the positive direction of the Y axis. As well, the θ_1 and θ_2 are the rotation angles of the first and second motor rotors, respectively, with the positive direction of the Y axis. The angle value formed by the clockwise rotation is set to be positive, and the angle value formed by the counterclockwise rotation is set to be negative.

Table 1. The parameters of the two-DOF FJMS.

Symbol	Description
L_1, L_2	Lengths of manipulator links (m)
L_{c1}, L_{c2}	Distances between center-of-mass positions and joints (m)
m_1, m_2	Masses of manipulator links (kg)
I_1, I_2	Rotational inertias of manipulator links (kg·m^2)
J_1, J_2	Rotational inertias of motor rotors (kg·m^2)
k_1, k_2	Spring-stiffness factors of flexible joints (N·m/rad)
τ_1, τ_2	Output torques of motors (N·m)
g	Gravitational acceleration (m/s^2)

At present, there are mainly two kinds of methods commonly used to establish the dynamic model for the system: the Lagrange equation method [41,42] and the Newton–Euler method [43]. In contrast to the latter method, the Lagrange equation method dramatically simplifies the complex dynamic equations due to the fact that it does not account for the internal binding forces of the system, and it allows the dynamic equations of the manipulator system to be expressed in a straightforward and concise manner. Therefore, the Lagrange equation method is chosen here in this paper.

The Lagrange equation method is based on the law of the conservation of energy by calculating the kinetic and potential energy of the system to accomplish the modeling. The system's Lagrange function ($L \in \mathbb{R}$) is defined as the difference between the kinetic energy ($K \in \mathbb{R}$) and the potential energy ($P \in \mathbb{R}$) of the system [44]:

$$L = K - P. \tag{1}$$

The Lagrange equation is as follows:

$$\begin{cases} \frac{d}{dt}\frac{\partial L}{\partial \dot{q}_i} - \frac{\partial L}{\partial q_i} = \tau_i \\ \frac{d}{dt}\frac{\partial K}{\partial \dot{q}_i} - \frac{\partial K}{\partial q_i} + \frac{\partial P}{\partial q_i} = \tau_i \end{cases} (i = 1, 2), \tag{2}$$

where q_i is the rotation angle of the joint, and τ_i is the torque of the actuator.

In the two-DOF FJMS, how the flexible joints are handled is critical. Based on Spong's simplified model, the flexible joint might be considered as a linear-torsion spring with zero inertia between the motor rotor and the link [13]. The simplified model of the flexible joint is shown in Figure 2, where k_i is this spring's stiffness factor. In this case, the motor rotor's rotation angle (θ_i) will not always equal the link's rotation angle (q_i) (i.e., $\theta_i \neq q_i$).

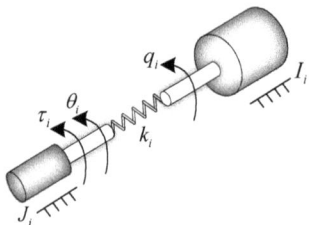

Figure 2. The simplified model of the flexible joint.

According to Lagrange's second type Equation (2), the dynamic equations of the two-DOF FJMS are identified as follows [45,46]:

$$\begin{cases} M(q)\ddot{q} + C(q,\dot{q})\dot{q} + G(q) = k(\theta - q) \\ J\ddot{\theta} + k(\theta - q) = \tau \end{cases}, \tag{3}$$

where $\theta = \begin{bmatrix} \theta_1 & \theta_2 \end{bmatrix}^T \in \mathbb{R}^{2\times 1}$, $q = \begin{bmatrix} q_1 & q_2 \end{bmatrix}^T \in \mathbb{R}^{2\times 1}$, $J = diag\{J_1, J_2\} \in \mathbb{R}^{2\times 2}$ is the diagonal and positive definite inertia matrix of the motors; $k = diag\{k_1, k_2\} \in \mathbb{R}^{2\times 2}$ is the simplified linear-torsion spring-stiffness-coefficient matrix of the flexible joints; $M(q) \in \mathbb{R}^{2\times 2}$ is the manipulator's symmetric and positive definite inertia matrix; $C(q,\dot{q})\dot{q} \in \mathbb{R}^{2\times 1}$ is a column vector incorporating the Coriolis force and the centrifugal force; $G(q) \in \mathbb{R}^{2\times 1}$

represents the vector of the gravity; $\tau = \begin{bmatrix} \tau_1 & \tau_2 \end{bmatrix}^T \in \mathbb{R}^{2\times 1}$ is the output torque of the motors. The specific forms of $M(q)$, $C(q,\dot{q})$, and $G(q)$ in Equation (3) are shown as follows:

$$\begin{cases} M(q) = \begin{bmatrix} \alpha_1 + \alpha_2 + 2\alpha_3 \cos q_2 & \alpha_2 + \alpha_3 \cos q_2 \\ \alpha_2 + \alpha_3 \cos q_2 & \alpha_2 \end{bmatrix} \\ C(q,\dot{q}) = \begin{bmatrix} -\dot{q}_2 \alpha_3 \sin q_2 & -(\dot{q}_1 + \dot{q}_2)\alpha_3 \sin q_2 \\ \dot{q}_1 \alpha_3 \sin q_2 & 0 \end{bmatrix} \\ G(q) = \begin{bmatrix} -\alpha_4 \sin q_1 - \alpha_5 \sin(q_1 + q_2) \\ -\alpha_5 \sin(q_1 + q_2) \end{bmatrix} \end{cases}, \qquad (4)$$

where $\alpha_1 = m_1 L_{c1}^2 + m_2 L_1^2 + I_1$, $\alpha_2 = m_2 L_{c2}^2 + I_2$, $\alpha_3 = m_2 L_1 L_{c2}$, $\alpha_4 = (m_1 L_{c1} + m_2 L_1)g$, and $\alpha_5 = m_2 g L_{c2}$.

2.2. LFT Technique

LFT was proposed by Redheffer scholars in 1960 and has been widely applied in the research of robust control, as well as in the research of the control of linear parameter-varying systems [47]. LFT is a powerful technique for representing the uncertainties in matrices and systems that is able to perform structural analyses for uncertain systems, and to directly represent systems with uncertainties in the form of state-space expressions. This method has the advantage of decoupling the systems into deterministic and uncertain parts, and it provides an effective tool to construct parameter-uncertain system models. Ultimately, it is possible to make explicit considerations for such uncertainties during the designing of the system controllers [48,49]. The LFT contains the lower LFT and the upper LFT, and the upper LFT structure is highlighted here.

Consider the complex matrix M, the partition form of which is in [50]:

$$M = \begin{bmatrix} M_{11} & M_{12} \\ M_{21} & M_{22} \end{bmatrix} \in \mathbb{C}^{(p_1+p_2)\times(q_1+q_2)}, \qquad (5)$$

where each matrix has the appropriate dimension, and $\delta_u \in \mathbb{C}^{q_1 \times p_1}$ is also a complex matrix. Assuming that there exists an inverse matrix of $(I - M_{11}\delta_u)$, then the upper LFT of the mapping corresponding to the matrix δ_u could be expressed as follows:

$$F_u(M, \delta_u) = M_{22} + M_{21}\delta_u(I - M_{11}\delta_u)^{-1}M_{12} : \mathbb{C}^{q_1 \times p_1} \to \mathbb{C}^{p_2 \times q_2}. \qquad (6)$$

The graphical representation of the upper LFT is shown in Figure 3, where M represents the known part of the system, the matrix δ_u represents all the uncertain components (including structural parameters, non-structural parameters, modeling uncertainties, etc.) with $\delta_u \in Y_\delta$, where $Y_\delta = \text{diag}\{\delta_1 I_1, \ldots, \delta_s I_s\}$ and $|\delta_i| \leq 1 (i = 1, \ldots, s)$. In addition, η_0 and v_0 represent, respectively, the auxiliary input and output of the system, and ω_0 and z_0 represent, respectively, the real input and output of the system.

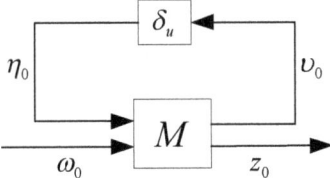

Figure 3. The graphical representation of the upper LFT.

The mathematical description of Figure 3 is as follows:

$$\begin{bmatrix} v_0 \\ z_0 \end{bmatrix} = M \begin{bmatrix} \eta_0 \\ \omega_0 \end{bmatrix} = \begin{bmatrix} M_{11} & M_{12} \\ M_{21} & M_{22} \end{bmatrix} \begin{bmatrix} \eta_0 \\ \omega_0 \end{bmatrix}, \quad (7)$$

$$\eta_0 = \delta_u v_0. \quad (8)$$

By means of LFT, the uncertain part of the system model is separated and the connection with the known exact model part is established, which is convenient to analyze and design the system controller effectively.

2.3. State Transformation Procedure of the Two-DOF FJMS

The equation $x = \begin{bmatrix} q_1 & \dot{q}_1 & q_2 & \dot{q}_2 & \theta_1 & \dot{\theta}_1 & \theta_2 & \dot{\theta}_2 \end{bmatrix}^T \in \mathbb{R}^{8\times 1}$ is selected as the state of the two-DOF FJMS, and $u = \begin{bmatrix} u_1 & u_2 \end{bmatrix}^T = \begin{bmatrix} \tau_1 & \tau_2 \end{bmatrix}^T \in \mathbb{R}^{2\times 1}$ is selected as the control input of this system. This manipulator system is known to be nonlinear according to Equations (3) and (4). As the angles and angular velocities of the first and second links are both close to zero in the attraction domain, the system may be approximately linearized around the equilibrium point, where the values of the variables are $q_1 = 0$, $q_2 = 0$, $\dot{q}_1 = 0$, and $\dot{q}_2 = 0$. The approximate linearization process via Taylor series expansion is performed as follows [33,34,37,39]:

$$\begin{cases} \cos q_1 \approx 1, \sin q_1 \approx q_1, \dot{q}_1 \approx 0 \\ \cos q_2 \approx 1, \sin q_2 \approx q_2, \dot{q}_2 \approx 0. \\ \sin(q_1 + q_2) \approx q_1 + q_2 \end{cases} \quad (9)$$

Remark 1. *The controlled system investigated in this paper is the two-DOF FJMS moving around the vertical equilibrium position, and the types of control problems are addressed via the control algorithms designed in this paper when the manipulator is moving in the vicinity of the equilibrium point. Because the dynamic equation of this manipulator is nonlinear in nature, it thus requires linearization about the equilibrium point. Hereby, the higher-order nonlinear terms are eliminated to attain the linear model by using Taylor series expansion. Of course, considering that there are certain conditions for linearizing the two-DOF FJMS using Taylor series expansion, we have given some constraints on the two joint angles to ensure that this manipulator system moves around the equilibrium point.*

Then, dynamic Equation (3) of the two-DOF FJMS is linearized and rewritten as a state-space expression with the following form:

$$\dot{x} = Ax + B_u u, \quad (10)$$

where the coefficient matrices A and B_u are as follows:

$$A = \begin{bmatrix} 0 & 1 & 0 & 0 & 0 & 0 & 0 & 0 \\ A_{21} & 0 & A_{23} & 0 & A_{25} & 0 & A_{27} & 0 \\ 0 & 0 & 0 & 1 & 0 & 0 & 0 & 0 \\ A_{41} & 0 & A_{43} & 0 & A_{45} & 0 & A_{47} & 0 \\ 0 & 0 & 0 & 0 & 0 & 1 & 0 & 0 \\ A_{61} & 0 & 0 & 0 & A_{65} & 0 & 0 & 0 \\ 0 & 0 & 0 & 0 & 0 & 0 & 0 & 1 \\ 0 & 0 & A_{83} & 0 & 0 & 0 & A_{87} & 0 \end{bmatrix}, B_u = \begin{bmatrix} 0 & 0 \\ 0 & 0 \\ 0 & 0 \\ 0 & 0 \\ 0 & 0 \\ b_6 & 0 \\ 0 & 0 \\ 0 & b_8 \end{bmatrix}, \quad (11)$$

where A_{ij} are represented as follows:

$$\begin{cases} A_{21} = \frac{\alpha_2\alpha_4 - \alpha_3\alpha_5 - k_1\alpha_2}{\alpha_1\alpha_2 - \alpha_3^2}, A_{23} = \frac{-\alpha_3\alpha_5 + k_2(\alpha_2+\alpha_3)}{\alpha_1\alpha_2 - \alpha_3^2} \\ A_{25} = \frac{k_1\alpha_2}{\alpha_1\alpha_2 - \alpha_3^2}, A_{27} = \frac{-k_2(\alpha_2+\alpha_3)}{\alpha_1\alpha_2 - \alpha_3^2} \\ A_{41} = \frac{\alpha_1\alpha_5 - \alpha_3\alpha_4 - \alpha_2\alpha_4 + \alpha_3\alpha_5 + k_1(\alpha_2+\alpha_3)}{\alpha_1\alpha_2 - \alpha_3^2} \\ A_{43} = \frac{(\alpha_1+\alpha_3)\alpha_5 - k_2(\alpha_1+\alpha_2+2\alpha_3)}{\alpha_1\alpha_2 - \alpha_3^2} \\ A_{45} = \frac{-k_1(\alpha_2+\alpha_3)}{\alpha_1\alpha_2 - \alpha_3^2}, A_{47} = \frac{k_2(\alpha_1+\alpha_2+2\alpha_3)}{\alpha_1\alpha_2 - \alpha_3^2} \\ A_{61} = \frac{k_1}{J_1}, A_{65} = \frac{-k_1}{J_1}, b_6 = \frac{1}{J_1} \\ A_{83} = \frac{k_2}{J_2}, A_{87} = \frac{-k_2}{J_2}, b_8 = \frac{1}{J_2} \end{cases} \quad (12)$$

3. Analysis of the LFT Uncertain System

The spring-stiffness coefficients are the key parameters of the flexible joints for the two-DOF FJMS, and the accuracy of their values plays a highly significant role in the overall controlled system. If there are fluctuations in the stiffness coefficients of the flexible joints, then this nominal manipulator system will become an uncertain dynamic system. The uncertainties of the k_1 and k_2 are described via the nominal values of the parameters themselves and their possible ranges of variation, as shown in the following equation:

$$\begin{cases} k_1 = \bar{k}_1(1 + W_{k_1}\delta_{k_1}) \\ k_2 = \bar{k}_2(1 + W_{k_2}\delta_{k_2})' \end{cases} \quad (13)$$

where $\bar{k}_1$ and $\bar{k}_2$ represent, respectively, the nominal values of k_1 and k_2; W_{k_1} and W_{k_2} are the normalized weighted coefficients of uncertainties; δ_{k_1} and δ_{k_2} are used to describe the fluctuation ranges of the corresponding parameters; and $|\delta_i| \leq 1 (i = k_1, k_2)$.

To handle the problem of parameter uncertainties, the LFT technique is used to separate the uncertain part and the definite part of the system. Through the upper LFT, the k_1 and k_2 in Equation (13) are converted into the upper linear fraction structure described in Equation (6):

$$\begin{cases} k_1 = \bar{k}_1(1 + W_{k_1}\delta_{k_1}) = \bar{k}_1 + W_{k_1}\delta_{k_1}(I - \delta_{k_1} \cdot 0)^{-1}\bar{k}_1 = F_u(M_{k_1}, \delta_{k_1}) \\ k_2 = \bar{k}_2(1 + W_{k_2}\delta_{k_2}) = \bar{k}_2 + W_{k_2}\delta_{k_2}(I - \delta_{k_2} \cdot 0)^{-1}\bar{k}_2 = F_u(M_{k_2}, \delta_{k_2})' \end{cases} \quad (14)$$

where

$$M_{k_1} = \begin{bmatrix} 0 & \bar{k}_1 \\ W_{k_1} & \bar{k}_1 \end{bmatrix}, M_{k_2} = \begin{bmatrix} 0 & \bar{k}_2 \\ W_{k_2} & \bar{k}_2 \end{bmatrix}. \quad (15)$$

The LFT uncertain model of the two-DOF FJMS considering parameter uncertainties and external disturbances can be obtained from Equations (10) and (14), as shown in Figure 4.

The uncertainties of the system considered during this research are categorized into internal and external uncertainties, where the parameter uncertainties are the internal uncertainties, and the external disturbances are the external uncertainties. Both of these have been considered and are presented in our system model. The approach of this research is capable of dealing with a class of control problems that consider the system's uncertainties. The reason why only the spring-stiffness coefficients are considered with parameter uncertainties is that they have somewhat more inaccuracy compared to the other parameters.

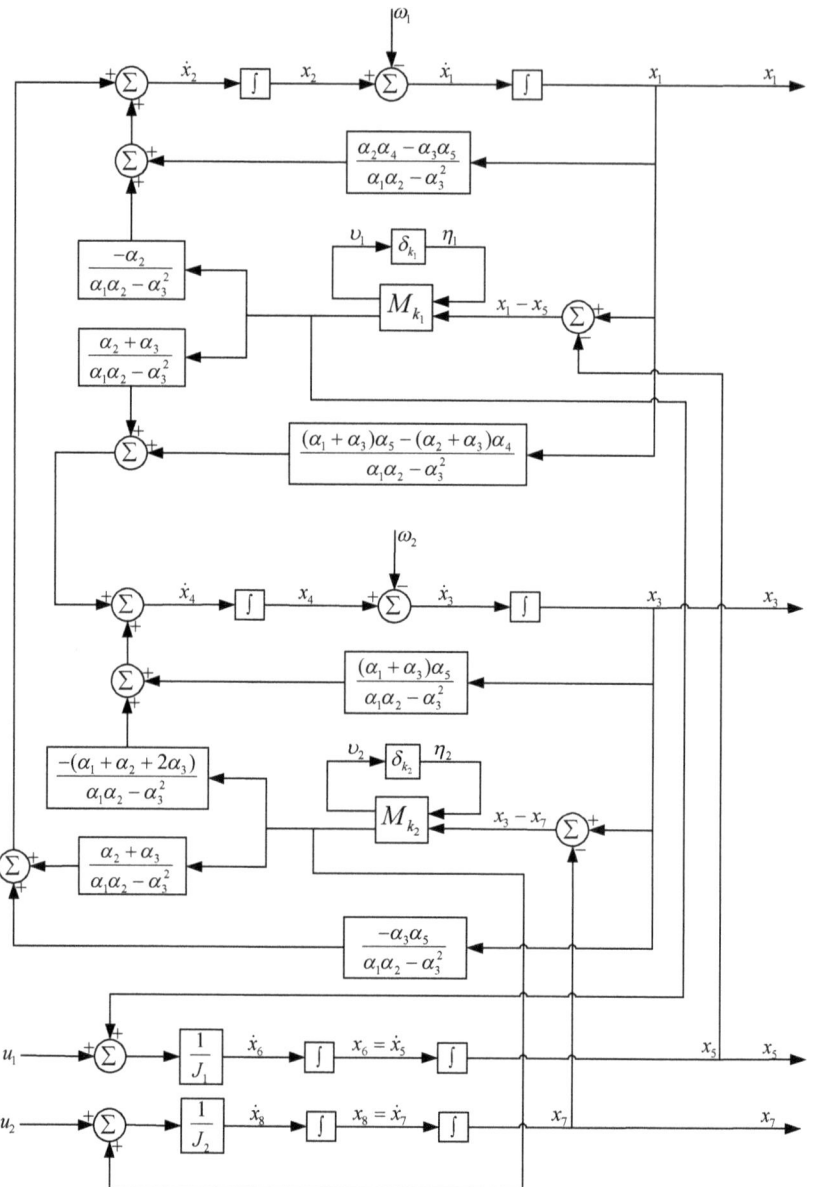

Figure 4. The LFT uncertain model of the two-DOF FJMS.

In the case of the real system, this manipulator system is naturally subject to angular constraints and drive-torque constraints. This is why a new variable (z_∞) is introduced to represent the constrained output of the uncertain system. The angular accelerations of the two joints are selected as the performance output of the system. Hence, the performance output and constrained output of this system are defined as follows:

$$\begin{cases} z_2 = \begin{bmatrix} \ddot{q}_1 & \ddot{q}_2 \end{bmatrix}^T = \begin{bmatrix} \dot{x}_2 & \dot{x}_4 \end{bmatrix}^T \\ z_\infty = \begin{bmatrix} \dfrac{u_1}{u_{1\max}} & \dfrac{u_2}{u_{2\max}} & \dfrac{q_1}{q_{1\max}} & \dfrac{q_2}{q_{2\max}} \end{bmatrix}^T \end{cases}. \quad (16)$$

It can be derived from Figure 4 that the state space and the corresponding mapping relationships between $v_i (i = 1,2)$ and $\eta_i (i = 1,2)$ are as follows:

$$\begin{bmatrix} \dot{x} \\ v \\ z_2 \\ z_\infty \end{bmatrix} = \begin{bmatrix} A & B_\eta & B_u & B_\omega \\ C_v & 0 & 0 & 0 \\ C_2 & D_{2\eta} & D_2 & 0 \\ C_\infty & 0 & D_\infty & 0 \end{bmatrix} \begin{bmatrix} x \\ \eta \\ u \\ \omega \end{bmatrix}, \quad (17)$$

$$\eta = \delta v = \begin{bmatrix} \delta_{k_1} & 0 \\ 0 & \delta_{k_2} \end{bmatrix} v, \quad (18)$$

where $\omega = \begin{bmatrix} \omega_1 & \omega_2 \end{bmatrix}^T \in \mathbb{R}^{2\times 1}$ is the external disturbance vector of the uncertain system; $\eta = \begin{bmatrix} \eta_1 & \eta_2 \end{bmatrix}^T \in \mathbb{R}^{2\times 1}$ and $v = \begin{bmatrix} v_1 & v_2 \end{bmatrix}^T \in \mathbb{R}^{2\times 1}$ are auxiliary vectors that represent, respectively, the uncertainty input and output vectors of this system; $\delta = diag\{\delta_{k_1}, \delta_{k_2}\} \in Y_\delta$ represents the uncertainty matrix; and Y_δ is the collection of uncertainties.

Equations (17) and (18) could be transformed into state-space equations in which the uncertain and definite components of this system are separated, as follows:

$$\begin{cases} \dot{x} = (A + \Delta A)x + (B_u + \Delta B_u)u + B_\omega \omega \\ z_2 = (C_2 + \Delta C_2)x + (D_2 + \Delta D_2)u \\ z_\infty = C_\infty x + D_\infty u \end{cases}, \quad (19)$$

where $A, B_u, B_\omega, C_2, D_2, C_\infty$, and D_∞ are the known-constant-coefficient matrices describing the nominal system model of this manipulator; $\Delta A, \Delta B_u, \Delta C_2$, and ΔD_2 are the uncertainty matrix functions of appropriate dimensions, representing the parameter uncertainties of the system model. Therefore, in order to extract the variables in the uncertainty matrices $\Delta A, \Delta B_u, \Delta C_2$, and ΔD_2 containing δ_{k_1} and δ_{k_2}, $\Delta A, \Delta B_u, \Delta C_2$, and ΔD_2 could be written in the form of a bounded norm based on Equations (17) and (18), as follows:

$$\begin{cases} \begin{bmatrix} \Delta A & \Delta B_u \end{bmatrix} = E_1 \delta \begin{bmatrix} F_1 & F_2 \end{bmatrix} \\ \begin{bmatrix} \Delta C_2 & \Delta D_2 \end{bmatrix} = E_2 \delta \begin{bmatrix} F_1 & F_2 \end{bmatrix} \end{cases}, \quad (20)$$

where E_1, E_2, F_1, and F_2 are uncertain matrices of appropriate dimensions in the following forms, respectively:

$$\begin{cases} E_1 = \begin{bmatrix} 0 & \frac{-\alpha_2 W_{k_1}}{\alpha_1\alpha_2-\alpha_3^2} & 0 & \frac{(\alpha_2+\alpha_3)W_{k_1}}{\alpha_1\alpha_2-\alpha_3^2} & 0 & \frac{W_{k_1}}{J_1} & 0 & 0 \\ 0 & \frac{(\alpha_2+\alpha_3)W_{k_2}}{\alpha_1\alpha_2-\alpha_3^2} & 0 & \frac{-(\alpha_1+\alpha_2+2\alpha_3)W_{k_2}}{\alpha_1\alpha_2-\alpha_3^2} & 0 & 0 & 0 & \frac{W_{k_2}}{J_2} \end{bmatrix}^T, F_2 = \begin{bmatrix} 0 & 0 \\ 0 & 0 \end{bmatrix} \\ E_2 = \begin{bmatrix} \frac{-\alpha_2 W_{k_1}}{\alpha_1\alpha_2-\alpha_3^2} & \frac{(\alpha_2+\alpha_3)W_{k_2}}{\alpha_1\alpha_2-\alpha_3^2} \\ \frac{(\alpha_2+\alpha_3)W_{k_1}}{\alpha_1\alpha_2-\alpha_3^2} & \frac{-(\alpha_1+\alpha_2+2\alpha_3)W_{k_2}}{\alpha_1\alpha_2-\alpha_3^2} \end{bmatrix}, F_1 = \begin{bmatrix} \bar{k}_1 & 0 & 0 & 0 & -\bar{k}_1 & 0 & 0 & 0 \\ 0 & 0 & \bar{k}_2 & 0 & 0 & 0 & -\bar{k}_2 & 0 \end{bmatrix} \end{cases}$$

$$(21)$$

4. Robust Model Predictive Control with Constraints

The solution to the optimization control problem addressed in this paper is to purposely design a controller that firstly ensures that this manipulator system maintains strong robustness and stability under the dual influence of parameter uncertainties and external disturbances, and secondly, that minimizes the $\mathcal{H}_\infty$ norm of system perturbation ω to performance output z_2 while ensuring that all the constraints of this system are satisfied. The final control goal is to design a controller to stabilize the two-DOF FJMS at the equilibrium point under the influence of a series of factors. In addition, the state feedback structure with $u = Kx$ is considered in the design process of the controller to ensure that a good control performance can be obtained. Therefore, the key point of our designed controller is how to calculate the feedback gain that satisfies the system constraints while guaranteeing the system performance.

4.1. Robust Constrained $\mathcal{H}_\infty$ Control

For the LFT uncertain system with constraints, this subsection focuses on the design of a robust constrained $\mathcal{H}_\infty$ controller to ensure the improvement in the $\mathcal{H}_\infty$ performance and the fulfillment of the input–output constraints. With the application of the LMI technique, the constrained $\mathcal{H}_\infty$ control problem can be converted into the convex optimization problem with LMIs as constraints to make it easier to solve. The lemmas about the LMIs used in this procedure are as follows:

Lemma 1 ([51,52]). *Suppose that* $S = \begin{bmatrix} M & N \\ N^T & L \end{bmatrix} \in \mathbb{R}^{(k+l)\times(k+l)}$ *is non-singular and its inverse matrix is recorded as* $S^{-1} = \begin{bmatrix} \tilde{M} & \tilde{N} \\ \tilde{N}^T & \tilde{L} \end{bmatrix} \in \mathbb{R}^{(k+l)\times(k+l)}$. *Then, the nonlinear matrix inequality*

$$L \geq 0, \begin{bmatrix} I \\ F \end{bmatrix}^T \begin{bmatrix} M & N \\ N^T & L \end{bmatrix} \begin{bmatrix} I \\ F \end{bmatrix} \leq 0, \tag{22}$$

is equivalent to the following LMI:

$$\tilde{M} \leq 0, \begin{bmatrix} -F^T \\ I \end{bmatrix}^T \begin{bmatrix} \tilde{M} & \tilde{N} \\ \tilde{N}^T & \tilde{L} \end{bmatrix} \begin{bmatrix} -F^T \\ I \end{bmatrix} \geq 0. \tag{23}$$

Lemma 2 ([53,54]). *Suppose that there are three constant matrices, E, F, and G, and four affine-function matrices, $K(\beta)$, $L(\beta)$, $M(\beta) = M(\beta)^T$, and $N(\beta)$, with the independent variable β. Which $L(\beta)$ can be decomposed into $Y^T U(\beta)^{-1} Y$ and $U(\beta) < 0$ is also affine depending on the variable β. Then, the nonlinear matrix inequality*

$$\begin{bmatrix} E \\ K(\beta) \\ F \end{bmatrix}^T \begin{bmatrix} M(\beta) & \begin{bmatrix} G & N(\beta) \end{bmatrix} \\ \begin{bmatrix} G^T \\ N(\beta)^T \end{bmatrix} & L(\beta) \end{bmatrix} \begin{bmatrix} E \\ K(\beta) \\ F \end{bmatrix} \leq 0, \tag{24}$$

is equivalent to the following LMI:

$$\begin{pmatrix} E^T M(\beta) E + E^T (GK(\beta) + N(\beta)F) + (GK(\beta) + N(\beta)F)^T E & \begin{bmatrix} K(\beta)^T & F^T \end{bmatrix} Y \\ Y^T \begin{bmatrix} K(\beta) \\ F \end{bmatrix} & -U(\beta) \end{pmatrix} \leq 0. \tag{25}$$

Lemma 3 ([55,56]). *Suppose that $S_0, S_1, \cdots, S_j \in \mathbb{R}^{n \times n}$ are the symmetric matrices. If there exists $\varphi_1, \varphi_2, \ldots, \varphi_j \geq 0$ such that $S_0 - \sum_{i=1}^{j} \varphi_i S_i > 0$ holds, then it is obtained as follows:*

$$\zeta^T S_0 \zeta > 0 \text{ for all } \zeta \neq 0 \text{ such that } \zeta^T S_i \zeta \geq 0 (i = 1, \ldots, j). \tag{26}$$

Due to the fact that the discrete-time model is applied to the conventional MPC, it is necessary to discretize the system (17), and the results are processed via the method of Equations (17)–(21), as follows:

$$\begin{cases} x(k+1) = (A_d + \Delta A_d)x(k) + (B_{ud} + \Delta B_{ud})u(k) + B_{\omega d}\omega(k) \\ z_2(k) = (C_{2d} + \Delta C_{2d})x(k) + (D_{2d} + \Delta D_{2d})u(k) \\ z_\infty(k) = C_{\infty d}x(k) + D_{\infty d}u(k) \end{cases}, \quad (27)$$

where $A_d = e^{AT_s}$, $\Delta A_d = E_{1d}\delta F_{1d}$, $E_{1d} = A^{-1}(e^{AT_s} - I)E_1$, $F_{1d} = F_1$, $B_{ud} = A^{-1}(e^{AT_s} - I)B_u$, $\Delta B_{ud} = E_{1d}\delta F_{2d}$, $F_{2d} = F_2$, $B_{\omega d} = A^{-1}(e^{AT_s} - I)B_\omega$, $C_{2d} = C_2$, $\Delta C_{2d} = E_{2d}\delta F_{1d}$, $E_{2d} = E_2$, $D_{2d} = D_2$, $\Delta D_{2d} = E_{2d}\delta F_{2d}$, $C_{\infty d} = C_\infty$, $D_{\infty d} = D_\infty$, and T_s is the sample period. In addition, $|z_{\infty i}(k)| \leq z_{\infty i,\max} = 1 (i = 1, 2, 3, 4)$ represents the four normalized constrained output values of this system. The state feedback control law $u(k) = Kx(k)$ is substituted into (27), and the closed-loop system can be written as follows:

$$\begin{cases} x(k+1) = A_\delta x(k) + B_{\omega d}\omega(k) \\ z_2(k) = C_\delta x(k) \\ z_\infty(k) = (C_{\infty d} + D_{\infty d}K)x(k) \end{cases}, \quad (28)$$

where $A_\delta = A_d + B_{ud}K + E_{1d}\delta(F_{1d} + F_{2d}K)$ and $C_\delta = C_{2d} + D_{2d}K + E_{2d}\delta(F_{1d} + F_{2d}K)$.

If there exists a non-negative value of λ such that the system (28) satisfies the dissipation inequality shown below:

$$x(k+1)^T H x(k+1) - x(k)^T H x(k) \leq \lambda^2 \|\omega(k)\|_2^2 - \|z_2(k)\|_2^2, \quad (29)$$

where $H \in \mathbb{R}^{8 \times 8}$ is a positive definite symmetric matrix, and if the system is steady when $k \to \infty$, then $\lim_{k \to \infty} x(k+1) = 0$ is true. This could be further deduced as follows:

$$\max_{\delta \in Y_\delta} \left(\sum_{k=0}^{\infty} \left(\|z_2(k)\|_2^2 - \lambda^2 \|\omega(k)\|_2^2 \right) \right) \leq x(0)^T H x(0), \quad (30)$$

where it is marked by the establishment of (30) that the $\mathcal{H}_\infty$ norm of this system from ω to z_2 is less than λ.

Replacing with the components of (29) by the closed-loop system (28) gives the following:

$$(A_\delta x(k) + B_{\omega d}\omega(k))^T H (A_\delta x(k) + B_{\omega d}\omega(k)) - x(k)^T H x(k) \leq \lambda^2 \|\omega(k)\|_2^2 - \|C_\delta x(k)\|_2^2. \quad (31)$$

The above inequality is organized and transformed into the quadratic form as follows:

$$\begin{bmatrix} x(k) \\ \omega(k) \end{bmatrix}^T \begin{bmatrix} A_\delta^T H A_\delta - H + C_\delta^T C_\delta & A_\delta^T H B_{\omega d} \\ B_{\omega d}^T H A_\delta & B_{\omega d}^T H B_{\omega d} - \lambda^2 I \end{bmatrix} \begin{bmatrix} x(k) \\ \omega(k) \end{bmatrix} \leq 0. \quad (32)$$

The inequality (32) is equivalent to the following:

$$\begin{bmatrix} I & A_\delta^T \\ 0 & B_{\omega d}^T \end{bmatrix} \begin{bmatrix} -H & 0 \\ 0 & H \end{bmatrix} \begin{bmatrix} I & 0 \\ A_\delta & B_{\omega d} \end{bmatrix} + \begin{bmatrix} 0 & C_\delta^T \\ I & 0 \end{bmatrix} \begin{bmatrix} -\lambda^2 I & 0 \\ 0 & I \end{bmatrix} \begin{bmatrix} 0 & I \\ C_\delta & 0 \end{bmatrix} \leq 0. \quad (33)$$

In order to separate δ from A_δ and C_δ, the inequality (33) is modified into the following form:

$$\begin{bmatrix} I & 0 \\ 0 & I \\ \delta(F_{1d}+F_{2d}K) & 0 \end{bmatrix}^{\mathrm{T}} \begin{bmatrix} I & 0 & 0 \\ A_d+B_{ud}K & B_{\omega d} & E_{1d} \end{bmatrix}^{\mathrm{T}} \begin{bmatrix} -H & 0 \\ 0 & H \end{bmatrix} \begin{bmatrix} I & 0 & 0 \\ A_d+B_{ud}K & B_{\omega d} & E_{1d} \end{bmatrix} \begin{bmatrix} I & 0 \\ 0 & I \\ \delta(F_{1d}+F_{2d}K) & 0 \end{bmatrix}$$
$$+ \begin{bmatrix} I & 0 \\ 0 & I \\ \delta(F_{1d}+F_{2d}K) & 0 \end{bmatrix}^{\mathrm{T}} \begin{bmatrix} 0 & I & 0 \\ C_{2d}+D_{2d}K & 0 & E_{2d} \end{bmatrix}^{\mathrm{T}} \begin{bmatrix} -\lambda^2 I & 0 \\ 0 & I \end{bmatrix} \begin{bmatrix} 0 & I & 0 \\ C_{2d}+D_{2d}K & 0 & E_{2d} \end{bmatrix} \begin{bmatrix} I & 0 \\ 0 & I \\ \delta(F_{1d}+F_{2d}K) & 0 \end{bmatrix} \leq 0$$
(34)

Then, with the application of the full-block multiplier technique [57–59], it could be deduced that the holding of (33) is equivalent to the existence of an invertible multiplier matrix ($\tilde{T}$) such that

$$\begin{bmatrix} \delta \\ I \end{bmatrix}^{\mathrm{T}} \begin{bmatrix} \tilde{T}_a & \tilde{T}_b \\ \tilde{T}_b^{\mathrm{T}} & \tilde{T}_c \end{bmatrix} \begin{bmatrix} \delta \\ I \end{bmatrix} \geq 0, \tag{35}$$

where $\tilde{T}_a < 0$. Due to the invertibility of the multiplier matrix ($\tilde{T}$), the application of Lemma 1 to (35) results in the following:

$$\begin{bmatrix} I \\ -\delta^{\mathrm{T}} \end{bmatrix}^{\mathrm{T}} \begin{bmatrix} T_a & T_b \\ T_b^{\mathrm{T}} & T_c \end{bmatrix} \begin{bmatrix} I \\ -\delta^{\mathrm{T}} \end{bmatrix} \leq 0 \text{ and } T = \tilde{T}^{-1} = \begin{bmatrix} T_a & T_b \\ T_b^{\mathrm{T}} & T_c \end{bmatrix}, \tag{36}$$

where T_c is a positive definite diagonal matrix and $T_a = -T_c$, T_b is a diagonal matrix, and all the elements of T_b are antisymmetric matrices. Moreover, there is one thing worth noting:

$$\begin{bmatrix} \delta & 0 \\ I & 0 \end{bmatrix}(F_{1d}+F_{2d}K) = \begin{bmatrix} 0 & 0 & I \\ F_{1d}+F_{2d}K & 0 & 0 \end{bmatrix} \begin{bmatrix} I & 0 \\ 0 & I \\ \delta(F_{1d}+F_{2d}K) & 0 \end{bmatrix}. \tag{37}$$

The following inequality could be further reasoned from (34)–(37):

$$\begin{bmatrix} I & 0 & 0 \\ A_d+B_{ud}K & B_{\omega d} & E_{1d} \\ 0 & I & 0 \\ C_{2d}+D_{2d}K & 0 & E_{2d} \\ 0 & 0 & I \\ F_{1d}+F_{2d}K & 0 & 0 \end{bmatrix}^{\mathrm{T}} \begin{bmatrix} -H & 0 & 0 & 0 & 0 & 0 \\ 0 & H & 0 & 0 & 0 & 0 \\ 0 & 0 & -\lambda^2 I & 0 & 0 & 0 \\ 0 & 0 & 0 & I & 0 & 0 \\ 0 & 0 & 0 & 0 & \tilde{T}_a & \tilde{T}_b \\ 0 & 0 & 0 & 0 & \tilde{T}_b^{\mathrm{T}} & \tilde{T}_c \end{bmatrix} \begin{bmatrix} I & 0 & 0 \\ A_d+B_{ud}K & B_{\omega d} & E_{1d} \\ 0 & I & 0 \\ C_{2d}+D_{2d}K & 0 & E_{2d} \\ 0 & 0 & I \\ F_{1d}+F_{2d}K & 0 & 0 \end{bmatrix} \leq 0. \tag{38}$$

Then, the inequality (38) is subjected to the simple elementary matrix transformation, resulting in the following:

$$\begin{bmatrix} I & 0 & 0 \\ 0 & I & 0 \\ 0 & 0 & I \\ \hline A_d+B_{udl}K & B_{\omega d} & E_{1d} \\ C_{2d}+D_{2d}K & 0 & E_{2d} \\ F_{1d}+F_{2d}K & 0 & 0 \end{bmatrix}^{\mathrm{T}} \begin{bmatrix} -H & 0 & 0 & 0 & 0 & 0 \\ 0 & -\lambda^2 I & 0 & 0 & 0 & 0 \\ 0 & 0 & \tilde{T}_a & 0 & 0 & \tilde{T}_b \\ \hline 0 & 0 & 0 & H & 0 & 0 \\ 0 & 0 & 0 & 0 & I & 0 \\ 0 & 0 & \tilde{T}_b^{\mathrm{T}} & 0 & 0 & \tilde{T}_c \end{bmatrix} \begin{bmatrix} I & 0 & 0 \\ 0 & I & 0 \\ 0 & 0 & I \\ \hline A_d+B_{ud}K & B_{\omega d} & E_{1d} \\ C_{2d}+D_{2d}K & 0 & E_{2d} \\ F_{1d}+F_{2d}K & 0 & 0 \end{bmatrix} \leq 0. \tag{39}$$

Because $\tilde{T}_a < 0$ and $H > 0$, the $diag\{-H, -\lambda^2 I, \tilde{T}_a\} < 0$ holds. Applying Lemma 1 to (39) and performing the elementary matrix transformation, the result would be as follows:

$$
\begin{bmatrix}
I & 0 & 0 \\
0 & I & 0 \\
0 & 0 & I \\
\hdashline
-(A_{dd}+B_{udl}K)^{\mathrm{T}} & -(C_{2d}+D_{2d}K)^{\mathrm{T}} & -(F_{1d}+F_{2d}K)^{\mathrm{T}} \\
-B_{\omega d}^{\mathrm{T}} & 0 & 0 \\
-E_{1d}^{\mathrm{T}} & -E_{2d}^{\mathrm{T}} & 0
\end{bmatrix}^{\mathrm{T}}
\begin{bmatrix}
N & 0 & 0 & 0 & 0 & 0 \\
0 & I & 0 & 0 & 0 & 0 \\
0 & 0 & T_c & 0 & 0 & T_b^{\mathrm{T}} \\
\hdashline
0 & 0 & 0 & -N & 0 & 0 \\
0 & 0 & 0 & 0 & -\lambda^{-2}I & 0 \\
0 & 0 & T_b & 0 & 0 & T_a
\end{bmatrix}
\begin{bmatrix}
* \\ * \\ *
\end{bmatrix} \geq 0, \quad (40)
$$

where $N = H^{-1}$. Due to $T_a < 0$ and $N > 0$, the $diag\{-N, -\lambda^{-2}I, T_a\} < 0$ holds. With the help of Lemma 2, the nonlinear matrix inequality (40) is equivalent to the specific form of LMI, as follows:

$$
\begin{bmatrix}
N & 0 & -E_{1d}T_b & -A_d - B_{ud}K & -B_{\omega d} & -E_{1d} \\
0 & I & -E_{2d}T_b & -C_{2d} - D_{2d}K & 0 & -E_{2d} \\
-T_b^{\mathrm{T}}E_{1d}^{\mathrm{T}} & -T_b^{\mathrm{T}}E_{2d}^{\mathrm{T}} & T_c & -F_{1d} - F_{2d}K & 0 & 0 \\
-(A_d + B_{ud}K)^{\mathrm{T}} & -(C_{2d} + D_{2d}K)^{\mathrm{T}} & -(F_{1d} + F_{2d}K)^{\mathrm{T}} & N^{-1} & 0 & 0 \\
-B_{\omega d}^{\mathrm{T}} & 0 & 0 & 0 & \lambda^2 I & 0 \\
-E_{1d}^{\mathrm{T}} & -E_{2d}^{\mathrm{T}} & 0 & 0 & 0 & -T_a^{-1}
\end{bmatrix} \geq 0. \quad (41)
$$

Then, the above inequality (41) is subjected to the congruence transformation with $diag\{I, I, I, -N, I, T_a\}$, and defining $R = KN$ results in the following:

$$
\begin{bmatrix}
N & 0 & -E_{1d}T_b & A_dN + B_{ud}R & -B_{\omega d} & -E_{1d}T_a \\
0 & I & -E_{2d}T_b & C_{2d}N + D_{2d}R & 0 & -E_{2d}T_a \\
-T_b^{\mathrm{T}}E_{1d}^{\mathrm{T}} & -T_b^{\mathrm{T}}E_{2d}^{\mathrm{T}} & T_c & F_{1d}N + F_{2d}R & 0 & 0 \\
(A_dN + B_{ud}R)^{\mathrm{T}} & (C_{2d}N + D_{2d}R)^{\mathrm{T}} & (F_{1d}N + F_{2d}R)^{\mathrm{T}} & N & 0 & 0 \\
-B_{\omega d}^{\mathrm{T}} & 0 & 0 & 0 & \lambda^2 I & 0 \\
-T_a^{\mathrm{T}}E_{1d}^{\mathrm{T}} & -T_a^{\mathrm{T}}E_{2d}^{\mathrm{T}} & 0 & 0 & 0 & -T_a
\end{bmatrix} \geq 0. \quad (42)
$$

In summary, if there are variables (R, N, λ^2) and multipliers (T_a, T_b, T_c) making the LMI (42) be established, then the $\mathcal{H}_\infty$ norm of the controlled system (27) from ω to z_2 must be less than λ under the action of the controller $K = RN^{-1}$ and $u(k) = Kx(k)$. Meanwhile, the above LMI also guarantees that $A_d + B_{ud}K$ is quadratically stable.

The treatment of the constraints existing in the system is discussed in this section. Firstly, an elliptic domain is defined as $\Omega(H, \rho_f) := \left\{ x(k)^{\mathrm{T}} H x(k) \leq \rho_f \right\}$. If $x(k) \in \Omega(H, \rho_f)$, then it could be deduced from the closed-loop system (28) as follows:

$$
|z_{\infty i}(k)|^2 = \left| e_i^{\mathrm{T}}(C_{\infty d} + D_{\infty d}K)x(k) \right|^2 \leq \max_{x(k) \in \Omega(H, \rho_f)} \left| e_i^{\mathrm{T}}(C_{\infty d} + D_{\infty d}K)x(k) \right|^2 (i = 1, 2, 3, 4), \quad (43)
$$

where $e_i (i = 1, 2, 3, 4)$ are the standard basis vectors in the four-dimensional space. If $\left| e_i^{\mathrm{T}}(C_{\infty d} + D_{\infty d}K)x(k) \right|^2 \leq z_{\infty i,\max}^2 (i = 1, 2, 3, 4)$, then $|z_{\infty i}(k)| \leq z_{\infty i,\max}(i = 1, 2, 3, 4)$ is valid, and it means that the constraints of the controlled system are satisfied. Let $S_0 = z_{\infty i,\max}^2 - \left| e_i^{\mathrm{T}}(C_{\infty d} + D_{\infty d}K)x(k) \right|^2$ and $S_1 = \rho_f - x(k)^{\mathrm{T}} H x(k)$. Applying Lemma 3, it is equivalent to the existence of $\varphi > 0$, which enables (44) to hold for all $\delta \in Y_\delta$:

$$
z_{\infty i,\max}^2 - x(k)^{\mathrm{T}} (e_i^{\mathrm{T}}(C_{\infty d} + D_{\infty d}K))^{\mathrm{T}} e_i^{\mathrm{T}}(C_{\infty d} + D_{\infty d}K) x(k) - \varphi \rho_f + \varphi x(k)^{\mathrm{T}} H x(k) \geq 0 (i = 1, 2, 3, 4). \quad (44)
$$

And let $\varphi = \dfrac{z_{\infty i,\max}^2}{\rho_f}$, then inequality (44) can be transformed into the following:

$$
\begin{bmatrix} I \\ e_i^{\mathrm{T}}(C_{\infty d} + D_{\infty d}K) \end{bmatrix}^{\mathrm{T}} \begin{bmatrix} -H & 0 \\ 0 & \dfrac{\rho_f}{z_{\infty i,\max}^2} \end{bmatrix} \begin{bmatrix} I \\ e_i^{\mathrm{T}}(C_{\infty d} + D_{\infty d}K) \end{bmatrix} \leq 0. \quad (45)
$$

Using the technique of the full-block multiplier again, the holding of the above inequality is equivalent to the existence of four invertible multiplier matrices ($\tilde{V}_i(i = 1,2,3,4)$), satisfying the following:

$$\begin{bmatrix} I & 0 \\ e_i^T(C_{\infty d} + D_{\infty d}K) & 0 \\ 0 & I \\ F_{1d} + F_{2d}K & 0 \end{bmatrix}^T \begin{bmatrix} -H & 0 & 0 & 0 \\ 0 & \frac{\rho_f}{z_{\infty i,max}^2} & 0 & 0 \\ 0 & 0 & \tilde{V}_{ai} & \tilde{V}_{bi} \\ 0 & 0 & \tilde{V}_{bi}^T & \tilde{V}_{ci} \end{bmatrix} \begin{bmatrix} I & 0 \\ e_i^T(C_{\infty d} + D_{\infty d}K) & 0 \\ 0 & I \\ F_{1d} + F_{2d}K & 0 \end{bmatrix} \leq 0, \quad (46)$$

where $\tilde{V}_{ai} < 0$. Then, the inequality (46) after the elementary matrix transformation is as follows:

$$\begin{bmatrix} I & 0 \\ 0 & I \\ \hline e_i^T(C_{\infty d} + D_{\infty d}K) & 0 \\ F_{1d} + F_{2d}K & 0 \end{bmatrix}^T \begin{bmatrix} -H & 0 & 0 & 0 \\ 0 & \tilde{V}_{ai} & 0 & \tilde{V}_{bi} \\ \hline 0 & 0 & \frac{\rho_f}{z_{\infty i,max}^2} & 0 \\ 0 & \tilde{V}_{bi}^T & 0 & \tilde{V}_{ci} \end{bmatrix} \begin{bmatrix} I & 0 \\ 0 & I \\ \hline e_i^T(C_{\infty d} + D_{\infty d}K) & 0 \\ F_{1d} + F_{2d}K & 0 \end{bmatrix} \leq 0. \quad (47)$$

Owing to $\tilde{V}_{ai} < 0$ and $H > 0$, the $diag\{-H, \tilde{V}_{ai}\} < 0$ holds. Applying Lemma 1 and performing the elementary matrix transformation, the inequality (47) is equivalent to the nonlinear matrix inequality shown below:

$$\begin{bmatrix} I & 0 \\ 0 & I \\ \hline -(C_{\infty d} + D_{\infty d}K)^T e_i & -(F_{1d} + F_{2d}K)^T \\ 0 & 0 \end{bmatrix}^T \begin{bmatrix} \frac{z_{\infty i,max}^2}{\rho_f} & 0 & 0 & 0 \\ 0 & V_{ci} & 0 & V_{bi}^T \\ \hline 0 & 0 & -H^{-1} & 0 \\ 0 & V_{bi} & 0 & V_{ai} \end{bmatrix} \begin{bmatrix} * \\ * \\ * \end{bmatrix} \leq 0. \quad (48)$$

where $V_{ci}(i = 1,2,3,4)$ are four positive definite diagonal matrices and $V_{ai} = -V_{ci}$, $V_{bi}(i = 1,2,3,4)$ are four diagonal matrices, and all elements of V_{bi} are antisymmetric matrices. Because $V_{ai} < 0$ and $H > 0$, the $diag\{-H^{-1}, V_{ai}\} < 0$ holds. Applying Lemma 2, the inequality (48) is equivalent to the following LMI:

$$\begin{bmatrix} \frac{z_{\infty i,max}^2}{\rho_f} & 0 & -e_i^T(C_{\infty d} + D_{\infty d}K) & 0 \\ 0 & V_{ci} & -F_{1d} - F_{2d}K & 0 \\ -(C_{\infty d} + D_{\infty d}K)^T e_i & -(F_{1d} + F_{2d}K)^T & H & 0 \\ 0 & 0 & 0 & -V_{ai}^{-1} \end{bmatrix} \geq 0. \quad (49)$$

Then, the inequality (49) is congruent transformed with $diag\{I, I, N, V_{ai}\}$, which results in the following:

$$\begin{bmatrix} \frac{z_{\infty i,max}^2}{\rho_f} & 0 & -e_i^T(C_{\infty d}N + D_{\infty d}R) & 0 \\ 0 & V_{ci} & -F_{1d}N - F_{2d}R & 0 \\ -(C_{\infty d}N + D_{\infty d}R)^T e_i & -(F_{1d}N + F_{2d}R)^T & N & 0 \\ 0 & 0 & 0 & -V_{ai} \end{bmatrix} \geq 0 (i = 1,2,3,4). \quad (50)$$

Therefore, if the variables that enable LMI (42) to be feasible also make LMI (50) hold, then $|z_{\infty i}(k)| \leq z_{\infty i,max}(i = 1,2,3,4)$, which indicates that all constraints are satisfied for the uncertainties considered by the system.

In summary, the robust constrained $\mathcal{H}_\infty$ controller with the state feedback structure is constructed by addressing the LMI optimization solution problem for the given $\rho_f > 0$, as shown below:

$$\min_{\lambda^2, N, R, T_a, T_b, T_c, \{V_{ai}, V_{ci}\}} \lambda^2 \text{ subject to (42) and (50).} \quad (51)$$

In order to reduce the number of independent variables for this controller and the online computational burden of the algorithm, we may set $T_b = 0$.

The method described above guarantees the $\mathcal{H}_\infty$ performance of the system with parameter uncertainties while satisfying the four constraints by addressing multiple LMIs. In addition, the trade-off between a good system performance and the satisfaction of the constraints can be achieved by defining the state elliptic domain ($\Omega(H, \rho_f)$) and selecting the appropriate controller parameter (ρ_f).

4.2. Robust Constrained Moving-Horizon $\mathcal{H}_\infty$ Control

The concept of the control algorithm designed above highlights the inherent compromise between ensuring all the constraints and enhancing the control performance of the uncertain system. However, there might be a few large perturbations that cannot be anticipated in advance in the actual system, and it is possible to guarantee the system performance only by increasing the value of ρ_f to expand the range of the elliptic domain $\Omega(H, \rho_f)$, yet the consequence of doing so would be extremely limited values of N and R, ultimately resulting in the larger value of the performance index (λ), which, in turn, decreases the performance of the system. In contrast, the pursuit of a better performance could be achieved by lowering the value of ρ_f. But if the controlled system is subjected to larger external perturbations, there is no guarantee that the system constraints can be satisfied. Therefore, how to modify the LMI optimization control problem based on (51) is the top priority.

The stationarity and strong conservativeness of ρ_f, H, and λ prompt us to incorporate the idea of the moving-horizon control of MPC to overcome the weaknesses of the current algorithm and to coordinate online the conflict between satisfying the constraints and improving the performance of the uncertain system. The moving-horizon control principle of predictive control is to address the objective optimization problem online at each sample moment, which is constantly renewed by the latest measurements of the controlled system, and the calculated control input is actioned on this system until the next sample moment.

The conflict between constraints and performance could be nicely settled by altering the range of the elliptic domain $\Omega(H, \rho_f)$ at any moment according to the extent of the disturbances to the uncertain system. Hence, the following LMI optimization problem is constantly refreshed with the latest state ($x(k)$) at each sample moment (k) and addressed:

$$\min_{\rho_k, \lambda_k^2, N_k, R_k, T_{ak}, T_{bk}, T_{ck}, \{V_{aik}, V_{cik}\}} \chi_1 \rho_k + \chi_2 \lambda_k^2 \text{ subject to,} \tag{52}$$

$$\begin{bmatrix} N_k & 0 & -E_{1d}T_{bk} & A_dN_k + B_{ud}R_k & -B_{\omega d} & -E_{1d}T_{ak} \\ 0 & I & -E_{2d}T_{bk} & C_{2d}N_k + D_{2d}R_k & 0 & -E_{2d}T_{ak} \\ -T_{bk}^T E_{1d}^T & -T_{bk}^T E_{2d}^T & T_{ck} & F_{1d}N_k + F_{2d}R_k & 0 & 0 \\ (A_dN_k + B_{ud}R_k)^T & (C_{2d}N_k + D_{2d}R_k)^T & (F_{1d}N_k + F_{2d}R_k)^T & N_k & 0 & 0 \\ -B_{\omega d}^T & 0 & 0 & 0 & \lambda_k^2 I & 0 \\ -T_{ak}^T E_{1d}^T & -T_{ak}^T E_{2d}^T & 0 & 0 & 0 & -T_{ak} \end{bmatrix} \geq 0, \tag{53}$$

$$\begin{bmatrix} \frac{z_{\infty i, \max}^2}{\rho_f} & 0 & -e_i^T(C_{\infty d}N_k + D_{\infty d}R_k) & 0 \\ 0 & V_{cik} & -F_{1d}N_k - F_{2d}R_k & 0 \\ -(C_{\infty d}N_k + D_{\infty d}R_k)^T e_i & -(F_{1d}N_k + F_{2d}R_k)^T & N & 0 \\ 0 & 0 & 0 & -V_{aik} \end{bmatrix} \geq 0 (i = 1,2,3,4), \tag{54}$$

$$\begin{bmatrix} \rho_k & x(k)^T \\ x(k) & N_k \end{bmatrix} \geq 0, \tag{55}$$

$$\begin{bmatrix} x(k)^T H_{k-1} x(k) + h_0 - h_{k-1} & x(k)^T \\ x(k) & N_k \end{bmatrix} \geq 0, \tag{56}$$

$$\rho_k \leq \rho_f, \tag{57}$$

where χ_1 and χ_2 are the weighting factors, which are used to adjust the weights between the minimization of the $\mathcal{H}_\infty$ norm from ω to z_2 and the minimization of the range of the elliptic domain $\Omega(H_k, \rho_k)$ while maintaining the constraints satisfied, thereby meeting the multifaceted requirements of the controlled system. The LMI (56) is an additional dissipation inequality that is implemented to ensure the dissipativity of the closed-loop system that is destroyed under the moving-horizon control scheme. Moreover, it is determined by the H_{k-1} and dissipation index h_{k-1} at the last moment, and the iterative updates of the H_k and h_k are calculated via $H_k := N_k^{-1}$ and $h_k := h_{k-1} - [x(k)^\mathrm{T} H_{k-1} x(k) - x(k)^\mathrm{T} H_k x(k)]$ with $h_0 := x(0)^\mathrm{T} H_0 x(0)$.

At the moment k, if the semi-definite programming (52) can be addressed online for a given ρ_f, resulting in ρ_k, λ_k, N_k, R_k, and several multipliers, then the feedback gain at the current moment is considered as $K(k) = R_k N_k^{-1}$, and thus the closed-loop control input of the controlled system can be specified as follows:

$$u(k) = K(k)x(k), \forall k \geq 0. \tag{58}$$

At each sample moment, the values of the variables R_k and N_k are obtained by solving the LMI optimization problem (52), where N_k is a positive definite symmetric matrix and $R_k = K(k)N_k$. Then, the feedback gain $K(k) = R_k N_k^{-1}$ at the current moment can be calculated after obtaining the values of the above variables. In addition, the state feedback structure is chosen in this paper during the controller design process, so the control input of the system is thus calculated with $u(k) = K(k)x(k)$.

The state $x(k)$ includes the information about the external disturbances to the system and the modeling mistakes caused by the parameter uncertainties. As a matter of fact, the $x(k)$ is used to calculate the value of the feedback gain $K(k)$ at the current moment, and it is also used as the feedback information of the closed-loop system. It should be noted here that the system state $x(k)$ in the control input $u(k) = K(k)x(k)$ and in the LMIs (55) and (56) are the state of the nonlinear system of the two-DOF FJMS. And it needs to be made clear that the purpose of linearizing the controlled system (3) is only to calculate the control feedback gain by solving the LMI optimization problem, and the feedback gain is subsequently used to calculate the control input of the two-DOF FJMS to act on the nonlinear system (3).

In addition to the introduction of the moving-horizon control strategy, another ingenious feature of the algorithm (52) is the treatment of ρ_k as an independent variable and as a portion of the objective optimization function. Moreover, the coupling between the system constraints and the performance index is separated by the constant ρ_f, which consequently makes the optimization problem easily solvable numerically.

For the algorithm described above, the feasibility of the optimization problem at every sample moment is crucial. However, the feasibility of the above online optimization algorithm would not be guaranteed in the case of strong disturbances increasing suddenly at some random moment. Accordingly, borrowing the idea of the scaling method, a non-negative number ($\sigma \geq 0$) is introduced in this paper. The purpose of this improvement is simply to diminish the conservativeness of the algorithm (52) and to enhance its feasibility so that it is capable of coping with larger external disturbances. The robust constrained moving-horizon $\mathcal{H}_\infty$ control optimization algorithm is as follows:

$$\min_{\rho_k, \lambda_k^2, N_k, R_k, T_{ak}, T_{bk}, T_{ck}, \{V_{aik}, V_{cik}\}} \chi_1 \rho_k + \chi_2 \lambda_k^2 \text{ subject to (53), (54), (55), (56) and,} \tag{59}$$

$$\rho_k \leq \rho_f (1 + \sigma). \tag{60}$$

Therefore, the robust constrained moving-horizon $\mathcal{H}_\infty$ control algorithm specifically addresses the LMI optimization problem (59) refreshed by the current moment state $x(k)$ at each sample moment, and if infeasibility occurs, then the value of σ is augmented and the optimization problem is recalculated. For one moment, the process of expanding the range of the elliptic domain is diagrammed in Figure 5.

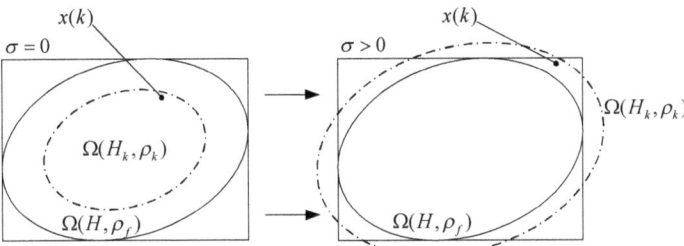

Figure 5. The diagram of expanding the elliptic domain for one moment in time.

In the theory of robust model predictive control, we resort to the LMI technique to transform the constrained $\mathcal{H}_\infty$ control problem into a convex optimization problem with LMIs as constraints. In this way, multiple objectives in the controller design requirements could be transformed into multiple LMIs that are used as constraints on the objective function. Secondly, the optimization problem is solved to obtain the optimal control input at the current moment, which is implemented on the nonlinear controlled system, and then the optimization problem is refreshed with the latest state and is resolved again at the next moment, and ultimately the system performance is improved, and the system requirements are fulfilled through continuous moving-horizon optimization control.

5. Properties of the Closed-Loop System

In this section, the closed-loop characteristics of this system are discussed by implementing the above optimization algorithm (59) on the controlled system. The closed-loop system is mainly confronted with two situations: when the external disturbances in the system are relatively small, the optimization algorithm (52) is feasible at each sample moment, and when the external disturbances are sudden or large, it is only necessary to enlarge the value of σ to guarantee the feasibility of the optimization algorithm (59) to achieve some degree of relaxation so that the controlled system can handle unpredictable and large disturbances.

Theorem 1. *Suppose the following:*

1. *At every moment, the semi-definite programming (52) based on the state $x(k)$ at the current moment has the results as $\rho_k, \lambda_k, N_k, R_k$, and several multipliers;*
2. *The performance optimization metric$\{\lambda_0, \lambda_1, \cdots, \lambda_{k-1}, \lambda_k\}$is bounded.*

Then, for all $\delta \in Y_\delta$, the closed-loop controlled system under the action of $u(k) = K(k)x(k)$would have the following properties:

1. *The constraints of the controlled system are all fulfilled;*
2. *Under the perturbations of external limiting energy, the state $x(k)$of the system will converge to zero when $k \to \infty$;*
3. *The dissipation inequality $\sum_{i=0}^{k}\left(\|z_2(i)\|_2^2 - \overline{\lambda}^2\|\omega(i)\|_2^2\right) \le x(0)^\mathrm{T}H_0x(0)$ is valid for any moment (k), where $\overline{\lambda} = \max\{\lambda_0, \lambda_1, \cdots, \lambda_{k-1}, \lambda_k\}$;*
4. *The $\mathcal{H}_\infty$ norm from the system perturbation ω to the performance output z_2 is always no greater than λ_∞, where $\lambda_\infty = \lim_{k \to \infty} \max\{\lambda_0, \lambda_1, \cdots, \lambda_{k-1}, \lambda_k\}$.*

Proof of Theorem 1. If there is an appropriate ρ_f such that LMIs (55) and (57) are valid, then it also means that the closed-loop system state $x(k)$ is within the elliptic domain $\Omega(H_k, \rho_k) = \left\{x(k)^\mathrm{T}H_kx(k) \le \rho_k\right\}$. For all $\delta \in Y_\delta$, the inequalities (43), (45), (50), and (54) are all equivalent to each other, which subsequently leads to $|z_{\infty i}(k)| \le z_{\infty i,\max}(i = 1, 2, 3, 4)$, and the first property is proved.

At any moment (k), if the optimization solution (ρ_k, λ_k, N_k, R_k and several multipliers) is valid by substitution in the LMIs (42) and (53), then the dissipative inequality (29) is expressed to work. Therefore, the following may be obtained:

$$x(k+1)^T H_k x(k+1) - \sum_{i=1}^{k}\left(x(i)^T H_i x(i) - x(i)^T H_{i-1} x(i)\right) - x(0)^T H_0 x(0) \leq \sum_{i=0}^{k}\left(\lambda_i^2 \|\omega(i)\|_2^2 - \|z_2(i)\|_2^2\right). \quad (61)$$

The (56) formed by the dissipation constraint satisfies $\sum_{i=1}^{k}\left(x(i)^T H_i x(i) - x(i)^T H_{i-1} x(i)\right) \leq 0$. Letting $\bar{\lambda} = \max\{\lambda_0, \lambda_1, \cdots, \lambda_{k-1}, \lambda_k\}$, the inequality (61) can be simplified as follows:

$$x(k+1)^T H_k x(k+1) - x(0)^T H_0 x(0) \leq \sum_{i=0}^{k}\left(\bar{\lambda}^2 \|\omega(i)\|_2^2 - \|z_2(i)\|_2^2\right). \quad (62)$$

Due to the fact that H_k is positive definite and $\{\lambda_0, \lambda_1, \cdots, \lambda_{k-1}, \lambda_k\}$ is bounded, then the third property is proved to be permanent. If there is finite energy of the perturbations to the controlled system, then it is obtained that $\sum_{i=0}^{\infty}\|z_2(i)\|_2^2 \leq x(0)^T H_0 x(0) + \lambda_\infty^2 \sum_{i=0}^{\infty}\|\omega(i)\|_2^2$ when the limit of the inequality (62) at $k \to \infty$ is considered, where $\lambda_\infty = \lim_{k\to\infty} \max\{\lambda_0, \lambda_1, \cdots, \lambda_{k-1}, \lambda_k\}$. Thus, the second property is evidenced by the above procedure. As for the last property, it is proved when the zero initial state is selected. □

Considering the circumstances of the optimization control algorithm (59) to be implemented, the following findings would be produced with less conservatism.

Theorem 2. *Suppose the following:*

1. *The LMI (53) and LMI (54) are all feasible;*
2. *The amplitude of the perturbations at any moment is not infinite;*
3. *The performance optimization metric $\{\lambda_0, \lambda_1, \cdots, \lambda_{k-1}, \lambda_k\}$ is bounded.*

Then, for all $\delta \in Y_\delta$, the controlled system with the effect of the robust constrained moving-horizon $\mathcal{H}_\infty$ controller would have the following properties:

1. *At every moment (k), there is $\left|e_i^T(C_{\infty d} + D_{\infty d}K(k))x(k)\right| \leq z_{\infty i,\max}(i=1,2,3,4)$, and this relationship is established to symbolize that the constraint requirements of this controller are fulfilled;*
2. *The last three properties of Theorem 1 are also present.*

Proof of Theorem 2. The feasibility of the optimization algorithm (59) at any sample moment is guaranteed by the introduced factor $\sigma \geq 0$. The control gain is calculated with $K(k) = R_k N_k^{-1}$, and the first property is clearly well established. The proofs of the remaining properties are analogous to those of Theorem 1. □

6. Simulation Results

In this section, the robust constrained moving-horizon $\mathcal{H}_\infty$ control algorithm is implemented for the dynamic model of the two-DOF FJMS, and the $\mathcal{H}_\infty$ performance of this controlled system is tested. The nominal values of the parameters of this manipulator system are shown in Table 2 [46], selecting the sample time as $T_s = 0.01$s, and discretizing the two-DOF FJMS. In order to accomplish the stabilization of this system at the equilibrium point under the influence of external disturbances and parameter uncertainties, the whole system should have certain constraints on the sizes of both the joint angles of the two-DOF FJMS: $q_{1,\max} = 0.15$ and $q_{2,\max} = 0.15$. The main reason for constraining the maximum value of the two joint angles to be so small is mainly because of the consideration that there are certain conditions for linearizing the two-DOF FJMS using Taylor series expansion. Moreover, due to the saturation of the actuators, the torque values as the control input are also constrained: $u_{1,\max} = 100$ and $u_{2,\max} = 50$. To simplify the calculation procedure, the normalized control torques and joint angles are chosen as the constrained output here, where the system constraints are bounded by $z_{\infty i,\max} = 1 (i = 1, 2, 3, 4)$.

Table 2. The nominal values of the parameters of the two-DOF FJMS.

Symbol	Values
L_1, L_2	0.5 m, 0.5 m
L_{c1}, L_{c2}	0.25 m, 0.25 m
m_1, m_2	20 kg, 10 kg
I_1, I_2	5.6 kg·m^2, 2.8 kg·m^2
J_1, J_2	6.183 kg·m^2, 0.858 kg·m^2
$\bar{k}_1, \bar{k}_2$	1000 N·m/rad, 1000 N·m/rad
g	9.81 m/s^2

As for the parameter uncertainties, what are considered here are the spring-stiffness coefficients shown in Equation (13), where the normalized weighted coefficients are $W_{k_1} = 0.2$ and $W_{k_2} = 0.2$. In order to verify the feasibility and robustness of the designed control algorithm, the real values of the two uncertain parameters are set to different values at different time periods, as shown below:

$$k_1 = \begin{cases} \bar{k}_1(1-0.2), & 0 \leq k \leq 30 \\ \bar{k}_1(1+0.2), & 30 \leq k \leq 70 \\ \bar{k}_1(1+0.6), & 70 \leq k \leq 150 \end{cases}, k_2 = \begin{cases} \bar{k}_2(1-0.2), & 0 \leq k \leq 30 \\ \bar{k}_2(1+0.2), & 30 \leq k \leq 70 \\ \bar{k}_2(1+0.6), & 70 \leq k \leq 150 \end{cases}, \quad (63)$$

The external disturbances to the controlled system are assumed to be as follows:

$$\omega_1 = \omega_2 = \begin{cases} \frac{\pi}{6}\sin(\frac{k\pi}{10}), & 0 \leq k \leq 20 \\ -\frac{\pi}{6}\sin(\frac{k\pi}{10}), & 60 \leq k \leq 80 \\ 0, & else \end{cases}, \quad (64)$$

For the design of the robust constrained moving-horizon $\mathcal{H}_\infty$ controller, the weight factors are selected as $\chi_1 = 0.1$ and $\chi_2 = 1$ to achieve a greater system performance and less energy consumption. Moreover, the size of the fixed elliptic domain $\Omega(H, \rho_f)$ is chosen as $\rho_f = 10$. The RCHC (the LMI optimization algorithm (52)) and RCMHHC (the LMI optimization algorithm (59)) were implemented on the two-DOF FJMS for simulation, and the comparative outputs of the experiments are shown in Figures 6–12.

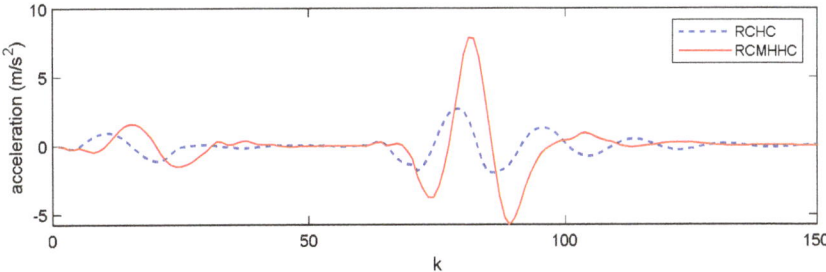

Figure 6. The performance output: the joint angular acceleration of q_1.

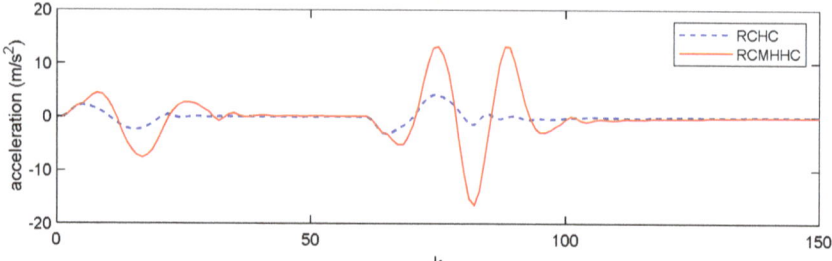

Figure 7. The performance output: the joint angular acceleration of q_2.

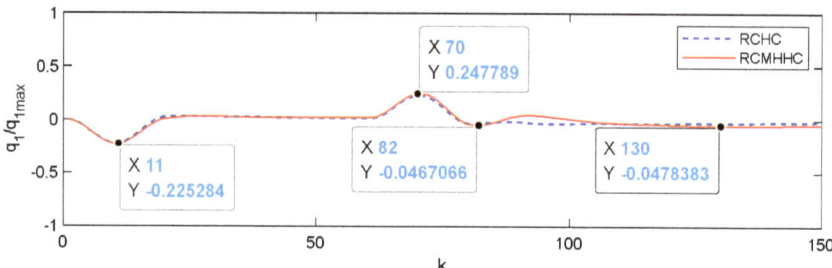

Figure 8. The constrained output: the normalized joint angle of q_1.

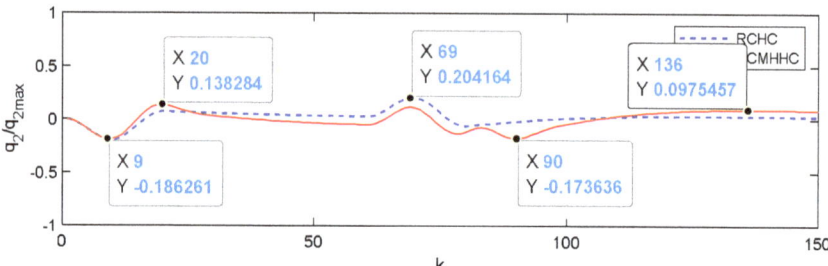

Figure 9. The constrained output: the normalized joint angle of q_2.

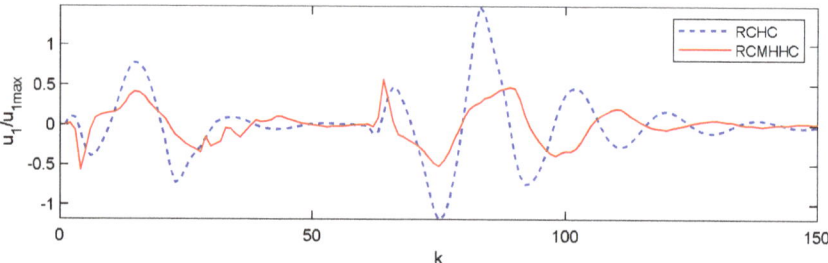

Figure 10. The control input: the normalized control torque of u_1.

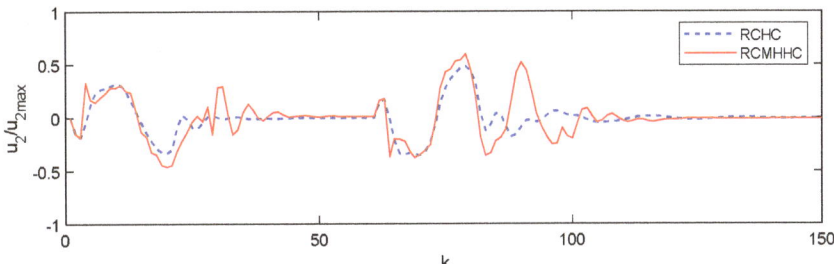

Figure 11. The control input: the normalized control torque of u_2.

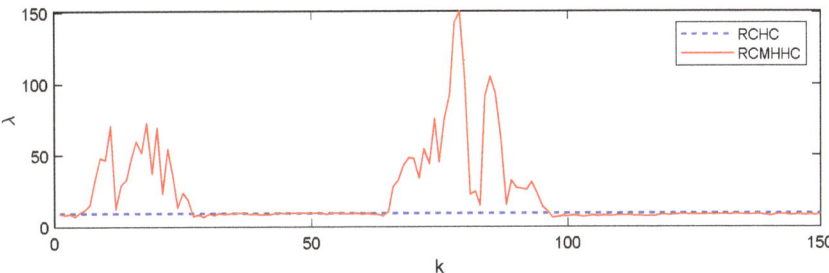

Figure 12. The performance indicator.

In Figure 10, it can be seen that the system under the action of RCHC has violated the control input constraint of u_1 at $74 \leq k \leq 77$ and $82 \leq k \leq 85$. The only way to make the system constraints satisfied is to increase the value of ρ_f to expand the elliptic domain. However, the result of doing so would make the performance of this system worse, which fully illustrates the disadvantages of RCHC. In contrast, the control input constraints of the controlled system are well satisfied with the action of RCMHHC. In addition, we can discover that both joint angles are not out of our constraints during the whole process, according to the simulation results in Figures 8 and 9. It should be noted here that the normalized values of the joint angles are shown in both images. Therefore, the actual maximum value of the two joint angles during the simulation is just $0.2478 \times 0.15 = 0.03717$ rad. In the later part of the simulation, the two joint angles of this system are basically stabilized around the X axis, where the output curve of q_2 may not be as smooth as that of q_1 in the later stages, but the actual value of the joint angle of q_2 at $k = 136$ is $0.0975 \times 0.15 = 0.014625$ rad. It is possible to determine that the designed controller is capable of stabilizing the nonlinear two-DOF FJMS at the vertical equilibrium position after acting on it. In Figures 6 and 7, the reason why the system performance output under the action of RCMHHC does not perform as well as RCHC at some moments is caused by the concession of the system performance to the unsatisfied constraints due to the impact of larger disturbances. Figure 12 displays the variation curves of the performance indices of both algorithms. After the comparative analysis of the above output curves, the significant advantage of RCMHHC can be found in the online reconciliation of the conflict between satisfying the system constraints and improving the control performance at any moment. The coordination mechanism of RCMHHC is to decrease the performance requirements when necessary to make sure that the hard constraints are satisfied, and to enhance the performance requirements in time when the controlled system is far from the boundary of the constraints.

Both algorithms work by solving the LMI optimization problem. All the LMIs in the LMI optimization problem are derived based on the linear system, and the problem is solved through the toolbox of MATLAB and is thus less time-consuming. The value of the feedback gain obtained by solving RCHC is fixed, and the LMI optimization problem only needs to be solved once. However, the idea of moving-horizon control in RCMHHC makes the state of the system appear in the LMIs, so that its feedback gain value is not fixed, and it is necessary to solve the LMI optimization problem at every sample moment. In contrast, the real-time property of RCMHHC may not be as good as that of RCHC, but RCMHHC can improve the $\mathcal{H}_\infty$ performance of the two-DOF FJMS well while ensuring that the system constraints are satisfied. In conclusion, the real-time property of solving the LMI optimization problem at each moment can be satisfied for conventional PCs.

7. Conclusions

For the two-DOF FJMS, this paper designs a robust constrained moving-horizon $\mathcal{H}_\infty$ controller to accomplish the control objectives of this system while considering external disturbances, parameter uncertainties, and input–output constraints simultaneously. After establishing the LFT uncertain system of the two-DOF FJMS, the semi-definite programming problem with LMIs as constraints is developed via the full-block multiplier technique, $\mathcal{H}_\infty$ control, and MPC, for which the control feedback gain of the two-DOF FJMS can be obtained after solving the LMI optimization problem. The feedback gain is subsequently used to calculate the control input of the controlled system to act on the nonlinear two-DOF FJMS under the state feedback structure. Based on the moving-horizon control principle of MPC, this LMI optimization problem is refreshed with the current state of the system at each sample moment and solved online, and so on, in a continuous iterative loop. The simulation of the designed controller implemented on the two-DOF FJMS shows that the proposed control algorithm is able to improve the system $\mathcal{H}_\infty$ performance while ensuring that the system constraints are satisfied, and it could coordinate online the conflict between both requirements.

Author Contributions: Conceptualization, R.L. and H.W.; methodology, G.Y.; software, H.W.; validation, R.L., H.W. and G.L.; formal analysis, R.L.; investigation, H.W.; resources, R.L.; data curation, L.J.; writing—original draft preparation, H.W.; writing—review and editing, R.L., H.W. and G.Y.; visualization, H.W.; supervision, G.L.; project administration, R.L.; funding acquisition, R.L. All authors have read and agreed to the published version of the manuscript.

Funding: This work was supported by the National Natural Science Foundation of China (Grant No. 62003233), the Fundamental Research Program of Shanxi Province (Grant Nos. 201901D211083 and 20210302124552), and the Science and Technology Innovation Project of Higher Education Institutions in Shanxi Province (Grant No. 2019L0236).

Data Availability Statement: Not applicable.

Conflicts of Interest: The authors declare no conflict of interest.

References

1. Zhang, W.; Yang, X.; Xu, Z.; Zhang, W.; Yang, L.; Liu, X. An Adaptive Fault-Tolerant Control Method for Robot Manipulators. *Int. J. Control Autom. Syst.* **2021**, *19*, 3983–3995. [CrossRef]
2. Phu, N.D.; Putov, V.V.; Su, C.T. Mathematical Models and Adaptive Control System of Rigid and Flexible 4-DOF Joint Robotic Manipulator with Executive Electric Drives. In Proceedings of the 2019 III International Conference on Control in Technical Systems (CTS), St. Petersburg, Russia, 30 October–1 November 2019.
3. Naidu, D.S. Singular Perturbation Analysis of a Flexible Beam Used in Underwater Exploration. *Int. J. Syst. Sci.* **2011**, *42*, 183–194. [CrossRef]
4. Nanos, K.; Papadopoulos, E.G. On the Dynamics and Control of Flexible Joint Space Manipulators. *Control Eng. Pract.* **2015**, *45*, 230–243. [CrossRef]
5. Nubert, J.; Köhler, J.; Berenz, V.; Allgöwer, F.; Trimpe, S. Safe and Fast Tracking on a Robot Manipulator: Robust MPC and Neural Network Control. *IEEE Robot. Autom. Lett.* **2020**, *5*, 3050–3057. [CrossRef]
6. Ahmadi, S.; Fateh, M.M. Control of Flexible Joint Robot Manipulators by Compensating Flexibility. *Iran. J. Fuzzy Syst.* **2018**, *15*, 57–71.
7. Wei, J.; Cao, D.; Wang, L.; Huang, H.; Huang, W. Dynamic Modeling and Simulation for Flexible Spacecraft with Flexible Jointed Solar Panels. *Int. J. Mech. Sci.* **2017**, *130*, 558–570. [CrossRef]
8. Zouari, L.; Abid, H.; Abid, M. Sliding Mode and PI Controllers for Uncertain Flexible Joint Manipulator. *Int. J. Autom. Comput.* **2015**, *12*, 117–124. [CrossRef]

9. Fateh, M.M. Robust Control of Flexible-Joint Robots Using Voltage Control Strategy. *Nonlinear Dyn.* **2012**, *67*, 1525–1537. [CrossRef]
10. Hassanzadeh, I.; Kharrati, H.; Bonab, J.R. Model Following Adaptive Control for a Robot with Flexible Joints. In Proceedings of the 2008 Canadian Conference on Electrical and Computer Engineering, Niagara Falls, ON, Canada, 4–7 May 2008.
11. Korayem, M.H.; Shafei, A.M.; Doosthoseini, M.; Absalan, F.; Kadkhodaei, B. Theoretical and Experimental Investigation of Viscoelastic Serial Robotic Manipulators with Motors at the Joints Using Timoshenko Beam Theory and Gibbs-Appell Formulation. *Proc. Inst. Mech. Eng. Part K J. Multi-Body Dyn.* **2015**, *230*, 37–51. [CrossRef]
12. Korayem, M.H.; Shafei, A.M.; Dehkordi, S.F. Systematic Modeling of a Chain of N-Flexible Link Manipulators Connected by Revolute–prismatic Joints Using Recursive Gibbs-Appell Formulation. *Arch. Appl. Mech.* **2014**, *84*, 187–206. [CrossRef]
13. Marino, R.; Spong, M. Nonlinear Control Techniques for Flexible Joint Manipulators: A Single Link Case Study. In Proceedings of the 1986 IEEE International Conference on Robotics and Automation, San Francisco, CA, USA, 7–10 April 1986.
14. Spong, M.W. Modeling and Control of Elastic Joint Robots. *Math. Comput. Model.* **1989**, *12*, 912. [CrossRef]
15. Sun, L.; Yin, W.; Wang, M.; Liu, J. Position Control for Flexible Joint Robot Based on Online Gravity Compensation with Vibration Suppression. *IEEE Trans. Ind. Electron.* **2018**, *65*, 4840–4848. [CrossRef]
16. Pan, Y.; Wang, H.; Li, X.; Yu, H. Adaptive Command-Filtered Backstepping Control of Robot Arms with Compliant Actuators. *IEEE Trans. Control Syst. Technol.* **2018**, *26*, 1149–1156. [CrossRef]
17. Rsetam, K.; Cao, Z.; Man, Z.; Mitrevska, M. Optimal Second Order Integral Sliding Mode Control for a Flexible Joint Robot Manipulator. In Proceedings of the IECON 2017-43rd Annual Conference of the IEEE Industrial Electronics Society, Beijing, China, 29 October–1 November 2017.
18. Qiu, B.; Guo, J.; Mao, M.; Tan, N. A Fuzzy-Enhanced Robust DZNN Model for Future Multi-Constrained Nonlinear Optimization with Robotic Manipulator Control. *IEEE Trans. Fuzzy Syst.* **2023**, 1–13. [CrossRef]
19. Xu, B.; Jiang, Q.; Ji, W.; Ding, S. An Improved Three-Vector-Based Model Predictive Current Control Method for Surface-mounted PMSM Drives. *IEEE Trans. Transp. Electrif.* **2022**, *8*, 4418–4430. [CrossRef]
20. Kali, Y.; Saad, M.; Benjelloun, K.; Fatemi, A. Discrete-Time Second Order Sliding Mode with Time Delay Control for Uncertain Robot Manipulators. *Robot. Auton. Syst.* **2017**, *94*, 53–60. [CrossRef]
21. Peng, Z.; Yan, W.; Huang, R.; Cheng, H.; Shi, K.; Ghosh, B.K. Event-Triggered Learning Robust Tracking Control of Robotic Systems with Unknown Uncertainties. *IEEE Trans. Circuits Syst. II Express Briefs* **2023**, *70*, 2540–2544. [CrossRef]
22. Ma, L.; Mei, K.; Ding, S.; Pan, T. Design of Adaptive Fuzzy Fixed-Time HOSM Controller Subject to Asymmetric Output Constraints. *IEEE Trans. Fuzzy Syst.* **2023**, 1–11. [CrossRef]
23. Peng, Z.; Luo, R.; Hu, J.; Shi, K.; Ghosh, B.K. Distributed Optimal Tracking Control of Discrete-Time Multiagent Systems Via Event-Triggered Reinforcement Learning. *IEEE Trans. Circuits Syst. I Regul. Pap.* **2022**, *69*, 3689–3700. [CrossRef]
24. Jiang, Z.H.; Higaki, S. Control of Flexible Joint Robot Manipulators Using a Combined Controller with Neural Network and Linear Regulator. *J. Syst. Control Eng.* **2011**, *225*, 798–806. [CrossRef]
25. Yan, Z.; Lai, X.; Meng, Q.; Wu, M. A Novel Robust Control Method for Motion Control of Uncertain Single-Link Flexible-Joint Manipulator. *IEEE Trans. Syst. Man Cybern. Syst.* **2021**, *51*, 1671–1678. [CrossRef]
26. Rsetam, K.; Cao, Z.; Man, Z. Cascaded-Extended-State-Observer-Based Sliding-Mode Control for Underactuated Flexible Joint Robot. *IEEE Trans. Ind. Electron.* **2020**, *67*, 10822–10832. [CrossRef]
27. He, W.; Yan, Z.; Sun, Y.; Ou, Y.; Sun, C. Neural-Learning-Based Control for a Constrained Robotic Manipulator with Flexible Joints. *IEEE Trans. Neural Netw. Learn. Syst.* **2018**, *29*, 5993–6003. [CrossRef] [PubMed]
28. Ma, H.; Zhou, Q.; Li, H.; Lu, R. Adaptive Prescribed Performance Control of a Flexible-Joint Robotic Manipulator with Dynamic Uncertainties. *IEEE Trans. Cybern.* **2022**, *52*, 12905–12915. [CrossRef]
29. Dong, F.; Han, J.; Chen, Y.H. Improved Robust Control for Multi-Link Flexible Manipulator with Mismatched Uncertainties. In Proceedings of the 2015 International Conference on Fluid Power and Mechatronics (FPM), Harbin, China, 5–7 August 2015.
30. Yim, J.G.; Yeon, J.S.; Park, J.H.; Lee, S.H.; Hur, J.S. Robust Control Using Recursive Design Method for Flexible Joint Robot Manipulator. In Proceedings of the 2007 IEEE International Conference on Robotics and Automation, Rome, Italy, 10–14 April 2007.
31. Abbas, H.S.; Cisneros, P.S.G.; Männel, G.; Rostalski, P.; Werner, H. Practical Model Predictive Control for a Class of Nonlinear Systems Using Linear Parameter-Varying Representations. *IEEE Access* **2021**, *9*, 62380–62393. [CrossRef]
32. Cisneros, P.S.G.; Sridharan, A.; Werner, H. Constrained Predictive Control of a Robotic Manipulator Using Quasi-LPV Representations. *Int. Fed. Autom. Control-Pap.* **2018**, *51*, 118–123. [CrossRef]
33. Do, T.T.; Vu, V.H.; Liu, Z. Linearization of Dynamic Equations for Vibration and Modal Analysis of Flexible Joint Manipulators. *Mech. Mach. Theory* **2022**, *167*, 104516. [CrossRef]
34. Lai, X.Z.; She, J.H.; Yang, S.X.; Wu, M. Comprehensive Unified Control Strategy for Underactuated Two-Link Manipulators. *IEEE Trans. Syst. Man Cybern. Part B (Cybern.)* **2009**, *39*, 389–398.
35. Spyrakos-Papastavridis, E.; Dai, J.S. Minimally Model-Based Trajectory Tracking and Variable Impedance Control of Flexible-Joint Robots. *IEEE Trans. Ind. Electron.* **2021**, *68*, 6031–6041. [CrossRef]
36. Richiedei, D.; Trevisani, A. Simultaneous Active and Passive Control for Eigenstructure Assignment in Lightly Damped Systems. *Mech. Syst. Signal Process.* **2017**, *85*, 556–566. [CrossRef]
37. Lai, X.Z.; Zhang, A.; She, J.H.; Wu, M. Motion Control of Underactuated Three-Link Gymnast Robot Based on Combination of Energy and Posture. *IET Control Theory Appl.* **2011**, *5*, 1484–1493. [CrossRef]

38. Lynch, A.G.; Vanderploeg, M.J. A Symbolic Formulation for Linearization of Multibody Equations of Motion. *J. Mech. Des.* **1995**, *117*, 441–445. [CrossRef]
39. Ghoreishi, A.; Nekoui, M.; Basiri, S. Optimal Design of LQR Weighting Matrices Based on Intelligent Optimization Methods. *Int. J. Intell. Inf. Process.* **2011**, *2*, 63–74.
40. Peng, Z.; Hu, J.; Cheng, H.; Huang, R.; Luo, R.; Zhao, P.; Ghosh, B.K. Tracking Control for Motion Constrained Robotic System Via Dynamic Event-Sampled Intelligent Learning Method. In Proceedings of the 2022 IEEE 61st Conference on Decision and Control (CDC), Cancun, Mexico, 6–9 December 2022.
41. Alam, W.; Mehmood, A.; Ali, K.; Javaid, U.; Alharbi, S.; Iqbal, J. Nonlinear Control of a Flexible Joint Robotic Manipulator with Experimental Validation. *Stroj. Vestn.* **2018**, *64*, 47–55.
42. Akhtaruzzaman, M.; Akmeliawati, R.; Yee, T.W. Modeling and Control of a Multi Degree of Freedom Flexible Joint Manipulator. In Proceedings of the 2009 Second International Conference on Computer and Electrical Engineering, Dubai, United Arab Emirates, 28–30 December 2009.
43. Bascetta, L.; Ferretti, G.; Scaglioni, B. Closed Form Newton-Euler Dynamic Model of Flexible Manipulators. *Robotica* **2017**, *35*, 1006–1030. [CrossRef]
44. Bilal, H.; Yin, B.; Kumar, A.; Ali, M.; Zhang, J.; Yao, J. Jerk-Bounded Trajectory Planning for Rotary Flexible Joint Manipulator: An Experimental Approach. *Soft Comput.* **2023**, *27*, 4029–4039. [CrossRef]
45. Hong, M.; Gu, X.; Liu, L.; Guo, Y. Finite Time Extended State Observer Based Nonsingular Fast Terminal Sliding Mode Control of Flexible-Joint Manipulators with Unknown Disturbance. *J. Frankl. Inst.* **2023**, *360*, 18–37. [CrossRef]
46. Kostarigka, A.K.; Doulgeri, Z.; Rovithakis, G.A. Prescribed Performance Tracking for Flexible Joint Robots with Unknown Dynamics and Variable Elasticity. *Automatica* **2013**, *49*, 1137–1147. [CrossRef]
47. Redheffer, R.M. On a Certain Linear Fractional Transformation. *J. Math. Phys.* **1960**, *39*, 269–286. [CrossRef]
48. Pasha, S.A.; Tuan, H.D.; Vo, B.N. Nonlinear Bayesian Filtering Using the Unscented Linear Fractional Transformation Model. *IEEE Trans. Signal Process.* **2010**, *58*, 477–489. [CrossRef]
49. Boukarim, G.E.; Chow, J.H. Modeling of Nonlinear System Uncertainties Using a Linear Fractional Transformation Approach. In Proceedings of the 1998 American Control Conference (ACC), Philadelphia, PA, USA, 26 June 1998.
50. Andrea, R.D.; Khatri, S. Kalman Decomposition of Linear Fractional Transformation Representations and Minimality. In Proceedings of the 1997 American Control Conference (ACC), Albuquerque, NM, USA, 6 June 1997.
51. Scherer, C.; Weiland, S. *Linear Matrix Inequalities in Control*, 2nd ed.; CRC Press: London, UK, 2011; pp. 123–142.
52. Herrmann, G.; Turner, M.C.; Postlethwaite, I. Linear Matrix Inequalities in Control. In *Mathematical Methods for Robust and Nonlinear Control: EPSRC Summer School*; Turner, M.C., Bates, D.G., Eds.; Springer: London, UK, 2007; Volume 367, pp. 123–142.
53. Chen, H. A Feasible Moving Horizon H∞ Control Scheme for Constrained Uncertain Linear Systems. *IEEE Trans. Autom. Control* **2007**, *52*, 343–348. [CrossRef]
54. Chen, H.; Scherer, C.W. Moving Horizon H∞ Control with Performance Adaptation for Constrained Linear Systems. *Automatica* **2006**, *42*, 1033–1040. [CrossRef]
55. Yaz, E.E. Linear Matrix Inequalities in System and Control Theory. *Proc. IEEE* **1998**, *86*, 2473–2474. [CrossRef]
56. Heemels, W.P.M.H.; Kundu, A.; Daafouz, J. On Lyapunov-Metzler Inequalities and S-Procedure Characterizations for the Stabilization of Switched Linear Systems. *IEEE Trans. Autom. Control* **2017**, *62*, 4593–4597. [CrossRef]
57. Scherer, C.W. Robust Mixed Control and Linear Parameter-Varying Control with Full Block Scalings. In *Advances in Linear Matrix Inequality Methods in Control*; Ghaoui, E.L., Niculescu, S.I., Eds.; Society for Industrial and Applied Mathematics: Philadelphia, PA, USA, 1999; pp. 187–207.
58. Scherer, C.W. A Full Block S-Procedure with Applications. In Proceedings of the 36th IEEE Conference on Decision and Control, San Diego, CA, USA, 12 December 1997.
59. Gyurkovics, É.; Takács, T. A Remark on Abstract Multiplier Conditions for Robustness Problems. *Syst. Control Lett.* **2009**, *58*, 276–281. [CrossRef]

Disclaimer/Publisher's Note: The statements, opinions and data contained in all publications are solely those of the individual author(s) and contributor(s) and not of MDPI and/or the editor(s). MDPI and/or the editor(s) disclaim responsibility for any injury to people or property resulting from any ideas, methods, instructions or products referred to in the content.

Review

Modeling and Control of Wide-Area Networks

Qiuzhen Wang [1] and Jiangping Hu [1,2,*]

[1] School of Automation Engineering, University of Electronic Science and Technology of China, Chengdu 611731, China; qzwang@std.uestc.edu.cn
[2] Yangtze Delta Region Institute (Huzhou), University of Electronic Science and Technology of China, Huzhou 313001, China
* Correspondence: hujp@uestc.edu.cn

Abstract: This paper provides a survey of recent research progress in mathematical modeling and distributed control of wide-area networks. Firstly, the modeling is introduced for two types of wide-area networks, i.e., coopetitive networks and cooperative networks, with the help of algebraic graph theory. Particularly, bipartite network topologies and cluster network topologies are introduced for coopetitive networks. With respect to cooperative networks, an intermittent clustered network modeling is presented. Then, some classical distributed control strategies are reviewed for wide-area networks to ensure some desired collective behaviors, such as consensus (or synchronization), bipartite consensus (or polarization), and cluster consensus (or fragmentation). Finally, some conclusions and future directions are summarized.

Keywords: wide-area network; clustered multi-agent systems; distributed control; swarm intelligence

MSC: 93C10; 93D05

Citation: Wang, Q.; Hu, J. Modeling and Control of Wide-Area Networks. *Mathematics* **2023**, *11*, 3984. https://doi.org/10.3390/math11183984

Academic Editor: Daniel-Ioan Curiac

Received: 28 July 2023
Revised: 9 September 2023
Accepted: 18 September 2023
Published: 19 September 2023

Copyright: © 2023 by the authors. Licensee MDPI, Basel, Switzerland. This article is an open access article distributed under the terms and conditions of the Creative Commons Attribution (CC BY) license (https://creativecommons.org/licenses/by/4.0/).

1. Introduction

With the development of science and technology, large-scale networks have gradually replaced local networks with simple structures and single functions, and thus, they have been increasingly used in industry and academia. Such large-scale networks are often called **wide-area networks** [1], which are generally composed of multiple groups or clusters to exhibit the sparsity of most complex networks [2]. The major characteristic of a wide-area network is the tight communication within clusters and sparse communication between clusters. This is particularly evident in post-disaster emergency communication systems [3–5], power grid optimization [6–8], internet of things applications [9–13], cloud computing [14–16], etc. In addition to the complex network topologies of wide-area networks, behavior emergence on such networks is also very fascinating. **Multi-agent systems** (MASs) have become a powerful approach to deal with the complexity and diversity of such large-scale networks. **Distributed control** in MASs is commonly used to study the underlying interaction mechanisms for the emergence of collective intelligent behaviors on wide-area networks. Desired behaviors and local interaction mechanisms are two distinctive features of distributed controls [17].

In general, wide-area networks can be divided into two categories, one of which is a class of networks with both cooperative and competitive interactions, known as **coopetitive networks** [18]. Cooperation and competition coexist in complex and subtle ways in natural evolution, human activities, and engineering applications [19–21]. For example, the group foraging of mixed species, which need to work together to explore the environment during foraging while also competing for limited resources [19]. The personal opinion is updated by taking the average of the neighbors' opinions [20], however, attempting to change someone else's opinion may also be seen as hostile or competitive. In engineering and military fields, friendly robots cooperate to capture and intercept enemy robots [21]. In terms of multi-objective convergence forms, bipartite consensus and cluster consensus are two typical types of emergent collective behavior in coopetitive networks [22–24]. Specifically, the former requires that the agents achieve a form of "modular consensus", while the latter is associated with a network topology partition that de-synchronizes the behavior among agents of different groups. Regarding bipartite consensus in networked MASs, structural balance is an indispensable concomitant of the network topology. Under this assumption, many studies have worked to understand the bipartite consensus of cooperative competitive networks across various contexts [25–28]. On the other hand, when achieving cluster consensus, acyclic partition and balanced couple partition are two typical network structures that ensure cluster consensus. These two communication topologies require the satisfying of the in-degree balance condition between different groups [29–31]. It is worth noting that in this type of clustered network, each subgraph only contains positive edges, i.e., cooperative relationships [32]. Recently, the concept of group-bipartite consensus has been introduced by combining acyclic partition and sign function [33]. The group-bipartite network topology removes the limitation that negative links are only allowed to exist between different groups, which is more in line with practical application scenarios. For example, there exist cooperative and competitive relationships between different provinces in terms of economics, politics, culture, etc., while different cities, regions, and enterprises within each province also cooperate and compete.

Life is not always a zero-sum game, where having a winner means having a loser. In fact, most successes do not come from competition, but from cooperation. Then, another type of wide-area network is called a **cooperative network**. For example, in post-disaster emergency communication networks where multiple subnetworks may become isolated [34], it is common for each group to be responsible for specific communication tasks, and to work quickly to collect, transmit, and process information. Intermittent communication may be established between groups. For simplicity, we refer to a MAS composed of multiple subnetworks as a cluster MAS (CMAS). For cooperative CMASs, consensus (or synchronization) emerges as a prominent collective behavior, whereby all agents ultimately converge to the same state. Bragagnolo et al. first studied a type of intermittent clustered network with a continuous–discrete communication mechanism in [35]. For instance, in a spatially clustered robot formation, robots that belong to the same cluster interact continuously. However, due to constraints such as limited energy and communication range, interactions between robots that are far apart are discrete. Until now, substantial research advancements have been achieved in the context of intermittent clustered networks, particularly focusing on synchronization or output synchronization [36–43].

Based on the above observations, this paper aims to provide a preliminary exploration of group behavior over wide-area networks and introduce some of the important research advances and application scenarios. Figure 1 shows a statistical analysis of several papers in Scopus between the years of 2010–2023 about four types of consensus, i.e., bipartite consensus, cluster consensus, group-bipartite consensus, and synchronization. The structure of this paper is as follows. Section 2 introduces two types of modeling methods for wide-area networks, i.e., coopetitive networks and cooperative networks. Section 3 reviews three collective behaviors on coopetitive networks, namely, bipartite consensus, cluster consensus, and group-bipartite consensus, and introduces some classical distributed control algorithms. Section 4 reviews recent distributed control in coopetitive networks. Section 5 summarizes the challenging issues in future relevant areas. Finally, the conclusion of this article is provided in Section 6.

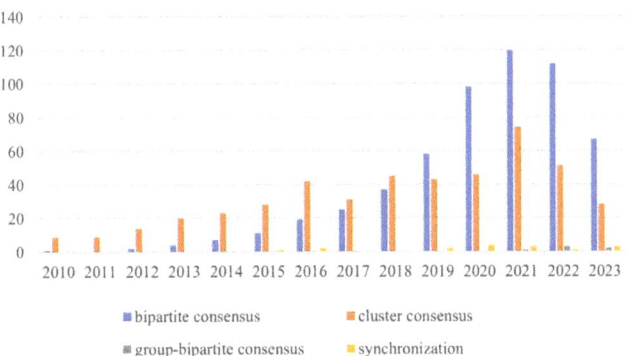

Figure 1. Statistical analysis of several research works related to four types of consensus in MASs published between 2010 and August 2023. Source: Scopus-indexed scientific papers on the field of MASs containing relevant keywords.

2. Notations and Modeling of Wide-Area Networks

Graph theory is a powerful tool for the modeling of wide-area networks. In this section, the modeling of coopetitive networks and cooperative networks over wide-area networks are introduced separately using graph theory. The symbols used throughout this article are listed in Table 1. The relationships between coopetitive networks and cooperative networks are shown in Table 2.

Table 1. Nomenclature.

Symbol	Definition
$\mathbb{R}_+$	The sets of non-negative real numbers
$\mathbb{N}$	The sets of non-negative integers
$\mathbb{R}^n$	The sets of n-dimensional real column vector space
$\mathbb{R}^{m \times n}$	The sets of $m \times n$-dimensional real matrix space
$\mathbb{C}^{m \times n}$	The sets of $m \times n$-dimensional complex matrix space
$diag\{\cdot\}$	The block diagonal matrix
$col(\cdot)$	The column vectors
$\|\cdot\|$	The Euclidean norm
$\otimes$	The Kronecker product for matrices
A^T	The transpose matrix of A
C^+	The generating inverse of matrix C
$\mathbf{0}$	The zero matrix with compatible dimensions
$\lambda(A)$	The eigenvalues of matrix A
$x(t_k^-)$	The left limit of $x(t_k)$
Hurwitz matrix	All eigenvalues of a matrix have negative real parts

2.1. Coopetitive Networks

This subsection employs graph theory to construct models for three distinct types of competitive networks: bipartite networks, cluster networks, and group-bipartite networks. The mathematical framework presented in this subsection provides a valuable tool for analyzing and modeling cooperative–competitive phenomena in both natural and human systems.

2.1.1. Bipartite Networks with Structural Balance Assumption

In bipartite networks, positive edges (represented by blue solid lines) and negative edges (represented by red dashed lines) are used to denote cooperative and competitive relationships between individuals, respectively. Cooperative relationships occur within clusters, while competition exists only between clusters. Herein, Figure 2 provides an intuitive illustration of the concept of bipartite networks, applied to describe healthcare coopetition networks [44]. In this network, there are two core hospitals, each modeled as a subnetwork. Within each subnetwork, hospitals cooperate by sharing healthcare resources, while simultaneously engaging in competition within the healthcare market across different subnetworks.

Table 2. Relationships among coopetitive networks and cooperative networks.

Network Types	Coopetitive Networks			Cooperative Networks
	Bipartite Networks	Cluster Networks	Group-Bipartite Networks	
Group number	2	more than 2		more than 2
Typical structure	structural balance	in-degree balance	acyclic partition, sign-balanced couple	a directed spanning tree
Intra-cluster communication	continuous, cooperation	continuous, cooperation	continuous, coopetition	continuous, cooperation
Inter-cluster communication	continuous, competition	continuous, coopetition	continuous, coopetition	**discrete**, cooperation
Global behavior	bipartite consensus	cluster consensus	group-bipartite consensus	synchronization

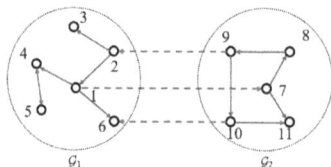

Figure 2. Bipartite networks. Between $\mathcal{G}_1$ and $\mathcal{G}_2$, only competitive relationships exist, while within each subcluster, only cooperative relationships are present [45].

Directed signed graphs are employed to describe this particular type of network.

Definition 1. *Directed signed graph.*

Given a digraph $\mathcal{G} = (\mathcal{V}, \mathcal{E}, \mathcal{A})$, where the set of nodes represents a collection of individuals, denoted as $\mathcal{V} = \{1, \ldots, N\}$; the set of directed edges is denoted as $\mathcal{E} \in \mathcal{V} \times \mathcal{V}$; $\mathcal{A} = [a_{ij}] \in \mathbb{R}^{N \times N}$ stands for the adjacency matrix of the graph, with element a_{ij} representing the strength of the interaction between node i and node j. The sign function $sgn(\bullet)$ is used to represent the coopetitive relationship between node i and node j, that is,

$$sgn(a_{ij}) = \begin{cases} 1, & a_{ij} > 0 \quad \text{friendly and cooperative} \\ 0, & a_{ij} = 0 \quad \text{no connection} \\ -1, & a_{ij} < 0 \quad \text{hostile and competitive} \end{cases}$$

To give the definition of the structural balance of signed graphs, we first introduce the concept of positive and negative cycles. Generally, the cycles in a signed graph contain both positive and negative edges. If the product of the weights a_{ij} of the edges in a cycle is positive, the cycle is called a positive cycle; otherwise, it is called a negative cycle.

Definition 2. *Structurally balanced.*

As presented in [45], if all the cycles in a signed graph $\mathcal{G}$ are positive cycles, then $\mathcal{G}$ is called structurally balanced; if there is at least one negative cycle, then $\mathcal{G}$ is called structurally unbalanced; if there are no cycles in $\mathcal{G}$, then it is called a vacuum-balanced graph. Different balanced structures are given in Figure 3, where the blue solid edges have positive weights, indicating a cooperative relationship; the red dashed edges have negative weights, indicating competitive relationships.

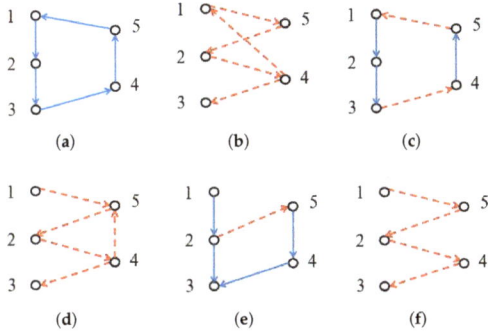

Figure 3. Structural balance, structural unbalance, and vacuum-balance of signed graphs [45]. (**a**) Structurally balanced cooperation network. (**b**) Structurally balanced competition network. (**c**) Structurally balanced coopetitive network. (**d**) Structurally unbalanced competition network. (**e**) Structurally unbalanced coopetitive network. (**f**) Vacuum-balanced competition network.

From a linear algebra perspective, if the signed graph exhibits structural balance, the set of nodes $\mathcal{V}$ can be partitioned into two groups: $\mathcal{V}_1 = \{1, \ldots, N_0\}$ and $\mathcal{V}_2 = \{N_0 + 1, \ldots, N\}$. The relationship is cooperative within subgroups, and competition exists only between two subgroups. **Bipartite consensus** is a typical feature of this type of network, i.e., the individuals in the two groups have the same absolute value of the final state, but opposite signs.

By assigning appropriate numbers to each individual, the adjacency matrix associated with the signed graph $\mathcal{G}$ can be transformed into the following block matrix form.

$$A = \begin{pmatrix} A_{11} & A_{12} \\ A_{21} & A_{22} \end{pmatrix}$$

where $A_{11} \in \mathbb{R}^{N_0 \times N_0}$ and $A_{22} \in \mathbb{R}^{(N-N_0) \times (N-N_0)}$ are non-negative matrices, while $A_{12} \in \mathbb{R}^{N_0 \times (N-N_0)}$ and $A_{21} \in \mathbb{R}^{(N-N_0) \times N_0}$ are non-positive matrices.

The Laplacian matrix $\mathcal{L}$ of a signed graph plays an important role in analyzing the collective behavior evolution of coopetitive systems, and its definition is as follows.

$$\mathcal{L} = \mathcal{D} - \mathcal{A} \tag{1}$$

where $\mathcal{D} = diag\{d_1, \ldots, d_N\}$ denotes the degree matrix, $d_i = \sum_{j \in N_i} |a_{ij}|$ with $N_i = \{j \mid (i,j) \in \mathcal{E}\}$.

2.1.2. Cluster Networks with In-Degree Balance Condition

In practice, systems need to be divided into multiple clusters to accomplish different tasks due to differences in performance and task requirements. Therein, individuals within a group are in a cooperative relationship, while individuals belonging to different subgroups can choose to compete or cooperate with each other [46]. For instance, a strategic alliance model in which companies cooperate and compete with each other in order to share research and development costs, mitigate risks, learn, and acquire complementary resources [22].

In cluster networks, the graph has a partition of the node set $\mathcal{V}$ that takes the form $\{\mathcal{V}_1, \ldots, \mathcal{V}_k\}$ such that $\mathcal{V}_i \neq \varnothing, \cup_{\ell=1}^{k} \mathcal{V}_\ell = \mathcal{V}$ and $\mathcal{V}_i \cap \mathcal{V}_j = \varnothing, i \neq j, i, j \in \{1, 2, \ldots, k\}$. Let $\mathcal{G}_\ell$ denote the underlying topology of cluster $\mathcal{V}_\ell, \ell = 1, \ldots, k$, i.e., $\mathcal{V}_\ell = \mathcal{V}(\mathcal{G}_\ell)$. Without loss of generality, the node set of each cluster can be represented by $\mathcal{V}(\mathcal{G}_\ell) = \left\{\sum_{j=0}^{\ell-1} n_j + 1, \cdots, \sum_{j=0}^{\ell} n_j\right\}, 1 \leq \ell \leq k$, where $n_0 = 0, \sum_{\ell=1}^{k} n_\ell = N$. For convenience, let $\bar{i}$

denote the index of the subset where node i belongs, that is, $i \in \mathcal{V}_{\bar{i}}$. Obviously, $1 \leq \bar{i} \leq k$. Nodes i and j are said to be in the same subgroup if $\bar{i} = \bar{j}$.

Assumption 1. *In-degree balance.*

$$\sum_{j \in \mathcal{V}(G_\ell)} a_{ij} = 0, \quad \forall i = 1, \ldots, N, i \in \mathcal{V} \backslash \mathcal{V}(G_\ell), \ell = 1, \ldots, k.$$

Two types of clustered networks satisfying Assumption 1, which have been studied by many scholars [46], are shown in Figure 4. One is Figure 4a with an acyclic partition structure, with the requirement that information from the former subcluster can be passed on to the latter subcluster, but information from the latter subcluster cannot be passed on to the former subcluster. In contrast, the one shown in Figure 4b is not an acyclic partition but a balanced couple partition, because the two subclusters can pass information to each other. **Cluster consensus** is typical for these examples, i.e., individuals within a group reach the same state while the states of individuals within different groups can end up being different.

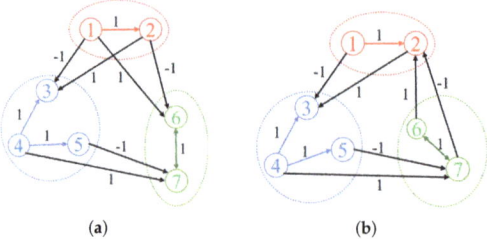

Figure 4. Two types of clustered networks that satisfy the in-degree balance condition. (**a**) Acyclic partition. (**b**) Balanced couple partition.

The Laplacian matrices $\mathcal{L} = [l_{ij}] \in \mathbb{R}^{N \times N}$ corresponding to $\mathcal{G}$ in Figure 4a,b have the following forms, respectively:

$$\mathcal{L} = [l_{ij}] = \begin{bmatrix} \mathcal{L}_{11} & \cdots & \mathbf{0}_{n_1 \times n_k} \\ \vdots & \ddots & \vdots \\ \mathcal{L}_{k1} & \cdots & \mathcal{L}_{kk} \end{bmatrix} \quad (2)$$

$$\mathcal{L} = [l_{ij}] = \begin{bmatrix} \mathcal{L}_{11} & \cdots & \mathcal{L}_{1k} \\ \vdots & \ddots & \vdots \\ \mathcal{L}_{k1} & \cdots & \mathcal{L}_{kk} \end{bmatrix} \quad (3)$$

where $\mathcal{L}_{ii}$ represents the information exchange within subgroup $\mathcal{G}_i$, and $\mathcal{L}_{ij}$ represents the information exchange from subgroup $\mathcal{G}_i$ to subgroup $\mathcal{G}_j$, with $i, j = 1, \ldots, k$.

2.1.3. Group-Bipartite Networks

Apart from occurring between different clusters, competition is possible between agents belonging to the same group in cluster consensus problems (see Figure 5). Consequently, a definition of a group-bipartite network has been proposed in the past year or two. The collective behavior of such networks is termed **group-bipartite consensus**, and can be used to describe some tasks of multi-objective symmetry. For example, in search and rescue missions, UAVs search the area and perform tasks in formations of symmetric patterns [47].

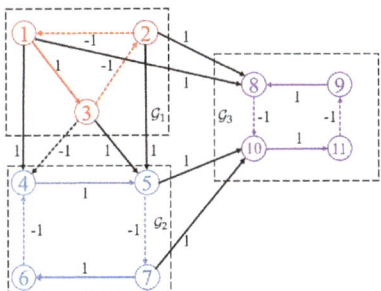

Figure 5. The network topology with an acyclic partition [33].

As described in [33], the group-bipartite network topology combines the structural balance properties of bipartite networks and the acyclic partition approach of cluster networks. Sign-balanced couple and the following assumptions are needed for the group-bipartite networks:

Assumption 2. *Each of $\mathcal{G}_i, i \in \{1, \ldots, k\}$ is structurally balanced.*

Assumption 3. *Each of $\mathcal{G}_i, i \in \{1, \ldots, k\}$ has a spanning tree.*

Under Assumption 2, the node set $\mathcal{V}_i$ of $\mathcal{G}_i (i = 1, \ldots, k)$ can be divided into two groups $\mathcal{V}_i^{(1)}$ and $\mathcal{V}_i^{(2)}$ satisfying $\mathcal{V}_i^{(1)} \cup \mathcal{V}_i^{(2)} = \mathcal{V}_i$, $\mathcal{V}_i^{(1)} \cap \mathcal{V}_i^{(2)} = \varnothing$. Moreover, a diagonal matrix $\Phi_i = diag\{\phi_{\sum_{j=0}^{l-1} n_j + 1}, \ldots, \phi_{\sum_{j=0}^{l} n_j}\}$ is defined, where $\phi_j = 1$ for $j \in \mathcal{V}_i^{(1)}$, and $\phi_j = -1$ for $j \in \mathcal{V}_i^{(2)}$.

On the other hand, since the partition $\{\mathcal{V}_1, \mathcal{V}_2, \ldots, \mathcal{V}_k\}$ is acyclic, a new form of Laplacian matrix $\mathcal{L} = [l_{ij}] \in \mathbb{R}^{N \times N}$ is constructed with a lower block triangular form, as (2), where

$$l_{ij} = \begin{cases} -a_{ij}, & i \neq j, \\ \sum_{j \notin \mathcal{V}_i} \phi_j a_{ij} + \sum_{j \in \mathcal{V}_i} |a_{ij}|, & i = j. \end{cases} \quad (4)$$

The modified Laplacian matrix $\mathcal{L}$ plays an important role in the consensus analysis of group-bipartite networks [33]. As a special case of cluster consensus, the collective behavior of such networks is termed **group-bipartite consensus**, and specifies multiple dual cluster consensus behaviors.

Figure 4 depicts the concept of group-bipartite networks. The node $\mathcal{V}$ in a group-bipartite network topology $\mathcal{G}$ is partitioned into three clusters $\mathcal{G}_1, \mathcal{G}_2$, and $\mathcal{G}_3$ with $\mathcal{V}_1 = \{1, 2, 3\}$, $\mathcal{V}_2 = \{4, 5, 6, 7\}$, and $\mathcal{V}_3 = \{8, 9, 10, 11\}$. It can be seen that coopetition exists within and between clusters. Clearly, each subcluster $\mathcal{G}_i, i = (1, 2, 3)$ is satisfied with the structural balance condition. Then, it can be obtained that $\phi_i = \begin{cases} 1, & i = 1, 3, 4, 5, 8, 9 \\ -1, & i = 2, 6, 7, 10, 11 \end{cases}$.

According to the definition of (4), $l_{44} = 0, l_{ii} = 1, i = (1, 2, 3, 5, 6, 7, 8, 9, 10, 11)$. Then, the corresponding $\mathcal{L}$ is

$$\mathcal{L} = \begin{pmatrix} 1 & 1 & 0 & 0 & 0 & 0 & 0 & 0 & 0 & 0 & 0 \\ 0 & 1 & 1 & 0 & 0 & 0 & 0 & 0 & 0 & 0 & 0 \\ -1 & 0 & 1 & 0 & 0 & 0 & 0 & 0 & 0 & 0 & 0 \\ -1 & 0 & 1 & 0 & 0 & 1 & 0 & 0 & 0 & 0 & 0 \\ 0 & -1 & -1 & -1 & 1 & 0 & 0 & 0 & 0 & 0 & 0 \\ 0 & 0 & 0 & 0 & 0 & 1 & -1 & 0 & 0 & 0 & 0 \\ 0 & 0 & 0 & 0 & 1 & 0 & 1 & 0 & 0 & 0 & 0 \\ -1 & -1 & 0 & 0 & 0 & 0 & 0 & 1 & -1 & 0 & 0 \\ 0 & 0 & 0 & 0 & 0 & 0 & 0 & 0 & 1 & 0 & 1 \\ 0 & 0 & 0 & 0 & -1 & 0 & -1 & 1 & 0 & 1 & 0 \\ 0 & 0 & 0 & 0 & 0 & 0 & 0 & 0 & 0 & -1 & 1 \end{pmatrix} \quad (5)$$

2.2. Cooperative Networks

In the subsection, an intermittent clustered network characterized only by cooperative connections is described. This network utilizes a combination of continuous and discrete communication mechanisms to facilitate interaction among agents. Specifically, continuous communication is used within clusters, and inter-cluster communication is discrete. Examples encompass spatially clustered robots, which exhibit a unique set of characteristics [48]. These robots are organized into clusters, wherein continuous interactions persist among the members of each cluster. However, owing to limitations posed by energy and communication resources, long-distance interactions are restricted to discrete exchanges.

Consider a clustered network $\mathcal{G} = (\mathcal{V}, \mathcal{E})$ modeled as a non-empty union consisting of several independently connected subnetworks or clusters $\mathcal{G}_k = (\mathcal{V}_k, \mathcal{E}_k)$ such that $\bigcup_{k=1}^{m} \mathcal{V}_k = \mathcal{V}$, $\mathcal{V}_k \cap \mathcal{V}_\tau = \emptyset$ for all $k, \tau \in \{1, \ldots, m\}, \tau \neq k$. Each subcluster $\mathcal{G}_k$, $k \in \{1, \ldots, m\}$ contains a specific agent called a leader $\ell_k \in \mathcal{V}_k$, and the remaining ones are called followers $f_\tau \in \mathcal{V}_k / \ell_k$. The instantaneous communication between clusters is executed by the leader, with its associated communication network represented as $\mathcal{G}_\ell = (\mathcal{I}, \mathcal{E}_\ell)$, where $\mathcal{I} = \{\ell_1, \ell_2, \ldots, \ell_m\}$ and $\mathcal{E}_\ell = \{e_{\ell(ij)} \mid (i,j) \in \mathcal{I} \times \mathcal{I}\}$. Correspondingly, the Laplacian matrices of $\mathcal{G}_k$ and $\mathcal{G}_\ell$ are denoted by $\mathcal{L}_k$ and $\mathcal{L}_\ell$, respectively. For simplicity, the first agent in each cluster is the leader. Thus, the node set of the cluster $\mathcal{G}_k, k \in \{1, \ldots, m\}$ is given by

$$\mathcal{V}_k = \{\ell_k, f_{o_{k-1}+2}, \ldots, f_{o_k}\}, \quad (6)$$

where $o_0 = 0$ and $o_m = N$. The cardinality of $\mathcal{G}_k$ is represented by $\|\mathcal{V}_k\| = n_k = o_k - o_{k-1}, \forall k \geq 1$, and $\sum_{k=1}^{m} n_k = N$.

Under this network topology, a global Laplace matrix is defined as follows:

$$\mathcal{L} = \begin{bmatrix} \mathcal{L}_1 & \cdots & 0_{n_1 \times n_k} \\ \vdots & \ddots & \vdots \\ 0_{n_k \times n_1} & \cdots & \mathcal{L}_k \end{bmatrix} \quad (7)$$

The objective of such cluster networks with continuous–discrete communications is **synchronization**. The following assumptions are required for intermittent clustered networks:

Assumption 4. *The subcluster $\mathcal{G}_k, k \in \{1, \ldots, m\}$ is strongly connected.*

Assumption 5. *The communication topology $\mathcal{G}_\ell$ formed between the leaders contains a directed spanning tree.*

To illustrate the notation (6), Figure 6 illustrates a CMAS that consists of seven agents grouped into two clusters. Agents 1 and 5 are the leaders in each cluster. Moreover, the nodes in $\mathcal{G}_k$ and $\mathcal{G}_\ell$ are represented as $\mathcal{V}_1 = \{\ell_1, f_2, f_3, f_4\}$, $\mathcal{V}_2 = \{\ell_2, f_6, f_7\}$, and $\mathcal{G}_\ell = \{\ell_1, \ell_2\}$. Then, the corresponding $\mathcal{L}$ is

$$\mathcal{L} = \begin{pmatrix} 2 & 0 & -1 & -1 & 0 & 0 & 0 \\ -1 & 2 & -1 & 0 & 0 & 0 & 0 \\ -1 & -1 & 2 & 0 & 0 & 0 & 0 \\ -1 & 0 & -1 & 2 & 0 & 0 & 0 \\ 0 & 0 & 0 & 0 & 2 & -1 & -1 \\ 0 & 0 & 0 & 0 & -1 & 1 & 0 \\ 0 & 0 & 0 & 0 & -1 & -1 & 2 \end{pmatrix} \tag{8}$$

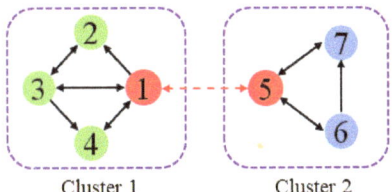

Figure 6. The intermittent clustered network of seven agents [40].

3. Distributed Control of Coopetitive Networks

This section focuses on the development of collective emergent behavior in coopetitive networks, including bipartite consensus, cluster consensus, and group-bipartite consensus. We review the subjects and issues pertaining to these consensuses that have been explored in recent years.

3.1. Bipartite Consensus

In 2012, Altafini et al. [49] first proposed a dynamic model of coopetitive networks based on a linear Laplace feedback design. It demonstrated that under the structural balance assumption, individuals form two groups with diametrically opposite final states, i.e., the absolute values of the individuals' final states are the same, but the signs are opposite. Such a dynamical behavior is referred to as **bipartite consensus**.

Hu et al. [45] investigated the modeling of the coexistence of cooperation and competition in social networks, along with the collective behaviors, under this modeling framework. Considering the influence of structural balance properties, This paper employed directed signed graphs to describe the interaction network of coopetitive relationships. The dynamics of each agent is modeled as

$$\dot{x}_i(t) = \sum_{j \in N_i} a_{ij}[x_j(t) - \text{sgn}(a_{ij})x_i(t)], i = 1, \ldots n \tag{9}$$

where the sign function $sign(a_{ij})$ equals 1 or -1, indicating the coopetitive relationship between agents. The above equation can be written in the following equivalent form:

$$\dot{x}(t) = -\mathcal{L}x(t) \tag{10}$$

where $x(t) = col(x_1(t), \ldots, x_N(t))$ is the state vector of all individuals, and $\mathcal{L}$ is the Laplacian matrix defined in (1). Three emergent collective behaviors (consensus, polarization or bipartite consensus, and fragmentation) were explored.

Case 1 (consensus): When the network is a purely cooperative network containing a spanning tree in Figure 3a, the state limits of all individuals satisfy $\lim_{t \to \infty} \|x_i(t) - x_j(t)\| = 0$. Figure 7 depicts the state evolution on this network.

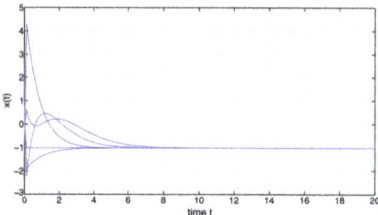

Figure 7. Consensus on cooperative networks.

Case 2 (bipartite consensus): When the network is Figure 3b, Figure 3c, or Figure 3f, i.e., the network is a vacuum-balanced or structurally balanced graph containing spanning trees, the final state of all individuals exhibits a bipartite consensus, that is, $\lim_{t\to\infty}\|x_i(t) - \mathrm{sgn}(a_{ij})x_j(t)\| = 0$. In Figure 8, the seven individuals eventually appear polarized, forming two limit states, 1 and -1.

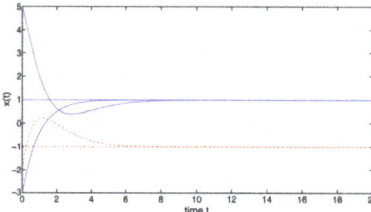

Figure 8. Bipartite consensus on structural balanced networks.

Case 3 (fragmentation): When the coopetitive network is a structurally unbalanced network, as shown in Figure 3d or Figure 3e, then the state of groups eventually splits, i.e., the states of all individuals converge to more than two limit states. Figure 9 depicts the state evolution on a structurally unbalanced network, where the blue solid lines and the red dotted lines respectively denote the trajectories of agents from two competitive subgroups. The three limit states are 1, 0, and -1.

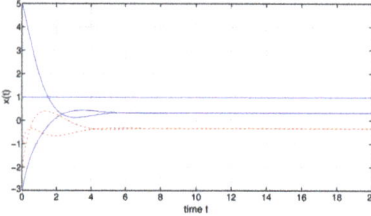

Figure 9. Group splits on structurally unbalanced networks.

From the above simulation results, it can be observed that for a coopetitive network, the structural balance condition is a necessary requirement to achieve bipartite consensus.

Immediately afterward, Proskurnikov et al. [50] extended Altafini's model for opinion dynamics in social networks and discussed the modulus consensus problem on time-varying directed signed graphs. This paper studied linear and nonlinear consensus protocols within a common framework, relaxing the constraints of strong connectivity and structural balance.

Ma et al. [51] investigated the bipartite consensus problem of first-order MASs on fixed directed graphs in the presence of measurement noise. This paper uncovered that even in the presence of measurement noise, achieving bipartite consensus on a signed graph with structural balance and a spanning tree was the minimum connectivity assumption required. Moreover, when the signed graph was structurally unbalanced, the state of the closed-loop system can achieve mean square convergence to zero as long as certain mild conditions are satisfied.

Guo et al. [52] discussed the bipartite consensus problem in MASs on directed signed graphs. This paper demonstrates the achievability of bipartite consensus on directed graphs with communication delays and strongly connected structurally balanced conditions. Additionally, the paper considered the case of structural balance, and proposed a method to pin a specific agent so that the system can achieve pinning bipartite consensus.

Shao et al. [53] investigated the bipartite consensus problem of MASs with second-order discrete-time dynamics under asynchronous settings. This paper took into account switching topologies and established an asynchronous distributed control protocol, while sufficient conditions were provided for achieving asynchronous bipartite consensus.

Recently, Wu et al. have worked constructively on the bipartite consensus of coopetitive networks with unknown disturbances or communication noise in [54–56].

In [54], the interventional bipartite consensus was addressed for high-order MASs with nonlinear unknown time-varying disturbances. Several adaptive estimators based on neural networks were proposed to estimate the nonlinear disturbances. The convergence of bipartite consensus was analyzed using the Lyapunov function method.

In [55], the mean square bipartite consensus problem was investigated in high-order MASs, incorporating modeling of coopetition interactions and communication noise. This paper proposed a distributed control strategy that was independent of the agent state matrix, and developed a novel randomized approximation strategy utilizing relative state information to suppress communication noise.

In [56], the bipartite consensus problem was explored for high-order MASs, considering scenarios with and without external systems (i.e., leaders). By designing distributed adaptive control laws and utilizing the linear parameterization approach to describe the time-varying characteristics of unknown disturbances, two fully distributed control strategies were proposed that did not rely on any global information. These strategies guaranteed the achievement of bipartite consensus MASs.

In addition, the output bipartite consensus of heterogeneous MASs was also discussed in [57,58].

Liang et al. [57] investigated the bipartite output synchronization problem for heterogeneous MASs with time-varying communication networks. This paper proposed a novel edge-based adaptive output feedback control strategy. By utilizing a sophisticated Lyapunov function, the convergence of the closed-loop system was analyzed.

Wu et al. [58] discussed the output bipartite consensus problem for heterogeneous MASs and further extended the results of [57]. In light of the unavailability of state information for leaders and unknown system matrices, new distributed estimators were designed for each of them.

The existing research works on bipartite consensus, along with additional details, are summarized in Table 3. Currently, there are some constraints on bipartite consensus, such as structural balance, spanning tree, or joint spanning tree. However, in the real world, the relationship between cooperation and competition can be very complex, and the network topology may not necessarily meet these conditions. Therefore, studying the dynamics of MASs on coopetition networks without structural constraints is of significant practical and theoretical importance.

3.2. Cluster Consensus

In numerous engineering applications, it is often necessary to classify individuals within MASs into multiple groups based on physical attributes or assigned tasks. Each of these groups is commonly referred to as a cluster. In response to this scenario, researchers have proposed a broader concept known as cluster consensus. In cluster consensus, individuals belonging to the same cluster are required to converge to the same constant value, while different clusters may not coincide with each other.

Table 3. A survey of the research undertaken on the bipartite consensus problem.

Reference	Year	Dynamics	Network Topology	Contribution
[49]	2012	First-order integrator	Static signed graphs	Giving the concept of bipartite consensus
[45]	2014	General linear dynamics	Directed signed graphs	Delving into three emergent collective behaviors (consensus, bipartite consensus, and fragmentation)
[50]	2015	First-order integrator	Time-varying topology	Extending Altafini's model into time-varying directed signed graphs
[51]	2017	First-order integrator	Fixed signed digraphs	Considering bipartite consensus under measurement noise
[52]	2018	First-order integrator	Directed signed graphs	Investigating pinning bipartite consensus under communication delays
[53]	2018	Second-order integrator	Switching topologies	Studying bipartite consensus under switching topologies
[54]	2016	High-order dynamics	Directed signed graphs	Addressing the interventional bipartite consensus with unknown disturbances
[55]	2018	General linear dynamics	Directed signed graphs	Investigating the mean square bipartite consensus with communication noise
[56]	2019	High-order dynamics	Directed signed graphs	Proposing fully distributed adaptive control laws
[57]	2020	Heterogeneous dynamics	Directed signed graphs	Proposing a novel edge-based adaptive output feedback control strategy
[58]	2023	Heterogeneous dynamics	Directed signed graphs	Designing the new distributed estimators for leader's unavailable information and unknown system matrix

In 2009, Yu and Wang [59] studied the cluster consensus problem of a first-order continuous MAS with two groups. In the scenario where information exchange was undirected, a novel consensus protocol was proposed to tackle the challenge of achieving group consensus. Several convergence conditions were established by leveraging principles from graph theory and matrix theory.

Subsequently, Yu et al. [60] incorporated topology switching and communication delays into a directed network and examined the problem of group consensus in MASs by introducing a double tree-form transformation. Certain necessary and/or sufficient conditions were derived for achieving group consensus.

Chen et al. [61] studied a first-order discrete-time CMAS with fixed and switching topologies, presenting a partitioning algorithm for strongly connected directed graphs. Then, by relying on the non-negative matrix analysis and Markov chain, two necessary conditions were presented for achieving cluster consensus. This paper can provide a response to the issues of determining clustering and ensuring consensus in MASs.

Thereafter, the group consensus problem with two groups was extended to a more general group consensus problem in [62]. The concepts of in-degree balance, out-degree balance, and balance pair were first proposed, and the condition of balance pair was used to restrict the coupling interaction between the clusters. These pioneering research works have attracted the attention of, and sparked discussions among, numerous scholars regarding cluster consensus with multiple clusters.

In [46], Qin et al. addressed the cluster consensus problem for linear MASs using an acyclic partitioning approach to rearrange all directed edges in acyclic directed graphs, and gave the following distributed feedback control protocol:

$$u_i(t) = K \left[\sum_{j \in N_i} a_{ij}(x_j(t) - x_i(t)) + d_i(s_{\bar{i}}(t) - x_i(t)) \right] \quad (11)$$

where K is the feedback matrix that needs to be designed, and $s_i, i = 1, \ldots, p$ are p particular solutions of a homogeneous system $\dot{s}(t) = As(t)$, such that $\lim_{t \to \infty} \|s_i(t) - s_j(t)\| \neq 0$, where $d_i > 0$ when agent i is pinned by s_i, otherwise $d_i = 0$. The results showed that for acyclic directed graphs, regardless of the intra- and inter-cluster coupling strength, the cluster consensus for general linear MASs can be achieved by designing the feedback control matrix K.

Later on, Yu et al. [63] introduced another partition structure for coopetitive networks, i.e., balanced coupled partition, to investigate the cluster consensus problem for linear MASs under pinning control. The results indicated that for fixed topology networks, if each cluster contained a directed spanning tree and the intra-cluster coupling strength was strong enough compared to the inter-cluster coupling strength, it was easy to design a feedback controller to make the system achieve cluster consensus; for switching topology networks, the necessary conditions to achieve cluster consensus were given.

Ma et al. [64] examined the consensus problem in first-order multi-agent systems (MASs) with fixed and directed topologies, taking into account time delays. This study introduced the novel concept of cluster-delay consensus. By employing graph theory, Lyapunov stability analysis, and matrix theory, the authors derived sufficient conditions to ensure the maintenance of cluster-delay consensus in MASs.

Chen et al. [65] studied the cluster consensus problem of MASs with heterogeneous dynamics. The issue was addressed by employing output regulation techniques and constructing state feedback controllers. This paper can be considered to extend the previous research that focused on the standard consensus problem in homogeneous MASs, allowing for different dynamics of agents in heterogeneous MASs.

Dong et al. [66] investigated the problem of cluster consensus for general linear MASs and proposed a novel intermittent output control strategy that enables effective cluster consensus control under nonperiodic operation. This paper provided a new approach and direction for addressing the problem of cluster consensus in MASs.

A summary of the existing research work on the cluster consensus problem is presented in Table 4. Cluster consensus control can enhance the scalability and availability of a system. However, it may also introduce increased complexity and latency, potentially leading to reduced performance. Therefore, in practical applications, it is necessary to carefully weigh the pros and cons and select an appropriate consensus control method based on specific circumstances.

Table 4. A survey of the research undertaken on the cluster consensus problem.

Reference	Year	Dynamics	Conditions	Contribution
[59]	2009	First-order integrator	Undirected topology	Solving the cluster consensus with two groups
[60]	2010	First-order integrator	Directed topology	Examining the problem of group consensus incorporated topology switching and communication delays
[61]	2011	First-order integrator	Fixed and switching topology	Proposing a cluster factorization algorithm for directed graphs
[62]	2012	First-order integrator	Directed topology	Studying the group consensus problem with multiple groups; proposing the concept of in-degree balance
[46]	2013	General linear dynamics	Fixed and switching topology	Proposing a acyclic partition; investigating the correlation between cluster consensus behavior and the coupling strength among agents
[63]	2014	General linear dynamics	Fixed and switching topology	Proposing a balanced coupled partition; giving the necessary conditions for achieving cluster consensus
[64]	2016	First-order integrator	Directed topology	Examining the cluster-delay consensus problem for nonlinear dynamics MASs
[65]	2017	Heterogeneous dynamics	Directed topology	Employing output regulation techniques to solve the cluster consensus problem
[66]	2022	General linear dynamics	Directed topology	Proposing a novel intermittent output control strategy in the cluster consensus problem

3.3. Group-Bipartite Consensus

Given the potential for competition between agents of the same group, a special class of cluster consensus, group-bipartite consensus has emerged over the past two years. The group-bipartite consensus is expected to accomplish multi-objective symmetric tasks, such as the formation of multiple symmetric shapes by UAVs simultaneously [67].

Liu et al. [33] combined group consensus and bipartite consensus, introduced the concept of group-bipartite consensus, and proposed a distributed control protocol to solve the group-bipartite consensus problem for first-order continuous MASs. The design protocol was as follows:

$$u_i(t) = \sum_{j \in \mathcal{V}_{\bar{i}}} a_{ij}[x_j(t) - \text{sgn}(a_{ij})x_i(t)] + \sum_{j \notin \mathcal{V}_{\bar{i}}} a_{ij}[x_j(t) - \phi_j x_i(t)] \quad (12)$$

The above equation can be written in the following equivalent form:

$$\dot{x}(t) = -\mathcal{L}x(t) \quad (13)$$

where $x(t) = col(x_1(t), \ldots, x_N(t))$ is the state vector of all individuals, and $\mathcal{L}$ is a new Laplacian matrix established to facilitate the control implementation and is defined in (4).

When the coopetitive network is a group-bipartite network, as shown in Figure 4, the trajectories of all agents satisfy the definition of group-bipartite consensus, that is, $\lim_{t \to \infty} \|x_i(t) - \alpha_{\bar{i}}\| = 0$ for $i \in \mathcal{V}_{\bar{i}}^{(1)}$ and $\lim_{t \to \infty} \|x_i(t) + \alpha_{\bar{i}}\| = 0$ for $i \in \mathcal{V}_{\bar{i}}^{(2)}$, where $i = 1, \ldots, N$ and $\alpha_{\bar{i}}$ are k constants. Figure 10 depicts that the trajectories of all agents gradually converge to a triple-bipartite final convergence state under the control protocol (12). Compared to Figure 8, Figure 10 introduces bipartite consensus as a foundation for a single group, and incorporates multiple groups to achieve bipartite consensus across multiple groups. This implies that group consensus and bipartite consensus are special cases of group-bipartite consensus.

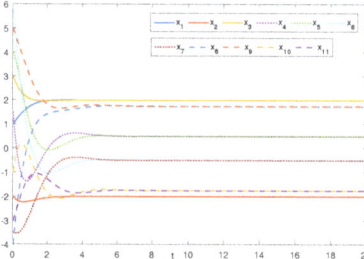

Figure 10. State evolution of the agents [33].

The group-bipartite consensus problem in a heterogeneous MAS composed of first-order integrators and second-order integrators was explored in [68]. By leveraging graph theory and Lyapunov stability methods, sufficient conditions and corresponding consensus protocols were derived for heterogeneous MASs under undirected communication topologies. In contrast to existing work [33], this paper eliminated the constraint of network topology, such as acyclic partitions. However, it only analyzed bipartite networks with two groups.

Thereafter, Liu et al. [69] discussed the problem of oscillatory group-bipartite consensus control in swarm robots with multiple oscillatory leaders. This paper modeled the robotic cluster using Euler–Lagrange equations and verified the proposed control method through two simulations conducted on two typical group-bipartite network topologies.

Wang et al. [67] investigated the problem of coordinated task control in a swarm of robots by introducing the concept of group-bipartite consensus in networked Euler–Lagrange systems. By utilizing the structure of acyclic partition network topologies, they proposed a static group-bipartite consensus control protocol and established geometric criteria to guarantee the achievement of multiple symmetric consensus in networked robot systems.

Recently, considering that most systems in reality are nonlinear, Lu et al. [70] investigated the finite-time problem of second-order nonlinear MASs with leaders. They generalized the system to a nonlinear form in order to achieve finite-time group-bipartite consensus. This paper constrained the nonlinear functions in the system with a semi-Lipschitz condition, which is derived from fundamental inequalities and the Lipschitz condition.

Table 5 presents a comprehensive overview of references related to the group-bipartite problem from the preceding three years. Although group-bipartite consensus shows potential in handling multi-objective symmetric tasks, the current research fails to consider the impact of various practical factors, such as communication delays, unobservable states, and unknown perturbations. Therefore, in order to bridge the gap between the inherent properties or constraints of consensus algorithms and practical models, further investigation is needed on consensus problems that involve these practical factors.

Table 5. A survey of the research undertaken on the group-bipartite consensus problem.

Reference	Year	Dynamics	Typical Structure	Contribution
[33]	2020	First-order integrator	Acyclic partition, sign-balanced couple	Giving the concept of group-bipartite consensus
[68]	2022	Heterogeneous dynamics	Structurally balanced	Solving the group-bipartite consensus problem in heterogeneous CMASs
[69]	2022	Euler–Lagrange systems	Acyclic partition, sign-balanced couple	Applying oscillatory group-bipartite consensus to swarm robots
[67]	2023	Networked robot systems	Acyclic partition	Proposing a static group-bipartite consensus control protocol
[70]	2023	Double integrator	Structurally balanced	Studying the finite-time group-bipartite consensus for nonlinear systems

4. Cooperative Control over Intermittent Clustered Networks

Although cooperation and competition are prevalent in nature and human society, cooperation is also widespread. Bragagnolo et al. [35] first proposed a cooperative clustered network with continuous intra-cluster communication and discrete inter-cluster communication. Two fundamental challenges can be posed for this kind of intermittent clustered network:

- **Problem 1:** Design the distributed consensus protocols such that the hybrid continuous–discrete CMASs can achieve global consensus behavior.
- **Problem 2:** Determine the characterization of the global consensus value.

The single-integrator dynamic model was expressed as (Bragagnolo et al. [35])

$$\begin{cases} \dot{x}(t) = -\mathcal{L}x(t), & \forall t \in \mathbb{R}_+ \setminus \mathcal{T} \\ x_l(t_k) = P_l x_l(t_k^-) & \forall t_k \in \mathcal{T} \\ x(0) = x_0 \end{cases} \tag{14}$$

where $x_l(t)$ is the set of leader states, $P_l \in \mathbb{R}^{m \times m}$ is a row stochastic matrix associated with $\mathcal{G}_\ell$, $\mathcal{T} = \{t_k \in \mathbb{R}_+ | t_k < t_{k+1}, \forall k \in \mathbb{N}, t_k \text{ reset time}\}$, and $\mathcal{L}$ is the Laplacian matrix defined in (7).

To address the first question, Bragagnolo et al. [35] proposed a quasi-periodic reset strategy and provided LMI conditions to ensure the global consensus index stability of subnetworks represented by directed and strongly connected graphs. As for the second question, the global consensus value depended on the initial conditions and the topology of the networks involved, including networks associated with clusters and networks associated with leaders. It is worth noting that the consensus value is independent of the reset sequence used for the leader state.

The simulation results in Figure 11 demonstrate that the leader's trajectory was non-smooth while the follower's trajectory was smooth, and the phenomenon of jumps at the moment of reset had no effect on the calculated consensus.

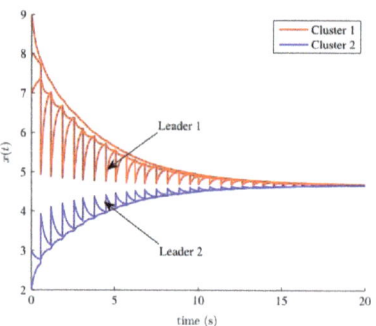

Figure 11. The state trajectories of the agents converging to the calculated consensus value [35].

Morarescu et al. [36] investigated the reset control problem for first-order time-varying CMASs. In contrast to [35], this study also took into account the possibility that the reset time could be triggered by certain events.

The consensus problem in cluster networks with general linear MASs is more challenging than the case of integrators, as pointed out by Pham et al. [37,38] and Wang et al. [40]:

$$\begin{cases} \dot{x}_i = Ax_i + Bu_i, \\ y_i = Cx_i, \end{cases} \quad t \in (t_{k-1}, t_k) \quad (15)$$

where $x_i \in \mathbb{R}^n$, $u_i \in \mathbb{R}^p$, and $y_i \in \mathbb{R}^q$ are the state, input, and output of the i-th $(i = 1, \ldots, N)$ agent, respectively.

To achieve global state consensus for (15), Pham et al. [37] proposed a reset state feedback control strategy using relative state measurements. Subsequently, due to physical or economical constraints in real systems, it is not possible to measure all states with sensors, so the states need to be observed. In [38], an output feedback control protocol based on an impulsive observer, more precisely a full-order state observer, was designed for linear CMASs:

$$\begin{cases} \dot{\hat{x}}_i = A\hat{x}_i + Bu_i + H(\hat{y}_i - y_i) + qL\xi_i, \\ \hat{y}_i = c\hat{x}_i, \\ u_i = pK\sum_{j=1}^{N} a_{ij}(\hat{x}_j - \hat{x}_i), \end{cases} \quad t \in (t_{k-1}, t_k) \quad (16)$$

where $\hat{x}_i \in \mathbb{R}^n$ is the observer state, $\hat{y}_i \in \mathbb{R}^n$ is the output of the observer, $H, L \in \mathbb{R}^{n \times q}$, $K \in \mathbb{R}^{p \times n}$ are the gain matrices, $p > 0, q > 0$ are the coupling gains, and ξ_i is the relative output measurement of the i-th agent defined by $\xi_i = \sum_{j=1}^{N} a_{ij}[(\hat{y}_j - \hat{y}_i) - (y_j - y_i)]$.

However, at the reset time, when it came to state updates between clusters, the state information that was not measurable at the prior time was exploited:

$$\begin{cases} x_{\ell_i}(t_k) = \sum_{j=1}^{m}(P_{l(ij)} \otimes I_n)x_{\ell_j}(t_k^-), x_{f_\tau}(t_k) = x_{f_\tau}(t_k^-), \\ \hat{x}_{\ell_i}(t_k) = \sum_{j=1}^{m}(P_{l(ij)} \otimes I_n)\hat{x}_{\ell_j}(t_k^-), \hat{x}_{f_\tau}(t_k) = \hat{x}_{f_\tau}(t_k^-), \end{cases} \quad t = t_k \quad (17)$$

where $x_{\ell_i}(t_k), x_{f_\tau}(t_k)$ and $\hat{x}_{\ell_i}(t_k), \hat{x}_{f_\tau}(t_k)$ represent the states and observer states of leaders and followers at time t_k, respectively.

To avoid such irrationality, Wang et al. [40] proposed a reset reduced-order observer-based output feedback control strategy as follows:

$$\begin{cases} \dot{v}_i = Fv_i + Gy_i + Hu_i, \\ u_i = cKQ_1\sum_{j=1}^{N}a_{ij}(y_i - y_j) + cKQ_2\sum_{j=1}^{N}a_{ij}(v_i - v_j), \end{cases} \quad t \in (t_{k-1}, t_k) \quad (18)$$

and

$$\begin{cases} y_{\ell_i}(t_k) = \sum_{j=1}^{m}(P_{l(ij)} \otimes I_q)y_{\ell_j}(t_k^-), y_{f_\tau}(t_k) = y_{f_\tau}(t_k^-), \\ v_{\ell_i}(t_k) = \sum_{j=1}^{m}(P_{l(ij)} \otimes I_{n-q})v_{\ell_j}(t_k^-), v_{f_\tau}(t_k) = v_{f_\tau}(t_k^-), \end{cases} \quad t = t_k \quad (19)$$

where $v_i \in \mathbb{R}^{n-q}$ is the observer state, $c > 0$ is the coupling strength, and $P_l \in \mathbb{R}^{m \times m}$ is a row stochastic matrix associated with $\mathcal{G}_\ell$. $F \in \mathbb{R}^{(n-q) \times (n-q)}$, $K \in \mathbb{R}^{p \times n}$, $G \in \mathbb{R}^{(n-q) \times q}$, $H \in \mathbb{R}^{(n-q) \times p}$, $Q_1 \in \mathbb{R}^{n \times q}$, and $Q_2 \in \mathbb{R}^{n \times (n-q)}$ are constant matrices designed according to Algorithm 1 in [40] below.

Algorithm 1: Output feedback control algorithm of the homogeneous MAS

Step 1: Hurwitz matrix F is selected to make its eigenvalues different from those of $\hat{A}$, where $\hat{A} = CAC^+$. Select the matrix G so that (F, G) is controllable.

Step 2: Find the unique solution T to the Sylvester equation $TC^+CA - FT = GC$, which satisfies that $\begin{bmatrix} C \\ T \end{bmatrix}$ is non-singular. If $\begin{bmatrix} C \\ T \end{bmatrix}$ is singular, return to step 1 and select G again, until it is non-singular. Then, $H = TC^+\hat{B}$ with $\hat{B} = CB$. Compute matrices Q_1 and Q_2 using $\begin{bmatrix} Q_1 & Q_2 \end{bmatrix} = \begin{bmatrix} C \\ T \end{bmatrix}^{-1}$.

Step 3: For a given positive-definite matrix $\hat{Q}$, solve the Riccati equation $\hat{A}^T P + P\hat{A} - P\hat{B}\hat{B}^T P = -\hat{Q}$ to obtain a positive-definite matrix P such that $\hat{K} = -\hat{B}^T P$.

Step 4: Select the coupling strength $c \geq \frac{1}{2\min_{\lambda_{\kappa,i} \neq 0}\{\text{Re}(\lambda_{\kappa,i})\}}$, where $\lambda_{\kappa,i}$ is the i-th nonzero eigenvalue of the Laplacian matrix L_κ.

In Algorithm 1, it is worth noting that the Hurwitz matrix F and the matrix T satisfying the Sylvester equation play the critical roles in solving the global output consensus value $\tilde{y}^*(t)$ in steps 1 and 2, respectively. In addition, the solution of the Riccati equation in step 3 and the determination of the coupling strength c in step 4 are sufficient conditions for the synchronization of the system described by Equation (15).

Regarding the first problem, due to the unavailability of states, the problem addressed by Wang et al. [40] was no longer the state consensus but the output consensus. To emphasize the second problem, as demonstrated by Theorem 3 in [40], the global output consensus value $\tilde{y}^*(t)$ was

$$\tilde{y}^*(t) - \frac{e^{\hat{A}t}(\phi^T Q \otimes I_q)Cx(0)}{\sum_{i=1}^{m}\phi_i} = 0, \quad \text{as } t \to \infty \quad (20)$$

Based on the simulation results, when the cooperative network is illustrated in Figure 5, the output convergence $y_i(t), i = 1, \ldots, 7$ of CMASs under Algorithm 1 and the reset reduced-order-based protocol (18)–(19) are displayed in Figure 11. It is evident from the results that the evolution of the two leaders abruptly changes at the reset time, whereas the followers exhibit a smooth evolution, ultimately achieving output consensus. Moreover, the consensus value computed by (20) is $\tilde{y}^* = \begin{bmatrix} -0.94 \\ -0.94 \end{bmatrix}$. It can be seen that the convergence of the system does not suffer from the chattering phenomenon during the reset time. From Figure 12, it is evident that the reduced-order observer, as compared to the reset full-order observer protocols in [40], exhibits a longer convergence time. This can be attributed to the reduction in computational redundancy and storage capacity in the system, which comes at the expense of a longer convergence time.

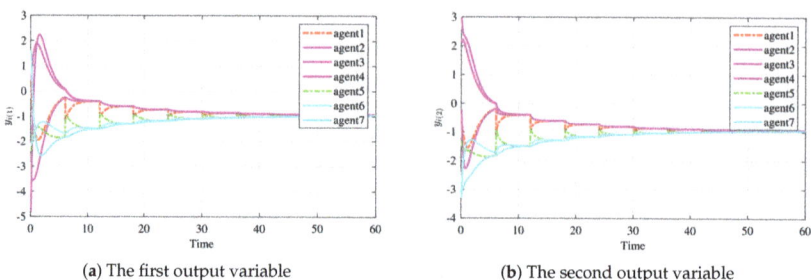

(a) The first output variable (b) The second output variable

Figure 12. Trajectories of output variables $y_i(t)$ with reset reduced-order observer [40].

Most recently, Wang et al. [42] investigated the output synchronization problem for heterogeneous CMASs with hybrid continuous–discrete dynamics as follows:

$$\begin{cases} \begin{cases} \dot{x}_i = A_i x_i + B_i u_i, \\ y_i = C_i x_i \end{cases} \quad t \in [t_{k-1}, t_k) \\ y_{\ell_i}(t_k) = \sum_{j=1}^m (P_{l(ij)} \otimes I_q) y_{\ell_j}(t_k^-), y_{f_\tau}(t_k) = y_{f_\tau}(t_k^-), \quad t = t_k \end{cases} \quad (21)$$

where $x_i \in \mathbb{R}^{n_i}$, $u_i \in \mathbb{R}^{p_i}$, and $y_i \in \mathbb{R}^q$ are the state, input, and output of the i-th ($i = 1, \ldots, N$) agent, respectively. Here $y_{\ell_i}(t_k)$, $y_{f_\tau}(t_k)$, and P_l have the same meanings as in (17). A_i, B_i, and C_i denote constant matrices with compatible dimensions and satisfy the following assumption.

Assumption 6. (A_i, B_i) *is stabilizable,* (A_i, C_i) *is detectable, and* C_i *is full rank with rank* q.

To address the output synchronization problem in heterogeneous CMASs, the ideas in [42] were divided into three steps:

The first step was to artificially create an internal reference model for each agent (as shown in Figure 13, with the same physical network structure as the heterogeneous CMASs) and design a reduced-order observer-based reset output feedback controller based on reset internal models as follows:

$$\begin{cases} \dot{\xi}_i = S\xi_i + K \sum_{j=1}^N a_{ij}(\xi_i(t) - \xi_j(t)), \\ \hat{y}_i = D\xi_i, \\ \dot{v}_i = F_i v_i + G_i y_i + H_i u_i, \\ u_i = K_{i1} Q_{i1} y_i + K_{i1} Q_{i2} v_i + K_{i2} \xi_i \end{cases} \quad t \in [t_{k-1}, t_k) \quad (22)$$

and at the reset time t_k,

$$\begin{cases} \xi_{\ell_i}(t_k) = \sum_{j=1}^m P_{l(ij)} \xi_{\ell_j}(t_k^-), \quad \xi_{f_\tau}(t_k) = \xi_{f_\tau}(t_k^-), \\ v_{\ell_i}(t_k) = v_{\ell_i}(t_k^-) + T_{\ell_i} C_{\ell_i}^+ [\sum_{j=1}^m P_{l(ij)} y_{\ell_j}(t_k^-) - y_{\ell_i}(t_k^-)], \\ v_{f_\tau}(t_k) = v_{f_\tau}(t_k^-), \end{cases} \quad (23)$$

where $\xi_i(t) \in \mathbb{R}^r$ and $\hat{y}_i(t) \in \mathbb{R}^q$ represent the internal reference model states and outputs, respectively. $v_i(t) \in \mathbb{R}^{n_i - q}$ denotes the reduced-order-observer states. $S \in \mathbb{R}^{r \times r}$ and $D \in \mathbb{R}^{q \times r}$ represent the state and output matrices of the internal reference models, respectively. $K \in \mathbb{R}^{r \times q}$, $F_i \in \mathbb{R}^{(n_i - q) \times (n_i - q)}$, $G_i \in \mathbb{R}^{(n_i - q) \times q}$, $H_i \in \mathbb{R}^{(n_i - q) \times p_i}$, and $T_{\ell_i} \in \mathbb{R}^{r \times q}$ are designed according to Algorithm 2 below.

Algorithm 2: Output feedback control algorithm of the heterogeneous MAS

Step 1: Select the Hurwitz matrix F_i to ensure that its eigenvalues are different from those of $\hat{A}_i$, where $\hat{A}_i = C_i A_i C_i^+$. Select the matrix G_i such that (F_i, G_i) is controllable.

Step 2: Determine the unique solution T_i to the Sylvester equation $T_i C_i^+ C_i A_i - F_i T_i = G_i C_i$ such that $\begin{bmatrix} C_i \\ T_i \end{bmatrix}$ is non-singular. If $\begin{bmatrix} C_i \\ T_i \end{bmatrix}$ is singular, return to Step 1 and select G_i again until it is non-singular. Compute matrices Q_{i1} and Q_{i2} using $\begin{bmatrix} Q_{i1} & Q_{i2} \end{bmatrix} = \begin{bmatrix} C_i \\ T_i \end{bmatrix}^{-1}$ and then select $H_i = T_i C_i^+ \hat{B}_i$, where $\hat{B}_i = C_i B_i$.

Step 3: Select the gain matrices K_{i1} such that $\hat{A}_i + \hat{B}_i K_{i1} C_i^+$ is Hurwitz, $K_{i2} = \Gamma_i - K_{i1} \Pi_i$, where the solution pairs (Π_i, Γ_i) with $\Pi_i \in \mathbb{R}^{n_i \times r}, \Gamma_i \in \mathbb{R}^{p_i \times r}$ depend on the following regulator equation:

$$\Pi_i S = A_i C_i^+ C_i \Pi_i + B_i \Gamma_i,$$
$$\Pi_i = C_i^+ D, i = 1, 2, \ldots, N. \tag{24}$$

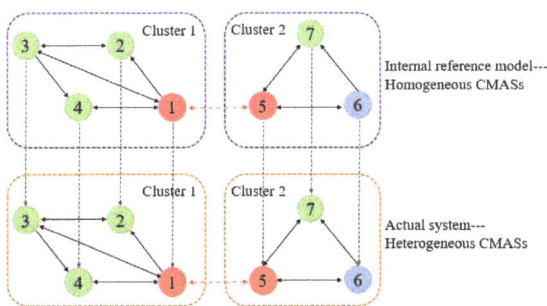

Figure 13. The heterogeneous CMASs in physical space and the artificially created homogeneous internal reference model [42].

In the implementation of Algorithm 2, all parameters are carefully selected to ensure the synchronization of heterogeneous CMASs (21) and are utilized during the convergence analysis. Additionally, the utilization of the generalized inverse of matrix C_i from [42] leads to the derivation of the solution pair (Π_i, Γ_i) using the regulator equation (24), which is based on the information provided by S, D, and (A_i, B_i, C_i).

The second step is to demonstrate that the homogeneous internal reference models achieve synchronization. With Figure 13 as a simulation example, Figure 14 shows the trajectories of state $\xi_i, i = 1, \ldots, 7$. It is clear that the internal reference model state achieves synchronization under Algorithm 2 and the hybrid communication mechanism (22)–(23).

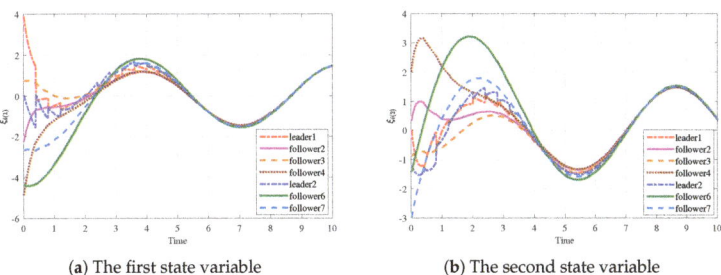

(a) The first state variable (b) The second state variable

Figure 14. State trajectories $\xi_i(t)$ of the reset internal models [42].

The third step is to let each agent track the corresponding internal reference model to obtain the output synchronization of the heterogeneous CMAS, that is, $\lim_{t\to\infty}\|y_i(t) - y_j(t)\|$ = 0. The trajectory of the output variable $y_i(t)$ can be seen to achieve synchronization, as shown in Figure 15.

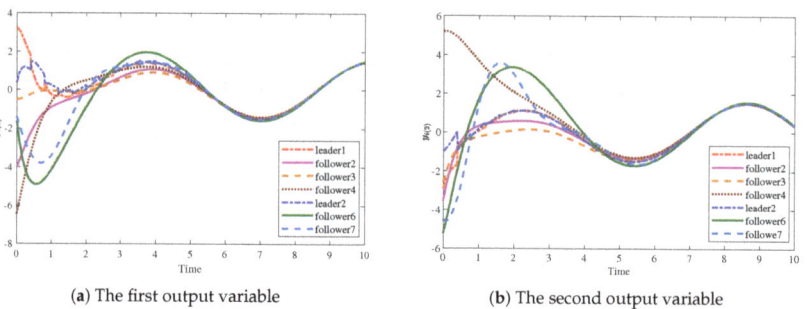

(a) The first output variable (b) The second output variable

Figure 15. Trajectories of output variables $y_i(t)$ [42].

To comprehensively investigate the disparity between consensus algorithms and the inherent attributes or constraints present in practical models, a thorough exploration of consensus issues incorporating real-world factors is imperative. These real-world factors encompass communication constraints (e.g., time delay and sampling perturbations), quantization, and state saturation, all of which can be considered vital constraint characteristics within practical models.

Recently, Pham et al. [39] focused on investigating the challenge of formation control in state-constrained clustered network systems. To tackle the issue arising from the combination of hybrid communication and state saturation, a highly resilient formation control protocol was proposed.

The output consensus problem for heterogeneous CMASs with disturbances was investigated in [41]. A new output consensus protocol was proposed in which each agent had an observer to reconstruct the state and disturbances, and a virtual reference model to take into account continuous intra-communication and discrete inter-communication.

Based on the above review, Table 6 lists the key characteristics of these studies, which explicitly show the advantages and disadvantages of the research questions. From Table 6, it is easy to see that although various cooperation control problems over intermittent cluster networks have been studied, there are still many critical issues to be further addressed in future work, such as the time sequence of inter-cluster communication being determined by event triggers rather than predetermined.

Table 6. Advantages and disadvantages of cooperation control problems over intermittent cluster networks.

Reference	Year	Dynamics	Advantages and Disadvantages		
			State Information	Communications Constraint	Reset Instants
[35]	2016	First-order integrator	✓	None	Predetermined
[36]	2016	First-order integrator	✓	Time-varying topology	Event-triggered
[37]	2019	General linear dynamics	✓	None	Predetermined
[38]	2020	General linear dynamics	×	None	Predetermined
[39]	2020	General linear dynamics	✓	State saturation	Predetermined
[40]	2021	General linear dynamics	×	None	Predetermined
[41]	2022	Heterogeneous dynamics	✓	Disturbances	Predetermined
[42]	2023	Heterogeneous dynamics	×	None	Predetermined

If state information can be available, it is marked by ✓, otherwise use ×.

5. Prospects for Future Research

An overview of the mathematical modeling and distributed control of wide-area networks has been presented. While some coordination controls of MASs over wide-area networks have been discussed in the literature, there are still several critical issues that need to be addressed in future research. In the following section, several relevant issues are suggested as potential directions for future investigations.

- Current research on coordination control of MASs has some structural constraints on communication topology. For example, bipartite consensus requires a structurally balanced condition, cluster consensus requires an in-degree balance assumption, and group-bipartite consensus needs to be structurally balanced, and requires acyclic partition, and sign-balanced couple. Furthermore, to achieve such dynamical behaviors, a spanning tree or joint spanning tree in the network topology is necessary. Therefore, investigating the dynamical behavior of MASs on more general unstructured constrained networks would be highly beneficial.
- In the existing results of cooperative control on intermittent clustered networks, the interaction moments between neighboring clusters are artificially predetermined. However, in order to enhance work efficiency and facilitate flexible information interactions, the inter-cluster communication often needs to change based on actual demand. Therefore, considering the incorporation of demand-based event-triggered impulsive control strategies for intermittent clustered networks is of great practical importance.
- Although group-bipartite consensus has the potential to address multi-objective symmetric tasks, it still faces challenges and limitations in practical applications. For instance, designing effective protocols to achieve group-bipartite consensus and addressing noise and faults in the systems that warrant further investigation.

In the last decade, numerous innovative frameworks and models have emerged in the study of wide-area networks for MASs, enhancing the literature on agent-based systems and expanding their potential applications. However, this field is still in its early stages and requires further refinement of theoretical results. In our review, we focus on group-bipartite networks, which have the potential to effectively handle complex tasks involving multiple levels and structures. However, these networks face various constraints related to topology, and communication constraints (e.g., delays, saturation, disturbances) have not been adequately addressed. To make group-bipartite networks more practical, future research should tackle these limitations. Cooperative intermittent clustered networks are particularly suitable for post-disaster communication networks and power grids due to their distinctive communication methods. Nevertheless, there are significant hurdles to overcome, such as predefined long-distance interactions. To address these issues, it is necessary to develop more flexible control algorithms that facilitate sparse communication between clusters.

6. Conclusions

MASs serve as important methodologies and tools for the analysis and modeling of complex systems. In this review, we categorize wide-area networks based on the interaction modes of cooperation or competition among agents. We provide mathematical modeling for each type, review the relevant literature, and summarize the key findings. Finally, we provide future research prospects. Exploring the clustering and sparsity of MASs can help explain real-world clustering phenomena and can be applied to practical engineering.

Author Contributions: Conceptualization, Q.W. and J.H.; methodology, Q.W. and J.H.; software, Q.W.; validation, Q.W. and J.H.; investigation, Q.W. and J.H.; writing—original draft, Q.W.; writing—review and editing, Q.W. and J.H.; visualization, Q.W.; supervision, J.H. All authors have read and agreed to the published version of the manuscript.

Funding: This work was partially supported by the National Key Research and Development Program of China under Grant 2022YFE0133100, by the National Natural Science Foundation of China under Grant No. 62103341, and by the Sichuan Science and Technology Program under Grant 2020YFSY0012.

Data Availability Statement: Not applicable.

Conflicts of Interest: The authors declare no conflict of interest.

References

1. Grossman, R.L.; Mazzucco, M.; Sivakumar, H.; Pan, Y.; Zhang, Q. Simple available bandwidth utilization library for high-speed wide area networks. *J. Supercomput.* **2005**, *34*, 231–242. [CrossRef]
2. Barabasi, A.-L. Scale-free networks: A decade and beyond. *Science* **2009**, *325*, 412–413. [CrossRef] [PubMed]
3. Tuna, G.; Nefzi, B.; Conte, G. Unmanned aerial vehicle-aided communications system for disaster recovery. *J. Netw. Comput. Appl.* **2014**, *41*, 27–36. [CrossRef]
4. Alvarez, F.; Almon, L.; Radtki, H.; Hollick, M. Bluemergency: Mediating post-disaster communication systems using the internet of things and bluetooth mesh. In Proceedings of the 2019 IEEE Global Humanitarian Technology Conference, Seattle, WA, USA, 17–20 October 2019; pp. 1–8. [CrossRef]
5. Do-Duy, T.; Nguyen, L.D.; Duong, T.Q.; Khosravirad, S.R.; Claussen, H. Joint optimisation of real-time deployment and resource allocation for UAV-aided disaster emergency communications. *IEEE J. Sel. Areas Commun.* **2021**, *39*, 3411–3424. [CrossRef]
6. Aly, A.A.; Felemban, B.F.; Mohammadzadeh, A.; Castillo, O.; Bartoszewicz, A. Frequency regulation system: A deep learning identification, type-3 fuzzy control and LMI stability analysis. *Energies* **2021**, *14*, 7801. [CrossRef]
7. Bostani, Y.; Jalilzadeh, S.; Mobayen, S.; Rojsiraphisal, T.; Bartoszewicz, A. Damping of subsynchronous resonance in utility DFIG-based wind farms using wide-area fuzzy control approach. *Energies* **2022**, *15*, 1787. [CrossRef]
8. Taghieh, A.; Mohammadzadeh, A.; Zhang, C.; Kausar, N.; Castillo, O. A type-3 fuzzy control for current sharing and voltage balancing in microgrids. *Appl. Soft. Comput.* **2022**, *129*, 109636. [CrossRef]
9. Tripathi, S.; Pandey, O.J.; Hegde, R.M. Energy-Efficient Data Transfer in Delay-Throughput Constrained Small-World LPWAN Using RISs. In Proceedings of the 2022 IEEE 8th World Forum on Internet of Things, Yokohama, Japan, 26 October 2022; pp. 1–6. [CrossRef]
10. Li, P.; Hu, J.; Qiu, L.; Zhao, Y.; Ghosh, B.K. A distributed economic dispatch strategy for power-water networks. *IEEE Trans. Control Netw. Syst.* **2022**, *9*, 356–366. [CrossRef]
11. Pandey, O.J.; Yuvaraj, T.; Paul, J.K.; Nguyen, H.H.; Gundepudi, K.; Shukla, M.K. Improving energy efficiency and qos of lpwans for iot using q-learning based data routing. *IEEE Trans. Cogn. Commun. Netw.* **2021**, *8*, 365–379. [CrossRef]
12. Chilamkurthy, N.S.; Pandey, O.J.; Ghosh, A.; Cenkeramaddi, L.R.; Dai, H.N. Low-power wide-area networks: A broad overview of its different aspects. *IEEE Access* **2022**, *10*, 81926–81959. [CrossRef]
13. Georgiou, O.; Raza, U. Low power wide area network analysis: Can LoRa scale? *IEEE Wirel. Commun. Lett.* **2017**, *6*, 162–165. [CrossRef]
14. Mishra, A.; Jain, R.; Durresi, A. Cloud computing: Networking and communication challenges. *IEEE Commun. Mag.* **2012**, *50*, 24–25. [CrossRef]
15. Borylo, P.; Lason, A.; Rzasa, J.; Szymanski, A.; Jajszczyk, A. Green cloud provisioning throughout cooperation of a WDM wide area network and a hybrid power IT infrastructure: A study on cooperation models. *J. Comput.* **2016**, *14*, 127–151. [CrossRef]
16. Persico, V.; Botta, A.; Marchetta, P.; Montieri, A.; Pescape, A. On the performance of the wide-area networks interconnecting public-cloud datacenters around the globe. *Comput. Netw.* **2017**, *112*, 67–83. [CrossRef]
17. Ballerini, M.; Cabibbo, N.; Candelier, R. Interaction ruling animal collective behavior depends on topological rather than metric distance: Evidence from a field study. *Proc. Natl. Acad. Sci. USA* **2008**, *105*, 1232–1237. [CrossRef] [PubMed]
18. Brandenburger, A.; Nalebuff, B. *Co-Opetition: A Revolution Mindset That Combines Competition and Cooperation*; Currency Doubleday: New York, NY, USA, 1996.
19. Alexander, R.D. The evolution of social behavior. *Annu. Rev. Ecol. Syst.* **1974**, *5*, 325–383. [CrossRef]
20. Dong, J.; Hu, J.; Zhao, Y.; Peng, Y. Opinion formation analysis for Expressed and Private Opinions (EPO) models: Reasoning private opinions from behaviors in group decision-making systems. *Expert Syst. Appl.* **2024**, *236*, 121292. [CrossRef]
21. Hu, J.; Wu, Y. Interventional bipartite consensus on coopetition networks with unknown dynamics. *J. Frankl. Inst.* **2017**, *354*, 4438–4456. [CrossRef]
22. Lee, J.; Park, S.H.; Ryu, Y.; Baik, Y.S. A hidden cost of strategic alliances under Schumpeterian dynamics. *Res. Policy* **2010**, *39*, 229–238. [CrossRef]
23. Liu, B.; Chen, T. Consensus in networks of multiagents with cooperation and competition via stochastically switching topologies. *IEEE Trans. Neural Netw.* **2008**, *19*, 1967–1937. [CrossRef]
24. Ma, J.; Hu, J. Safe consensus control of cooperative-competitive multi-agent systems via differential privacy. *Kybernetika* **2022**, *58*, 426–439. [CrossRef]
25. Wu, Y.; Liu, L.; Hu, J.; Feng, G. Adaptive antisynchronization of multilayer reaction–diffusion neural networks. *IEEE Trans. Neural Netw. Learn. Syst.* **2017**, *29*, 807–818. [CrossRef] [PubMed]

26. Hu, J.; Wu, Y.; Liu, L.; Feng, G. Adaptive bipartite consensus control of high-order multiagent systems on coopetition networks. *Int. J. Robust Nonlinear Control* **2018**, *28*, 2868–2886. [CrossRef]
27. Wu, Y.; Hu, J.; Zhao, Y.; Ghosh, B.K. Adaptive scaled consensus control of coopetition networks with high-order agent dynamics. *Int. J. Control* **2021**, *94*, 909–922. [CrossRef]
28. Ning, B.; Han, Q.; Zuo, Z. Bipartite consensus tracking for second-order multiagent systems: A time-varying function-based preset-time approach. *IEEE Trans. Autom. Control* **2020**, *66*, 2739–2745. [CrossRef]
29. An, B.; Liu, G.; Tan, C. Group consensus control for networked multi-agent systems with communication delays. *ISA Trans.* **2018**, *76*, 78–87. [CrossRef] [PubMed]
30. Chen, K.; Wang, J.; Zeng, X.; Zhang, Y.; Lewis, F.L. Cluster output regulation of heterogeneous multi-agent systems. *Int. J. Control* **2020**, *93*, 2973–2981. [CrossRef]
31. Qin, J.; Fu, W.; Shi, Y.; Gao, H.; Kang, Y. Leader-following practical cluster synchronization for networks of generic linear systems: An event-based approach. *IEEE Trans. Neural Netw. Learn. Syst.* **2018**, *30*, 215–224. [CrossRef]
32. Zhang, T.; Li, H.; Liu, J.; Lu, D.; Xie, S.; Luo, J. Distributed multiple-bipartite consensus in networked Lagrangian systems with cooperative–competitive interactions. *Nonlinear Dyn.* **2021**, *106*, 2229–2244. [CrossRef]
33. Liu, J.; Li, H.; Ji, J.; Luo, J. Group-bipartite consensus in the networks with cooperative-competitive interactions. *IEEE Trans. Circuits Syst. II-Express Briefs* **2020**, *67*, 3292–3296. [CrossRef]
34. Deepak, G.C.; Ladas, A.; Sambo, Y.A.; Pervaiz, H.; Politis, C.; Lmran, M.A. An overview of post-disaster emergency communication systems in thefuture networks. *IEEE Wirel. Commun.* **2019**, *26*, 132–139. [CrossRef]
35. Bragagnolo, M.C.; Morărescu, I.C.; Daafouz, J.; Riedinger, P. Reset strategy for consensus in networks of clusters. *Automatica* **2016**, *65*, 53–63. [CrossRef]
36. Morărescu, I.C.; Martin, S.; Girard, A.; Muller-Gueudin, A. Coordination in networks of linear impulsive agents. *IEEE Trans. Autom. Control* **2016**, *61*, 2402–2415. [CrossRef]
37. Pham, T.V.; Messai, N.; Manamanni, N. Consensus of multi-agent systems in clustered networks. In Proceedings of the 2019 18th European Control Conference, Naples, Italy, 25–28 June 2019; pp. 1085–1090.
38. Pham, T.V.; Messai, N.; Manamanni, N. Impulsive observer-based control in clustered networks of linear multi-agent systems. *IEEE Trans. Netw. Sci. Eng.* **2020**, *7*, 1840–1851. [CrossRef]
39. Pham, T.V.; Messai, N.; Nguyen, D.H.; Manamanni, N. Robust formation control under state constraints of multi-agent systems in clustered networks. *Syst. Control Lett.* **2020**, *140*, 104689. [CrossRef]
40. Wang, Q.; Zhao, Y.; Hu, J. Reset output feedback control of cluster linear multi-agent systems. *J. Frankl. Inst.* **2021**, *358*, 8419–8442. [CrossRef]
41. Pham, T.V.; Nguyen, Q.T.T.; Messai, N.; Manamanni, N. Output consensus design in clustered networks of heterogeneous linear MASs under disturbances. *Eur. J. Control* **2023**, *69*, 100726. [CrossRef]
42. Wang, Q.; Hu, J.; Wu, Y.; Zhao, Y. Output synchronization of heterogeneous multi-agent systems over intermittent clustered networks. *Inf. Sci.* **2023**, *619*, 263–275. [CrossRef]
43. Shi, Y.; Hu, J.; Wu, Y.; Ghosh, B.K. Intermittent output tracking control ofheterogeneousmulti-agent systems over wide-area clustered communicationnetworks. *Nonlinear Anal.-Hybrid Syst.* **2023**, *50*, 101387. [CrossRef]
44. Peng, T.J.A.; Bourne, M. The coexistence of competition and cooperation between networks: Implications from two Taiwanese healthcare networks. *Brit. J. Manag.* **2009**, *20*, 377–400. [CrossRef]
45. Hu, J.; Zheng, W. Emergent collective behaviors on coopetition networks. *Phys. Lett. A* **2014**, *378*, 1787–1796. [CrossRef]
46. Qin, J.; Yu, C. Cluster consensus control of generic linear multi-agent systems under directed topology with acyclic partition. *Automatica* **2013**, *49*, 2898–2905. [CrossRef]
47. Hassija, V.; Saxena, V.; Chamola, V. Scheduling drone charging for multi-drone network based on consensus time-stamp and game theory. *Comput. Commun.* **2020**, *149*, 51–61. [CrossRef]
48. Pavlopoulos, G.A.; Secrier, M.; Moschopoulos, C.N.; Soldatos, T.G.; Kossida, S.; Aerts, J.; Schneider, R.; Bagos, P.G. Using graph theory to analyze biological networks. *BioData Min.* **2011**, *4*, 1–27. [CrossRef] [PubMed]
49. Altafini, C. Consensus problems on networks with antagonistic interactions. *IEEE Trans. Autom. Control* **2012**, *58*, 935–964. [CrossRef]
50. Proskurnikov, A.V.; Matveev, A.S.; Cao, M. Opinion dynamics in social networks with hostile camps: Consensus vs. polarization. *IEEE Trans. Autom. Control* **2015**, *61*, 1524–1536. [CrossRef]
51. Ma, C.; Qin, Z.; Zhao, Y. Bipartite consensus of integrator multiagent systems with measurement noise. *IET Contr. Theory Appl.* **2017**, *11*, 3313–3320. [CrossRef]
52. Guo, X.; Lu, J.; Alsaedi, A.; Alsaadi, F.E. Bipartite consensus for multi-agent systems with antagonistic interactions and communication delays. *Physica A* **2018**, *495*, 488–497. [CrossRef]
53. Shao, J.; Shi, L.; Zhang, Y.; Cheng, Y. On the asynchronous bipartite consensus for discrete-time second-order multi-agent systems with switching topologies. *Neurocomputing* **2018**, *316*, 105–111. [CrossRef]
54. Wu, Y.; Hu, J.; Zhang, Y.; Zeng, Y. Interventional consensus for high-order multi-agent systems with unknown disturbances on coopetition networks. *Neurocomputing* **2016**, *194*, 126–134. [CrossRef]
55. Hu, J.; Wu, Y.; Li, T.; Ghosh, B.K. Consensus control of general linear multiagent systems with antagonistic interactions and communication noises. *IEEE Trans. Autom. Control* **2018**, *64*, 2122–2127. [CrossRef]

56. Wu, Y.; Zhao, Y.; Hu, J. Bipartite consensus control of high-order multiagent systems with unknown disturbances. *IEEE Trans. Syst. Man Cyber.* **2019**, *49*, 2189–2199. [CrossRef]
57. Liang, Q.; Wu, Y.; Hu, J.; Zhao, Y. Bipartite output synchronization of heterogeneous time-varying multi-agent systems via edge-based adaptive protocols. *J. Frankl. Inst.* **2020**, *357*, 12808–12824. [CrossRef]
58. Wu, Y.; Liang, Q.; Zhao, Y.; Hu, J.; Xiang, L. Distributed estimation-based output consensus control of heterogeneous leader-follower systems with antagonistic interactions. *Sci. China-Inf. Sci.* **2023**, *66*, 139204. [CrossRef]
59. Yu, J.; Wang, L. Group consensus of multi-agent systems with undirected communication graphs. In Proceedings of the 2019 7th Asian Control Conference, Hong Kong, China, 27–29 August 2009; pp. 105–110.
60. Yu, J.; Wang, L. Group consensus in multiagent systems with switching topologies and communication delays. *Syst. Control Lett.* **2010**, *59*, 340–348. [CrossRef]
61. Chen, Y.; Lü, J.; Han, F.; Yu, X. On the cluster consensus of discrete-time multi-agent systems. *Syst. Control Lett.* **2011**, *60*, 517–523. [CrossRef]
62. Yu, J.; Wang, L. Group consensus of multi-agent systems with directed information exchange. *Int. J. Syst. Sci.* **2012**, *43*, 334–348. [CrossRef]
63. Yu, C.; Qin, J.; Gao, H. Cluster synchronization in directed networks of partial-state coupled linear systems under pinning control. *Automatica* **2014**, *50*, 2341–2349. [CrossRef]
64. Ma, Z.; Wang, Y.; Li, X. Cluster-delay consensus in first-order multi-agent systems with nonlinear dynamics. *Nonlinear Dyn.* **2016**, *83*, 1303–1310. [CrossRef]
65. Chen, K.; Wang, J.; Zhang, Y.; Lewis, F.L. Cluster consensus of multi-agent systems with heterogeneous dynamics. In Proceedings of the 2017 29th Chinese Control and Decision Conference, Chongqing, China, 28–30 May 2017; pp. 1517–1522.
66. Dong, S.; Chen, G.; Liu, M.; Wu, Z. Intermittent cluster consensus control of multiagent systems from a static/dynamic output approach. *IEEE Trans. Syst. Man Cyber.* **2022**, *52*, 7727–7736. [CrossRef]
67. Wang, Z.; Li, H.; Liu, J.; Zhang, T.; Ma, X.; Xie, S.; Luo, J. Static group-bipartite consensus in networked robot systems with integral action. *Int. J. Adv. Robot. Syst.* **2023**, *20*, 17298806231177148. [CrossRef]
68. Liu, C.; Li, R.; Liu, B. Group-bipartite consensus of heterogeneous multi-agent systems over signed networks. *Physica A* **2022**, *592*, 126712. [CrossRef]
69. Liu, J.; Xie, S.; Li, H. Oscillatory group-bipartite consensus in a swarm of robots with multiple oscillatory leaders. *IEEE Trans. Control Netw. Syst.* **2022**, *10*, 124–133. [CrossRef]
70. Lu, R.; Wu, J.; Zhan, X.; Yan, H. Finite-time group-bipartite consensus tracking for second-order nonlinear multi-agent systems. *Neurocomputing* **2023**, *545*, 126283. [CrossRef]

Disclaimer/Publisher's Note: The statements, opinions and data contained in all publications are solely those of the individual author(s) and contributor(s) and not of MDPI and/or the editor(s). MDPI and/or the editor(s) disclaim responsibility for any injury to people or property resulting from any ideas, methods, instructions or products referred to in the content.

Article

Bilingual–Visual Consistency for Multimodal Neural Machine Translation

Yongwen Liu [1,*], Dongqing Liu [2] and Shaolin Zhu [1]

[1] College of Software Engineering, Zhengzhou University of Light Industry, Zhengzhou 450001, China; zhushaolin@tju.edu.cn

[2] National Engineering Laboratory for Internet Medical Systems and Applications, Zhengzhou University, Zhengzhou 450052, China; liudongqing@zzu.edu.cn

* Correspondence: yongwen.liu@zzuli.edu.cn

Abstract: Current multimodal neural machine translation (MNMT) approaches primarily focus on ensuring consistency between visual annotations and the source language, often overlooking the broader aspect of multimodal coherence, including target–visual and bilingual–visual alignment. In this paper, we propose a novel approach that effectively leverages target–visual consistency (TVC) and bilingual–visual consistency (BiVC) to improve MNMT performance. Our method leverages visual annotations depicting concepts across bilingual parallel sentences to enhance multimodal coherence in translation. We exploit target–visual harmony by extracting contextual cues from visual annotations during auto-regressive decoding, incorporating vital future context to improve target sentence representation. Additionally, we introduce a consistency loss promoting semantic congruence between bilingual sentence pairs and their visual annotations, fostering a tighter integration of textual and visual modalities. Extensive experiments on diverse multimodal translation datasets empirically demonstrate our approach's effectiveness. This visually aware, data-driven framework opens exciting opportunities for intelligent learning, adaptive control, and robust distributed optimization of multi-agent systems in uncertain, complex environments. By seamlessly fusing multimodal data and machine learning, our method paves the way for novel control paradigms capable of effectively handling the dynamics and constraints of real-world multi-agent applications.

Keywords: multi-modal neural machine translation; bilingual-visual harmony; visual annotation

MSC: 68-02

Citation: Liu, Y.; Liu, D.; Zhu, S. Bilingual–Visual Consistency for Multimodal Neural Machine Translation. *Mathematics* **2024**, *12*, 2361. https://doi.org/10.3390/math12152361

Academic Editors: Jiangping Hu and Zhinan Peng

Received: 31 May 2024
Revised: 3 July 2024
Accepted: 22 July 2024
Published: 29 July 2024

Copyright: © 2024 by the authors. Licensee MDPI, Basel, Switzerland. This article is an open access article distributed under the terms and conditions of the Creative Commons Attribution (CC BY) license (https://creativecommons.org/licenses/by/4.0/).

1. Introduction

The complex and uncertain environments of multi-agent systems, along with inaccurate system dynamics, present significant challenges for effective modeling, control, and optimization. In multimodal neural machine translation (MNMT), the concept of visual annotation, which captures the essence of content in bilingual sentence pairs, has gained much attention [1–8]. Visual annotations are typically represented as images or videos that depict the main concepts and actions described in the corresponding text. This method uses an extra encoder to turn visual annotations into visual representations, effectively conveying the content of the source sentence [9]. These visual representations are then integrated into the decoder along with the source sentence representation [10]. This process enriches the context vector, which evolves over time and contributes to the step-by-step generation of the target translation. The successful integration of visual information has led to the development of the bi-encoder-to-decoder framework, which simultaneously translates the source sentence and its visual annotation into a target sentence with the same meaning, opening up new possibilities in MNMT [11,12].

While MNMT has demonstrated the ability to extract valuable translation cues from visual information, thereby enhancing the context vector's role in generating the target translation through auto-regressive decoding, current approaches primarily focus on aligning the visual annotation with the source language. This narrow focus fails to fully address the broader concept of multimodal consistency, which encompasses not only source–visual alignment but also target–visual and bilingual–visual alignment. Visual annotation captures the meaning of both the source sentence (e.g., English) and the corresponding target sentence (e.g., German), a phenomenon known as target–visual consistency. Exploiting this consistency allows for the MNMT model to extract valuable contextual information from the visual input, leading to a more informed and accurate translation process. Furthermore, the visual annotation reflects the semantic content of both the source and target sentences, introducing the concept of bilingual–visual consistency. To fully leverage the potential of visual information, MNMT should aim to replicate this bilingual alignment, promoting semantic harmony between the parallel sentences and the visual annotation. Incorporating both target–visual and bilingual–visual consistencies enables MNMT to fully exploit the rich information contained in visual annotations, leading to significant improvements in translation quality. By leveraging these two forms of consistency, MNMT can access a wealth of contextual information, resulting in more accurate, nuanced, and coherent translations.

Despite the potential benefits of incorporating target–visual and bilingual–visual consistencies, current MNMT methods often treat visual data as supplementary information rather than integrating them deeply into the core translation process. This limitation arises because these methods fail to fully exploit the potential of visual information, resulting in suboptimal contextual integration and less coherent translations. To address this issue, there is a need for novel MNMT approaches that deeply integrate visual information into the translation process, leveraging target–visual and bilingual–visual consistencies to improve translation quality and coherence.

To address the limitations of current MNMT approaches and fully leverage the potential of visual information, we propose a novel multimodal consistency approach that effectively utilizes target–visual consistency (TVC) and bilingual–visual consistency (BiVC) derived from visual annotations. For TVC, we employ an attention layer to extract future context from the visual annotation under the supervision of the ground-truth future target textual context, forming a multimodal target context. This extracted feature is then fed into a masked self-attention module to learn a target representation summarizing both past and future context information, enabling the model to capture long-range dependencies and generate more coherent translations. To promote BiVC, we introduce a bilingual–visual consistency loss term to guide the training of MNMT, encouraging semantic agreement between the learned bilingual sentence representations and the visual representation. The main contributions of this work are as follows:

- New target–visual consistency approach: We propose a new method to leverage future context cues from visual data, addressing the limitations of auto-regressive decoders and enabling the model to generate more accurate and coherent translations.
- Bilingual–visual consistency: We introduce a new loss term that guides the learning of semantically aligned textual and visual representations, fostering a tighter semantic integration and improving the overall quality of the translations.
- Performance evaluation:Through extensive experiments on widely used multimodal translation datasets, such as Multi30k English-to-French/German/Czech [13] and Flickr30kEnt-JP Japanese-to-English [14], we demonstrate that our approach achieves significant performance improvements over strong baselines and sets new state-of-the-art results.

By deeply integrating visual information into the translation process, our approach not only enhances the accuracy and fluency of translations but also opens up new possibilities for multimodal communication. This work paves the way for future research that can harness the rich contextual cues provided by visual data, ultimately leading to the development of more advanced and human-like language processing systems capable of understanding and translating complex, multimodal content.

2. Related Work

MNMT encompasses the translation of a target sentence alongside pertinent non-linguistic cues, such as visual information [4,15–17]. A notable approach, introduced by [18], involves a latent variable model that intricately intertwines visual information and textual features, forming a robust foundation for MNMT. This pioneering work delved into the complex interplay between visual and textual modalities, showcasing the substantial benefits of incorporating visual cues into the translation process. Another noteworthy contribution, as discussed by [19], explored MNMT by incorporating visual information as an additional spatiotemporal context to facilitate the translation of a source sentence into the target language. Their approach dynamically emphasized key words within the source sentence and integrated essential spatiotemporal cues from images into the decoder, enabling the generation of the target sentence. While their method effectively leveraged visual information to guide the decoding process, it is important to note that they did not directly encode image features or explicitly model the varying importance of different modalities. Furthermore, the work by [12] introduced a multimodal approach based on the transformer architecture [20]. Their approach induced hidden representations of images from the text, guided by image-aware attention mechanisms. This innovative methodology laid the groundwork for a more comprehensive integration of textual and visual information, enhancing the model's capacity to understand and generate translations that faithfully capture the essence of both modalities. Taken together, these seminal works highlight the growing importance of integrating visual cues into the MNMT framework and propose diverse strategies to effectively harness the complementary nature of textual and visual contents, ultimately leading to significant improvements in translation quality.

In the realm of MNMT, the concept of multimodal consistency revolves around the synchronization of visual and textual information to convey the same underlying semantics. An influential study by [2] integrated global visual features into an encoder–decoder framework, leveraging an attention-based recurrent neural network (RNN). This work laid the foundation for subsequent approaches, such as [17,21,22], which harnessed global visual information to establish simultaneous neural machine translation (NMT). These methods effectively utilize visual cues to complement the incomplete textual modality during the decoding process, demonstrating the potential of multimodal consistency in enhancing the robustness and efficiency of MNMT systems. Furthermore, visual information has been leveraged as a pivot to facilitate the creation of a shared multilingual visual–semantic embedding space in various approaches. For instance, Ref. [23] highlighted the significance of visual information in enhancing alignments within the latent language spaces, emphasizing the shared physical perceptual nature of visual cues across different languages. This insight has important implications for the development of more effective multilingual MNMT models that can better capture cross-lingual semantic correspondences. Additionally, Ref. [24] put forth a technique that employed visual agreement regularization during training to foster bilingual representations by aligning source-to-target and target-to-source models, further underscoring the critical role of multimodal consistency in improving the quality and coherence of translations. Moreover, LSTM networks have been widely used in NMT and have shown robust performance in various tasks. They effectively handle sequential data and maintain long-term dependencies through their gating mechanisms. While LSTMs are effective, they tend to be less efficient in capturing long-range dependencies compared to transformers. Additionally, LSTMs rely on sequential processing, which can be a bottleneck for training speed and scalability. The self-attention mechanism in transformers allows for more flexible and context-aware representations, which are crucial for handling the complexities of multimodal inputs. Transformers are more scalable and efficient for large datasets, a critical factor given the size of our training data.

Drawing inspiration from these advancements, our study harnesses the power of multimodal consistency in two key aspects to push the boundaries of MNMT performance. Firstly, we utilize multimodal consistency to enable our model to capture future contexts from visual cues, introducing a novel approach to enhance target context modeling and generate more accurate and fluent translations. Secondly, we employ this consistency to ensure semantic coherence between bilingual parallel sentences and the anchored visual annotation, proposing a new training objective that encourages the model to learn more robust and semantically aligned representations. By deeply integrating multimodal consistency into the core of our MNMT framework, we aim to unlock the full potential of visual information and set a new state of the art in the field.

3. Background of Multimodal Transformer

In this section, we introduce an advanced multimodal transformer framework for MNMT [12], which has achieved state-of-the-art performance on the Multi30k multimodal translation task. Unlike the classical transformer framework, this model incorporates a multimodal self-attention mechanism to encode both textual and visual information, learning a visually aware representation of the source sentence that serves as input to the decoder for generating the target translation word-by-word.

3.1. Multimodal Self-Attention

Given an input textual sentence of length J, represented as $\mathbf{X}^{text} = (\mathbf{x}_1, \ldots, \mathbf{x}_J)$, the traditional self-attention mechanism, denoted as ATT_s, computes a new representation $\mathbf{H}^{text} = (\mathbf{h}_1, \mathbf{h}_2, \ldots, \mathbf{h}_J)$. This mechanism dynamically weights the importance of each word within the sentence when computing the representation of each word. The traditional self-attention mechanism ATT_s projects each word $\mathbf{x}_i$ into Query ($\mathbf{x}_i W_e^Q$), Key ($\mathbf{x}_j W_e^K$), and Value ($\mathbf{x}_j W_e^V$) spaces using layer-specific trainable matrices W_e^Q, W_e^K, and $W_e^V \in \mathbb{R}^{d_{model} \times d_{model}}$, where d_{model} is the dimension of the word embedding. The attention score for each word pair (i, j) is computed using the scaled dot-product:

$$\text{score}_{ij} = \frac{(\mathbf{x}_i W_e^Q)(\mathbf{x}_j W_e^K)^\top}{\sqrt{d_{model}}}. \tag{1}$$

Next, the new representation $\mathbf{h}_i$ for each word $\mathbf{x}_i$ is computed as the weighted sum of the value projections:

$$\mathbf{h}_i = \sum_{j=1}^{J} \alpha_{ij} (\mathbf{x}_j W_e^V), \tag{2}$$

where $\alpha_{ij} = \text{softmax}(\text{score}_{ij})$ and softmax is applied to obtain the attention weights, ensuring that they sum up to 1.

Formally, this process is expressed as

$$\mathbf{h}_i = \text{ATT}_s(\mathbf{x}_i, \mathbf{X}^{text}) = \sum_{j=1}^{J} \text{softmax}\left(\frac{(\mathbf{x}_i W_e^Q)(\mathbf{x}_j W_e^K)^\top}{\sqrt{d_{model}}}\right)(\mathbf{x}_j W_e^V). \tag{3}$$

This model enhances focus on relevant parts of the input sequence by assigning higher weights to more significant words based on their contextual relationships.

In contrast to the traditional self-attention mechanism that processes only textual modality, the multimodal self-attention mechanism, denoted as MATT, seamlessly integrates visual information into the text processing framework. Guided by image-aware attention, this mechanism adaptively combines textual and visual inputs to enhance the representational power of the model. Formally, the inputs consist of two modalities: *text* represented by $\mathbf{X}^{text} \in \mathbb{R}^{J \times d_{model}}$, and *image* represented by $\mathbf{X}^{image} \in \mathbb{R}^{N \times d_{model}}$. These are concatenated into a single input $\mathbf{B}^{multi} = \{\mathbf{X}^{text} : \mathbf{X}^{image} \in \mathbb{R}^{(J+N) \times d_{model}}\}$, simplified as $\mathbf{B}^{multi} = (\mathbf{b}_1, \mathbf{b}_2, \cdots, \mathbf{b}_M)$, where $M = J + N$. Each multimodal input $\mathbf{b}_m$ and text $\mathbf{x}_j$ is projected into Query, Key, and Value spaces. Similar to the discussion of ATT, the process of MATT is formally expressed as

$$\mathbf{z}_m = \text{MATT}_s(b_m, \mathbf{X}^{text}) = \sum_{j=1}^{J} \text{softmax}\left(\frac{(\mathbf{b}_m W_e^Q)(\mathbf{x}_j W_e^K)^\top}{\sqrt{d_{model}}}\right)(\mathbf{x}_j W_e^V). \quad (4)$$

This results in a visually-informed representation $\mathbf{Z}^{multi} = (\mathbf{z}_1, \mathbf{z}_2, \cdots, \mathbf{z}_M)$, where $\mathbf{Z}^{multi} \in \mathbb{R}^{(J+N) \times d_{model}}$ can effectively capture the nuances of both textual and visual inputs.

3.2. Auto-Regressive Decoder

The auto-regressive decoder in MNMT generates target words sequentially, conditioned on previously generated words and multimodal inputs. Given the previously generated target words $\mathbf{y}_{<t} = (\mathbf{y}_1, \mathbf{y}_2, \ldots, \mathbf{y}_{t-1})$, the decoder computes the target representation $\mathbf{s}_t$ using the target attention mechanism ATT_t, as defined in Equation (5):

$$\mathbf{s}_t = \text{ATT}_t(\mathbf{q}_t, \mathbf{y}_{<t}) = \sum_{k=1}^{t-1} \text{softmax}\left(\frac{(\mathbf{q}_t W_d^Q)(\mathbf{y}_k W_d^K)^\top}{\sqrt{d_{model}}}\right)(\mathbf{y}_k W_d^V), \quad (5)$$

where $\mathbf{q}_t$ represents the target hidden state at time step t, and $W_d^V, W_d^Q, W_d^K \in \mathbb{R}^{d_{model} \times d_{model}}$ are trainable parameter matrices specific to the decoder. $\mathbf{s}_t$ captures the dependencies among the previously generated target words and guides the decoding process.

Next, the decoder employs the context attention module ATT_c to compute the context vector $\mathbf{c}_t$, which integrates multimodal context information $\mathbf{Z}^{multi}$:

$$\mathbf{c}_t = \text{ATT}_c(\mathbf{s}_t, \mathbf{Z}^{multi}) = \sum_{m=1}^{M} \text{softmax}\left(\frac{(\mathbf{s}_t W_c^Q)(\mathbf{z}_m W_c^K)^\top}{\sqrt{d_{model}}}\right)(\mathbf{z}_m W_c^V), \quad (6)$$

where $W_c^V, W_c^Q, W_c^K \in \mathbb{R}^{d_{model} \times d_{model}}$ are additional trainable matrices. The context vector $\mathbf{c}_t$ is then passed through a feed-forward neural network to compute the probability distribution over the next target word $\hat{y}_t$:

$$P(\hat{y}_t | \mathbf{y}_{<t}, \mathbf{X}^{text}, \mathbf{X}^{image}) = \text{softmax}(\mathbf{W}_o \tanh(\mathbf{W}_w \mathbf{c}_t)), \quad (7)$$

where $\mathbf{W}_o$ and $\mathbf{W}_w$ are learnable parameters. This formulation ensures that each target word prediction is conditioned on both previous target words and the multimodal input $\mathbf{B}^{multi}$.

Training the MNMT model θ involves maximizing the log-likelihood of the correct translation sequence $\mathbf{Y}$ given textual and visual inputs $\mathbf{X}^{text}$ and $\mathbf{X}^{image}$:

$$\arg\max_{\theta} \sum_{t=1}^{T} \log P(y_t | \mathbf{y}_{<t}, \mathbf{X}^{text}, \mathbf{X}^{image}; \theta), \quad (8)$$

where T is the length of the target sequence. This objective is commonly optimized using cross-entropy loss, ensuring that the model learns to generate accurate translations based on both textual and visual contexts.

4. Multimodal Consistency-Based MNMT

In this section, we first propose to extract the target future context information from the visual annotation using the target–visual consistency for enhancing the dependent-time target representation, which is abbreviated as TVC. We then use the bilingual–visual consistency to guide the training of MNMT, thereby encouraging the semantic agreement between the learned bilingual sentences and the pivoted visual annotation, which is abbreviated as BiVC. Figure 1 shows an overview of the proposed multimodal consistency-based MNMT.

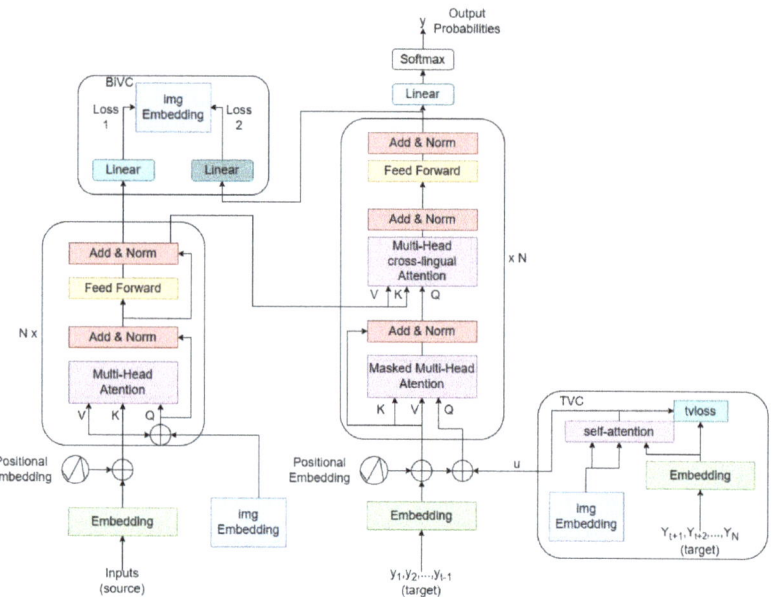

Figure 1. An overview of our method.

4.1. Target–Visual Consistency-Enhanced Target Representation

Auto-regressive decoders in NMT and MNMT are known to struggle with effectively modeling future target context during generation [25–29]. However, as discussed in Section 1, visual annotations encapsulate semantic information of both source and target sentences. We leverage this target–visual consistency to extract prospective target–side contextual cues from the visual input, thereby mitigating the aforementioned decoder limitation.

We first extract a visual feature related to the target future context under the supervision of the ground-truth target future textual context. This is achieved through an attention mechanism over the visual annotation:

$$\mathbf{u}_t = \text{ATT}_{image}(\mathbf{q}_t, \mathbf{X}^{image}) = \sum_{n=1}^{N} \text{softmax}(\frac{(\mathbf{q}_t W_r^Q)(\mathbf{r}_n W_r^K)^\top}{\sqrt{d_{model}}})(\mathbf{r}_n W_r^V), \quad (9)$$

where $\mathbf{u}_t \in \mathbb{R}^{1 \times d_{model}}$ is the extracted visual feature, and $\mathbf{q}_t$ is the query vector based on previously generated target words. To ensure $\mathbf{u}_t$ captures the desired target future context, we introduce an L1 regularization loss to minimize the mean absolute error between $\mathbf{u}_t$ and the ground-truth future target words $\mathbf{y}_t^{future}$:

$$tvloss = \text{L1loss}(\mathbf{u}_t, \text{Linear}_{future}(\mathbf{y}_t^{future})), \quad (10)$$

where Linear_{future} reduces the dimension of $\mathbf{y}_t^{future}$ to match $\mathbf{u}_t$. Both $\mathbf{q}_t$ and $\mathbf{u}_t$ are then fed into a masked multimodal self-attention module to learn an enriched target representation $\mathbf{s}'_t$:

$$\hat{\mathbf{s}}_t = \text{MATT}_t([\mathbf{q}_t, \mathbf{u}_t], \mathbf{y}^{past}) = \sum_{k=1}^{k<t} \text{softmax}(\frac{([\mathbf{q}_t, \mathbf{u}_t] W_d^Q)(\mathbf{y}_k W_d^K)^\top}{\sqrt{d_{model}}})(\mathbf{y}_k W_d^V). \quad (11)$$

Here, $\mathbf{s}'_t \in \mathbb{R}^{1 \times d_{model}}$ encodes both the target past context from previously generated words and the target future context from the visual annotation. This enriched representation $\mathbf{s}'_t$ is then used to compute the context vector c'_t via cross-attention with the source representation Z^{multi}:

$$c'_t = \text{ATT}_c(s'_t, Z^{multi}) = \sum_{m=1}^{M} \text{softmax}(\frac{(s'_t W_c^Q)(z_m W_c^K)^\top}{\sqrt{d_{model}}})(z_m W_c^V). \qquad (12)$$

Finally, c'_t is used to predict the current target word $\hat{y}_t$:

$$P(\hat{y}_t|y< t, X^{text}, X^{image}) \propto \exp(W_o \tanh(W_w c'_i)). \qquad (13)$$

The training objective is revised to maximize the conditional translation probability while minimizing the target–visual consistency loss *tvloss*:

$$\mathcal{J}(\phi) = \arg\max_{\phi} \text{celoss}(Y|X^{text}, X^{image}) - \text{tvloss}(Y, X^{image}). \qquad (14)$$

By explicitly modeling target–visual consistency during training and inference, our approach can effectively leverage future context cues from visual data, overcoming the inherent limitations of auto-regressive decoders and leading to more informed and coherent target translations.

4.2. Bilingual–Visual Consistency-Guided MNMT

While traditional text-only NMT models rely on source–target consistency, MNMT introduces an additional dimension, i.e., bilingual–visual consistency. This concept posits that the visual annotation should coherently represent the content described in both source and target sentences. However, existing MNMT approaches often overlook this crucial aspect, potentially underutilizing the rich information present in visual annotations.

To address this limitation, we introduce a bilingual–visual consistency loss term to guide our MNMT model training. This encourages semantic agreement between the learned bilingual sentence representations and the pivotal visual annotation representation. Given the source textual representation $Z^{multi} \in \mathbb{R}^{J \times d_{model}}$ and the target context representation $C' = (c'_1, c'_2, \cdots, c'_T) \in \mathbb{R}^{T \times d_{model}}$, we first project them to match the dimension of the visual representation X^{image}:

$$\mathcal{Z}^{multi} = \text{Linear}_s(Z^{multi}, X^{image}) \qquad (15)$$

$$\mathcal{C}' = \text{Linear}_s(C', X^{image}) \qquad (16)$$

Then, we compute the mean absolute error (or L1Loss) between the converted bilingual sentence representations $\{\mathcal{Z}^{multi}, \mathcal{C}'\}$ and the pivoted visual representation $\mathcal{X}^{image}$:

$$bivloss(X^{text}, Y^{text}, X^{image}) = \text{L1loss}_{s2i}(\mathcal{Z}^{multi}, X^{image}) + \text{L1loss}_{t2i}(\mathcal{C}', X^{image}), \qquad (17)$$

where L1loss$_{s2i}$ focuses on the mean absolute error (or L1Loss) between $\mathcal{Z}^{multi}$ and X^{image}, and L1loss$_{t2i}$ focuses on the mean absolute error (or L1Loss) between $\mathcal{C}'$ and X^{image} When the values of $bivloss(X^{text}, Y^{text}, X^{image})$ are smaller, the semantic consistency between the learned bilingual sentences and the pivoted visual annotation is higher. To obtain a bilingual–visual consistency-guided MNMT model φ, the training objection maximizes the conditional translation probability over the training dataset $\{[X^{text}, X^{image}, Y]\}$ as follows:

$$\mathcal{L}(\varphi) = \arg\max_{\varphi} \{celoss(Y|X^{text}, X^{image}) - bivloss(X^{text}, Y^{text}, X^{image})\}. \qquad (18)$$

By explicitly encouraging the alignment of learned bilingual sentence representations with visual annotations during training, our approach effectively captures the underlying semantic relationships across modalities. This bilingual–visual consistency serves as an inductive bias, guiding the model to learn more coherent and semantically aligned representations, thereby enhancing translation performance. The proposed TVC and BiVC components synergize, with TVC leveraging visual data to enrich target representations and BiVC ensuring a tight semantic coupling between bilingual text and visual information. Together, they enable our MNMT approach to fully exploit the complementary strengths of textual and visual modalities, overcoming the limitations of previous methods. Our multimodal consistency framework introduces a novel paradigm for MNMT, moving beyond traditional techniques that treat visual data as merely supplementary. Instead, our approach deeply integrates visual information into the core translation process, enabling more informed, contextually rich, and semantically coherent translations.

5. Experimental Setup

5.1. Dataset and Setup

In this section, we introduce the dataset and evaluation metrics, and provide the detailed experimental settings. We conducted experiments on four language pairs from two widely used multimodal translation datasets, including Multi30k [13] for English-to-German (En-De), English-to-French (En-Fr), and English-to-Czech (En-Cs), and Flickr30kEnt-JP for Japanese-to-English (En-Ja) [14]. The Multi30k dataset contains 29K bilingual parallel sentence pairs with visual annotation, 1K validation instances, and 1K test instances. Flickr30kEnt-JP contains Japanese translations of the first two original English captions for each image in the Flickr30k [30] Entities dataset. We used the Test2017 and Test2016 datasets for the evaluation of the English–German task. Additionally, we used the Test2016 and Test2017 test sets to evaluate the proposed methods on the English-to-Czech and English-to-French tasks, respectively. All sentences were preprocessed by tokenizing and normalizing the punctuation using the Moses Toolkit [31]. To tokenize Japanese, we used the MeCab version 0.996 (http://taku910.github.io/mecab, accessed on 18 February 2023).

For evaluating the translation performance, we used two widely used automatic evaluation metrics, BLEU [32] and METEOR [33]. We employed the transformer [20] as the underlying architecture to design our model. Each encoder and decoder of the model has 6-layer stacked self-attention networks, 8 heads, 1024 hidden units, and 2048 feed-forward filter size. We used the Adam optimizer with a minibatch size of 64. For the learning rate, we used the default configuration of the transformer. Specifically, the size of the word embedding was set to 256 dimensions, and embeddings were learned from scratch. We extracted global image features using ResNet-50. The spatial features were $14 \times 14 \times 1024$-dimensional vectors, which are representations of local spatial regions of the image. We trained the model for 20 epochs and set the warmup steps to 8000. During the training, the attention dropout and residual dropout were $p = 0.1$. An extra linear layer was utilized to project all visual features into 256 dimensions.

For text preprocessing, we tokenized the text data using the Moses tokenizer and performed sentence segmentation to ensure consistency in text length. Byte Pair Encoding (BPE) was applied to handle rare words and improve vocabulary efficiency. For image preprocessing, we resized images to a fixed resolution to maintain consistency across the dataset and normalized the pixel values to have zero mean and unit variance. Data augmentation techniques, such as random cropping and horizontal flipping, were used to increase the robustness of the model. In terms of aligning text and images, each sentence was aligned with its corresponding image based on the dataset annotations, ensuring accuracy by cross-referencing with the dataset documentation and performing manual checks on a subset of the data. The main hyperparameters used in our experiments are shown in Table 1 below.

Table 1. Hyperparameter settings.

Hyperparameter	Value	Description
Embedding Size	512	Size of the word embeddings.
Hidden Size	512	Size of the hidden layers in the network.
No. of Layers	6	Number of layers in the encoder and decoder.
Attention Heads	8	Number of attention heads.
Dropout Rate	0.1	Dropout rate used to prevent overfitting.
Learning Rate	0.0001	Initial learning rate for the optimizer.
Batch Size	64	Number of samples per batch.
Optimizer	Adam	Optimizer used for training the model.
Weight Decay	0.01	Weight decay factor.
Gradient Clipping	1.0	Maximum norm for gradient clipping.
Epochs	50	Number of training epochs.
BiVC Weight	0.5	Weight for the bilingual visual consistency loss.
TVC Weight	0.5	Weight for the temporal visual consistency loss.

5.2. Baselines

We compare our proposed approach with the following representative and competitive baselines:

- DMMT [4]: This method proposes distilling translations to solve the problem where visual information is only used by a second-stage decoder.
- IMG [2]: This approach uses global features extracted from visual information using a pre-trained convolutional neural network. These global image features are then incorporated into the translation model.
- SMMT [18]: This method models the interaction between visual and textual features through a latent variable, which is then used in the target-language decoder to predict image features.
- EMMT [12]: This approach introduces a new attention mechanism to learn the representations of images based on textual information, avoiding the encoding of irrelevant visual information into latent representations.
- VMMT [24]: This method employs a visual agreement regularized training on source-to-target and target-to-source models to obtain bilingual representations.

These baselines represent diverse approaches in MNMT, ranging from feature incorporation and latent variable modeling to attention mechanisms and regularization techniques. By comparing our method against these competitive baselines, we aim to demonstrate the effectiveness of our multimodal consistency approach in leveraging visual information for improved translation quality.

6. Results and Discussions

6.1. Main Results

Table 2 presents the primary results for our proposed methods and comparison methods on the En-De Test2016 and Test2017 test sets. The findings underscore the significant performance gains achieved by our EMMT+TVC, EMMT+BiVC, and EMMT+TVC+BiVC models over the baseline EMMT model, highlighting the effectiveness of TVC and BiVC in leveraging visual annotation to enhance MNMT performance.

Table 2. BLEU and METEOR scores for the proposed methods compared to benchmark methods on the Multi30k En-De Test2016 and Test2017 test sets. Results are averaged over five training runs.

Methods	Test16		Test17	
	BLEU	METEOR	BLEU	METEOR
Only-text NMT	35.61	53.6	23.8	45.3
Existing MNMT systems				
DMMT [4]	36.9	54.5	-	-
IMG [2]	37.3	55.1	-	-
SMMT [18]	37.5	55.8	26.1	49.9
EMMT [12]	38.5	55.7	-	-
VMMT [24]	-	-	29.3	51.2
Our MNMT systems (±std)				
EMMT	38.61 ± 0.5	56.2 ± 0.5	28.00 ± 0.6	51.1 ± 0.5
+TVC	40.71 ± 0.5	58.6 ± 0.3	29.11 ± 0.5	51.9 ± 0.4
+BiVC	39.11 ± 0.6	57.8 ± 0.4	28.53 ± 0.7	51.6 ± 0.5
+TVC+BiVC	41.27 ± 0.5	59.2 ± 0.4	29.70 ± 0.6	52.2 ± 0.5

Specifically, our experiments show that the EMMT+TVC model consistently outperformed the EMMT+BiVC model on both Test2016 and Test2017 test sets. This suggests that extracting target future context information from visual annotations contributes more effectively to translation quality than solely enforcing semantic agreement between bilingual sentences and visual annotations. Furthermore, the EMMT+TVC+BiVC model achieved higher BLEU scores compared to both the EMMT+TVC and EMMT+BiVC models on Test2016 and Test2017 test sets. This demonstrates that combining target–visual consistency and bilingual–visual consistency offers synergistic benefits, resulting in additional improvements in MNMT performance.

Our analysis indicates that the superior performance of the proposed method can be attributed to several key factors. The handling of specific linguistic phenomena is significantly improved; our model excels in translating sentences with ambiguous or context-dependent terms, as the visual context helps disambiguate such terms, leading to more accurate translations. For instance, in sentences with polysemous words, the visual context provides additional cues that help the model choose the correct meaning. Furthermore, the proposed method shows uniform performance across various sentence types, including simple declarative sentences, complex sentences with multiple clauses, and sentences with idiomatic expressions. The integration of visual information enhances the model's contextual understanding, which is particularly beneficial for translating descriptive texts where visual elements play a crucial role. Examples from our experiments show that sentences describing scenes, objects, or actions are translated more accurately when visual context is incorporated.

6.2. Evaluation of Semantic Agreement via Bilingual–Visual Consistency Loss

The bilingual–visual consistency loss (bivloss) is a pivotal element in our proposed approach, designed to promote semantic coherence between bilingual parallel sentences and visual annotations. To assess its impact, we conducted a comprehensive analysis of bivloss scores alongside corresponding BLEU, METEOR, and TER (Translation Edit Rate) scores for the baseline EMMT model, the EMMT+BiVC model, and the EMMT+TVC+BiVC model on the En-De Test2016 and Test2017 test sets. Table 3 presents compelling evidence that integrating the bivloss term significantly enhances model performance. On both the Test2016 and Test2017 sets, the EMMT+BiVC model achieved markedly lower bivloss scores compared to the baseline EMMT model. Specifically, on Test2016, the EMMT+BiVC model recorded a bivloss score of 13.89, surpassing the baseline EMMT's score of 15.01. Similarly, on Test2017, the EMMT+BiVC model achieved a bivloss score of 11.08, significantly lower than the baseline EMMT's 16.77. Importantly, these reductions in bivloss were accompanied by improved BLEU and METEOR scores. On Test2016, the EMMT+BiVC model achieved a BLEU score of 39.11, outperforming the baseline EMMT's score of 38.61, and a METEOR score of 57.8 compared to the baseline's 56.2. This trend persisted on Test2017, where the EMMT+BiVC model scored 28.53 BLEU and 51.6 METEOR compared to the baseline's 28.00 and 51.1, respectively.

Table 3. Semantic agreement metrics between bilingual sentence representations and visual representations on the Multi30k En-De Test2016 and Test2017 test sets. Results are averaged over 5 training runs.

Methods	Test2016				Test2017			
	bivloss	BLEU	METEOR	TER	bivloss	BLEU	METEOR	TER
EMMT	15.01	38.61	56.2	38.7	16.77	28.00	51.1	47.2
+TVC	8.13	40.71	58.6	36.8	9.02	29.11	51.9	45.0
+BiVC	13.89	39.11	57.8	37.5	11.08	28.53	51.6	46.7
+BiVC+TVC	6.88	41.27	59.2	36.1	5.93	29.70	52.2	44.2

Moreover, the TER scores provide additional insights into the translation quality. On Test2016, the EMMT+BiVC model achieved a TER of 37.5, improving over the baseline's 38.7. On Test2017, the TER for the EMMT+BiVC model was 46.7, compared to the baseline's 47.2. These findings underscore that encouraging bilingual–visual consistency through the bivloss term effectively aligns bilingual sentence representations with visual annotations, resulting in more coherent and higher-quality translations, as evidenced by the enhanced BLEU, METEOR, and TER scores. Furthermore, the EMMT+TVC+BiVC model, incorporating both TVC and BiVC approaches, achieved even lower bivloss scores compared to the EMMT+BiVC model. Specifically, on Test2016, the EMMT+TVC+BiVC model achieved a bivloss score of 6.88, further improving alignment between bilingual sentences and visual annotations. On Test2017, this score reduced to 5.93, indicating substantial progress in enhancing semantic coherence.

While the improvement in bivloss scores for the EMMT+TVC+BiVC model was moderate compared to the gain in BLEU scores over the EMMT+BiVC model, the TVC component played a pivotal role in enhancing translation quality. For instance, on Test2016, despite the bivloss score decreasing from 13.89 to 6.88, the EMMT+TVC+BiVC model achieved a BLEU score of 41.27, surpassing the EMMT+BiVC model's 39.11 BLEU. A similar trend was observed on Test2017, where the EMMT+TVC+BiVC model's BLEU score of 29.70 outperformed the EMMT+BiVC model's 28.53 BLEU, despite a smaller reduction in bivloss (from 11.08 to 5.93). Additionally, the METEOR and TER scores highlight the comprehensive improvement achieved by the EMMT+TVC+BiVC model. On Test2016, the METEOR score increased to 59.2, and the TER improved to 36.1. On Test2017, the METEOR score reached 52.2, and the TER was 44.2. These results highlight that while the bilingual–visual consistency loss effectively aligns textual and visual representations, the target–visual consistency introduced by the TVC component plays a critical role in enhancing overall translation quality. By leveraging visual annotations to extract future target context, the EMMT+TVC+BiVC model mitigates autoregressive decoder limitations, thereby generating more informed and coherent translations.

6.3. Learning Curves of Loss and BLEU Scores for Multimodal Consistency-Based MNMT

To investigate the effect of multimodal consistency on MNMT, we analyze the learning curves of loss scores for both the baseline EMMT and the EMMT+TVC+BiVC models. We focus on the En-De development set for loss curves and on the En-De Test2016 and Test2017 test sets for BLEU score curves.

The baseline EMMT model employs the standard cross-entropy loss, denoted as $celoss_{EMMT}$ in Equation (8). In contrast, the EMMT+TVC+BiVC model integrates additional loss components: a target–visual consistency loss $tvloss_{+TVC+BiVC}$ (Equation (10)) and a bilingual–visual consistency loss $bivloss_{+TVC+BiVC}$ (Equation (18)), alongside the standard cross-entropy loss $celoss_{+TVC+BiVC}$.

Figure 2a illustrates the learning curves of these loss components for the EMMT+TVC+BiVC model on the En-De development set. The $celoss_{+TVC+BiVC}$ curve shows a consistent downward trend, indicating effective optimization of the standard cross-entropy loss throughout training, crucial for generating accurate target translations. Notably, both $tvloss_{+TVC+BiVC}$ and $bivloss_{+TVC+BiVC}$ exhibit decreasing trends over time. $tvloss_{+TVC+BiVC}$, responsible for aligning target-side context with visual annotations, initially starts higher but steadily decreases as training progresses. Similarly, $bivloss_{+TVC+BiVC}$, aimed at maintaining semantic consistency between bilingual sentence representations and visual data, also shows a steady decline. These converging trends across all three loss components suggest that the EMMT+TVC+BiVC model effectively optimizes both standard translation objectives and multimodal consistency goals, enhancing overall model coherence and performance.

Figure 2b presents the learning curves of BLEU score, highlighting that the EMMT+TVC+BiVC model consistently outperforms the baseline EMMT model on both En-De Test2016 and Test2017 test sets throughout the training epochs. Starting with higher BLEU scores, the EMMT+TVC+BiVC model demonstrates continuous improvement, underscoring the benefits of multimodal consistency approaches. Importantly, the gap in BLEU scores between the EMMT+TVC+BiVC model and the baseline EMMT model widens over time, indicating that integrating multimodal consistency objectives not only boosts immediate performance but also facilitates more effective model learning. This alignment across textual and visual modalities leads to superior translation quality compared to traditional approaches.

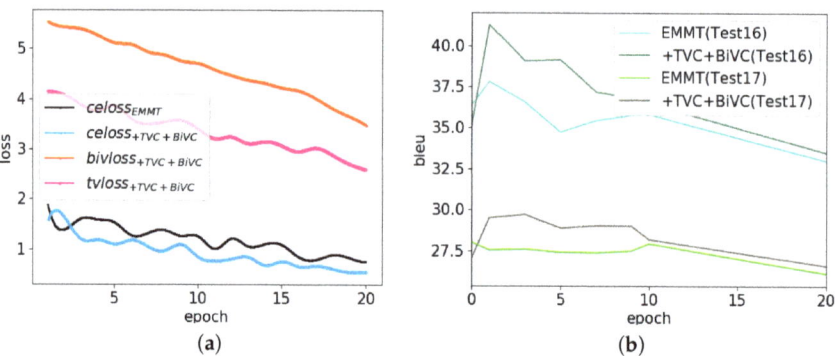

Figure 2. Learning curves and BLEU scores comparison for baseline EMMT and EMMT+TVC+BiVC models. (**a**) Learning curves comparison; (**b**) BLEU scores comparison. It shows the learning curves of loss scores for the baseline EMMT model and the EMMT+TVC+BiVC model on the En-De development set. It also presents the BLEU scores of both models on the En-De Test2016 and Test2017 test sets. The results are averaged over 5 training runs.

6.4. Ablation Study for Visual Annotation

To better understand the role and importance of visual annotation in our proposed multimodal consistency approaches, we conducted a series of ablation experiments. Specifically, we evaluated the performance of the baseline EMMT model and our EMMT+BiVC, EMMT+TVC, and EMMT+TVC+BiVC models using random image annotations instead of ground-truth image annotations. The results presented in Table 4 provide valuable insights into the significance of visual information in our models.

Table 4. Results of an ablation study comparing the baseline EMMT model and our enhanced models using random image annotations on the En-De Test2016 and Test2017 test sets. Results are averaged over 5 training runs.

Methods	Test2016				Test2017			
	Truth Image		Random Image		Truth Image		Random Image	
	BLEU	METEOR	BLEU	METEOR	BLEU	METEOR	BLEU	METEOR
EMMT	38.61	56.2	35.62	53.24	28.00	51.1	27.47	50.9
+BiVC	39.11	57.8	34.59	52.11	28.53	51.6	27.14	50.41
+TVC	40.71	58.6	35.71	52.59	29.11	51.9	26.39	49.74
+BiVC+TVC	41.27	59.2	34.77	51.78	29.70	52.2	26.78	50.26

Firstly, the results clearly demonstrate that all models, including the baseline EMMT and our proposed variants, perform significantly better when using ground-truth image annotations compared to random image annotations. On the En-De Test2016 test set, the BLEU score of the EMMT model drops from 38.61 with ground-truth images to 35.62 with random images. A similar pattern is observed on the En-De Test2017 test set, where the BLEU score decreases from 28.00 to 27.47 when using random images. This indicates that visual annotation is a crucial component contributing to the overall performance of multimodal machine translation models, with the content of the visual annotation playing a pivotal role in alignment with textual context.

Furthermore, we observe that with random visual annotations, our proposed models (EMMT+BiVC, EMMT+TVC, and EMMT+TVC+BiVC) actually perform worse than the baseline EMMT model in terms of both BLEU and METEOR scores. For example, on the En-De Test2016 test set, the EMMT+BiVC model scores 34.59 BLEU with random images, lower than the baseline EMMT's 35.62 BLEU. A similar trend is seen on the En-De Test2017 test set, where the EMMT+BiVC model's BLEU of 27.14 is inferior to the baseline EMMT's 27.47 BLEU. This suggests that when visual annotations are not aligned with textual content, our multimodal consistency approaches, which heavily leverage visual information, extract more noise than useful signal. In contrast, the baseline EMMT model, relying primarily on textual information and using visual input as supplementary information, is less affected by modal mismatches.

However, when ground-truth image annotations are used, the situation reverses. Our proposed models, EMMT+BiVC, EMMT+TVC, and EMMT+TVC+BiVC, consistently outperform the baseline EMMT model across both the En-De Test2016 and Test2017 test sets. Specifically, the EMMT+TVC+BiVC model achieves the highest BLEU scores of 41.27 on Test2016 and 29.70 on Test2017, significantly outperforming the baseline EMMT's 38.61 and 28.00 BLEU, respectively. This demonstrates that when visual annotations are accurate and well-aligned with textual content, our multimodal consistency approaches effectively leverage this information to enhance translation quality. The target–visual consistency and bilingual–visual consistency components enable our models to extract richer contextual cues from visual data and maintain tighter semantic coherence between textual and visual modalities, resulting in superior translation performance.

6.5. Impact of Different Loss Functions on Performance

We employed the smooth L1 loss as the primary loss function for our proposed multimodal consistency approaches. However, given the importance of selecting an appropriate loss function in deep learning models, we conducted further investigations to explore the impact of using different loss functions within our framework. Table 5 presents the results of our experiments evaluating the performance of our models when trained with various loss functions, including the L2 loss, KLDiv loss, BCEWithLogits loss, HingeEmbedding loss, and the L1 loss used in our main approach.

Table 5. Results of the impact of different loss functions on BLEU scores.

	En-De Test2016	En-De Test2017	En-Fr Test2017	En-Cs Test2016
L1loss	41.27	29.70	50.46	32.87
L2loss	39.21	30.02	50.21	31.06
KLDivloss	34.75	27.13	50.08	31.78
BCEWithLogitloss	36.31	27.01	50.57	31.24
HingeEmbeddingloss	35.78	27.45	50.06	26.14

We observe that the L1 loss consistently yields the most promising performance across the majority of evaluated scenarios. For instance, on the En-De Test2016 test set, the model trained with the L1 loss achieves a BLEU score of 41.27, outperforming the other loss functions by a significant margin. Specifically, the L2 loss-based model scores 39.21 BLEU, the KLDiv loss-based model scores 34.75 BLEU, the BCEWithLogits loss-based model scores 36.31 BLEU, and the HingeEmbedding loss-based model scores 35.78 BLEU. This trend persists on the En-De Test2017 test set, where the L1 loss-based model achieves a BLEU score of 50.46, surpassing the scores of the L2 loss-based model (50.21 BLEU), the KLDiv loss-based model (50.08 BLEU), the BCEWithLogits loss-based model (50.57 BLEU), and the HingeEmbedding loss-based model (50.06 BLEU). The superior performance of the L1 loss-based model is also evident in other language pair tasks. On the En-Fr Test2017 test set, the L1 loss-based model achieves a BLEU score of 32.87, outperforming the L2 loss-based model (31.06 BLEU), the KLDiv loss-based model (31.78 BLEU), the BCEWithLogits loss-based model (31.24 BLEU), and the HingeEmbedding loss-based model (26.14 BLEU).

Similarly, on the En-Cs Test2016 test set, the L1 loss-based model scores 29.70 BLEU, while the L2 loss-based model scores 30.02 BLEU, the KLDiv loss-based model scores 27.13 BLEU, the BCEWithLogits loss-based model scores 27.01 BLEU, and the HingeEmbedding loss-based model scores 27.45 BLEU. These results clearly demonstrate the effectiveness of the L1 loss function in the context of our proposed multimodal consistency approach for machine translation. Known for its robustness and ability to handle outliers, the L1 loss proves particularly suitable for aligning textual and visual representations, as well as capturing target–visual and bilingual–visual consistencies. In contrast, other loss functions such as the L2 loss, KLDiv loss, BCEWithLogits loss, and HingeEmbedding loss do not perform as well in our experiments. The L2 loss, being more sensitive to outliers, may struggle to provide the necessary guidance for the model to learn desired multimodal representations. The KLDiv loss, designed for probabilistic distributions, may not be the optimal choice for tasks involving the alignment of structured textual and visual features. Additionally, the BCEWithLogits loss and HingeEmbedding loss, typically used for classification tasks, appear less suitable for the multimodal translation problem compared to the L1 loss.

6.6. Universality of Multimodal Consistency

To assess the universality and broader applicability of our proposed multimodal consistency approaches, we conducted experiments across multiple language pairs beyond the English–German (En-De) task, which was the focus of our main analysis. Specifically, we evaluated the performance of the baseline EMMT model and our EMMT+TVC, EMMT+BiVC, and EMMT+TVC+BiVC models on the English–French (En-Fr), English–Czech (En-Cs), and English–Japanese (En-Ja) multimodal translation tasks. The results presented in Table 6 provide valuable insights into the generalizability and effectiveness of our multimodal consistency techniques. Firstly, the results demonstrate that our proposed approaches consistently outperform the baseline EMMT model across all the evaluated language pairs. On the En-Fr task, the EMMT+TVC+BiVC model achieves a BLEU score of 32.87, which is a 1.50 point improvement over the baseline EMMT's 31.37 BLEU. Similarly, on the En-Cs task, the EMMT+TVC+BiVC model scores 29.70 BLEU compared to 28.00 BLEU for the baseline EMMT. Even on the more distant language pair of En-Ja, the

EMMT+TVC+BiVC model outperforms the baseline by 1.08 BLEU points, scoring 45.73 BLEU versus the EMMT's 44.65 BLEU. These findings suggest that the core principles underlying our multimodal consistency approaches, namely, TVC and BiVC, are universal and effectively applicable to a diverse range of multimodal translation tasks beyond the initial En-De setup.

Table 6. Performance of multimodal consistency models across different multimodal language pairs. Results are averaged over 5 training runs.

Methods	En-Fr			En-Cs			En-Ja		
	METEOR	BLEU	bivloss	METEOR	BLEU	bivloss	METEOR	BLEU	bivloss
EMMT	67.59	48.42	13.62	52.56	31.37	16.01	60.36	44.65	14.82
+TVC	68.07	49.69	9.60	53.86	32.13	8.36	62.07	44.97	8.08
+BiVC	68.89	49.16	13.05	53.14	31.95	15.92	61.21	44.13	14.05
+TVC+BiVC	70.17	50.46	5.14	54.39	32.87	6.97	62.73	45.73	6.71

Interestingly, the magnitude of improvement achieved by our proposed models varies across the different language pairs. The EMMT+TVC+BiVC model shows the most significant BLEU score improvements of 2.04 and 1.50 points on the more linguistically similar En-Fr and En-Cs tasks, respectively. In contrast, the improvement on the more distant En-Ja task is relatively smaller at 1.08 BLEU points. This pattern indicates that multimodal consistency approaches may be particularly beneficial when the target language is more closely related to the source language, leveraging visual annotations to strengthen semantic coherence across bilingual parallel sentences. Even for the more distant language pair of En-Ja, our multimodal consistency techniques still outperform the baseline, highlighting their broad applicability. In addition to BLEU score improvements, we also analyze the bivloss scores for different models and language pairs. Consistent with findings from the En-De experiments, the EMMT+TVC+BiVC model consistently achieves the lowest bivloss scores across the En-Fr, En-Cs, and En-Ja tasks, indicating its effectiveness in aligning learned bilingual sentence representations with visual annotations. For example, on the En-Fr task, the bivloss score for the EMMT+TVC+BiVC model is 6.71, significantly lower than the 14.05 and 14.82 scores for the EMMT+BiVC and baseline EMMT models, respectively. Similar trends are observed in the En-Cs and En-Ja tasks, further supporting the ability of our multimodal consistency approaches to foster tighter semantic integration between textual and visual modalities.

6.7. Impact of Different Dataset Sizes

To evaluate the impact of dataset size on our proposed model, we conducted experiments using different portions of the full dataset. We created subsets of the original dataset with varying sizes: 25%, 50%, 75%, and 100% of the full dataset. Each subset was used to train our model separately, ensuring that the training conditions remained consistent across different dataset sizes in Test2017. The performance of our model was evaluated using BLEU, METEOR, and TER metrics for each subset. The results are summarized in Table 7 below:

Table 7. Three performance metrics for different dataset sizes.

Dataset Size	BLEU	METEOR	TER
25%	22.9	25.2	0.53
50%	31.5	42.3	0.49
75%	33.2	48.6	0.47
100%	34.5	52.1	0.45

When using only 25% of the dataset, our model's performance was significantly lower across all metrics. This indicates that a smaller dataset limits the model's ability to learn effectively from the available data, resulting in poorer translation quality. Training with 50% of the dataset showed a noticeable improvement in performance, though it still lagged behind the results achieved with the full dataset. Using 75% of the dataset further improved the model's performance, bringing it closer to the results obtained with the full dataset. The best performance was achieved with the full dataset, confirming the importance of a larger dataset for training robust and accurate translation models.

6.8. Discussions

Our approach dynamically integrates textual and visual information while maintaining bilingual consistency, making it adaptable across diverse datasets. The flexibility of our self-attention mechanism and visual integration module enables an effective processing of various textual and visual inputs. Future research will involve experiments with additional datasets and languages to validate the robustness and versatility of our approach. Despite the significant improvements in translation quality, our model has some limitations. Visual context ambiguities, low-quality images, and increased computational complexity can adversely affect performance. The model's generalization to other domains or languages remains to be fully explored. Future work includes extending our approach to incorporate other modalities, such as audio, optimizing the model for real-time translation scenarios, enhancing robustness to ambiguous or irrelevant visual contexts, and adapting the model to different domains and languages. By addressing these limitations and exploring new directions, we can further advance multimodal neural machine translation, making it more versatile and applicable to a wider range of real-world scenarios.

7. Conclusions

In this study, we present a novel multimodal consistency approach that advances the state-of-the-art in MNMT. Our approach synergistically combines two complementary facets: TVC and BiVC. The integration of target–visual consistency enables our MNMT model to extract valuable target-side contextual cues from the visual annotation. By effectively leveraging the future context information, our model can generate more accurate and coherent target translations, overcoming the inherent limitations of autoregressive decoders. Simultaneously, the bilingual–visual consistency acts as a guiding force, steering our MNMT model to maintain a tight semantic alignment between the learned bilingual sentence representations and the corresponding visual annotation. This ensures that the textual and visual modalities are tightly coupled, further enhancing the translation quality. The synergistic combination of these two multimodal consistency components propels our approach beyond the capabilities of prior MNMT techniques. Extensive empirical evaluations on diverse multimodal translation tasks, including English–German, English–French, English–Czech, and English–Japanese, demonstrate the effectiveness and universality of our approach. Notably, our models achieve new state-of-the-art benchmarks across these language pairs, underscoring the aptitude of our multimodal consistency framework in harnessing the complementary strengths of textual and visual information. This significant performance improvement highlights the pivotal role that multimodal consistency plays in advancing the field of MNMT. Future work involves further exploration of multimodal consistency within the MNMT framework. We aim to uncover additional dimensions of multimodal coherence and investigate their impact on translation quality. Furthermore, we intend to extend the applicability of our proposed approach to other multimodal language tasks, unlocking its potential across a broader spectrum of real-world applications. By seamlessly integrating textual and visual modalities through the lens of multimodal consistency, our work paves the way for a new paradigm in MNMT. This visually aware, data-driven framework represents a significant advancement, positioning it as a valuable tool for intelligent language understanding and generation in complex, multimodal environments.

Author Contributions: Conceptualization, D.L.; methodology, S.Z.; writing—review & editing, Y.L. All authors have read and agreed to the published version of the manuscript.

Funding: The present research was supported by the National Natural Science Foundation of China (Grant No.62276188) and the Natural Science Foundation of Henan Province (Grant No.242300420677).

Data Availability Statement: Data are contained within the article.

Acknowledgments: We would like to thank the anonymous reviewers for their insightful comments.

Conflicts of Interest: The authors declare no conflicts of interest.

References

1. Specia, L.; Frank, S.; Sima'an, K.; Elliott, D. A Shared Task on Multimodal Machine Translation and Crosslingual Image Description. In Proceedings of the First Conference on Machine Translation: Volume 2, Shared Task Papers, Berlin, Germany, 11–12 August 2016; pp. 543–553. [CrossRef]
2. Calixto, I.; Liu, Q. Incorporating Global Visual Features into Attention-based Neural Machine Translation. In Proceedings of the 2017 Conference on Empirical Methods in Natural Language Processing, Copenhagen, Denmark, 9–11 September 2017; pp. 992–1003. [CrossRef]
3. Hewitt, J.; Ippolito, D.; Callahan, B.; Kriz, R.; Wijaya, D.T.; Callison-Burch, C. Learning Translations via Images with a Massively Multilingual Image Dataset. In Proceedings of the 56th Annual Meeting of the Association for Computational Linguistics (Volume 1: Long Papers), Melbourne, Australia, 15–20 July 2018; pp. 2566–2576. [CrossRef]
4. Ive, J.; Madhyastha, P.; Specia, L. Distilling Translations with Visual Awareness. In Proceedings of the 57th Annual Meeting of the Association for Computational Linguistics, Florence, Italy, 28 July–2 August 2019; pp. 6525–6538. [CrossRef]
5. Zhang, Z.; Chen, K.; Wang, R.; Utiyama, M.; Sumita, E.; Li, Z.; Zhao, H. Neural Machine Translation with Universal Visual Representation. In Proceedings of the International Conference on Learning Representations, Addis Ababa, Ethiopia, 26–30 April 2020.
6. Yin, Y.; Meng, F.; Su, J.; Zhou, C.; Yang, Z.; Zhou, J.; Luo, J. A Novel Graph-based Multi-modal Fusion Encoder for Neural Machine Translation. In Proceedings of the 58th Annual Meeting of the Association for Computational Linguistics, Online, 5–10 July 2020; pp. 3025–3035. [CrossRef]
7. Wang, X.; Thomason, J.; Hu, R.; Chen, X.; Anderson, P.; Wu, Q.; Celikyilmaz, A.; Baldridge, J.; Wang, W.Y. (Eds.) *Advances in Language and Vision Research, Proceedings of the First Workshop on Advances in Language and Vision Research, Online, 9 July 2020*; The Association for Computational Linguistics: Stroudsburg, PA, USA, 2020.
8. Berahmand, K.; Daneshfar, F.; Salehi, E.S.; Li, Y.; Xu, Y. Autoencoders and their applications in machine learning: A survey. *Artif. Intell. Rev.* **2024**, *57*, 28. [CrossRef]
9. Zhu, S.; Li, S.; Xiong, D. VisTFC: Vision-guided target-side future context learning for neural machine translation. *Expert Syst. Appl.* **2024**, *249*, 123411. [CrossRef]
10. Zhu, S.; Li, S.; Lei, Y.; Xiong, D. PEIT: Bridging the Modality Gap with Pre-trained Models for End-to-End Image Translation. In Proceedings of the 61st Annual Meeting of the Association for Computational Linguistics (Volume 1: Long Papers), Toronto, ON, Canada, 9–14 July 2023; pp. 13433–13447.
11. Calixto, I.; Liu, Q.; Campbell, N. Doubly-Attentive Decoder for Multi-modal Neural Machine Translation. In Proceedings of the 55th Annual Meeting of the Association for Computational Linguistics (Volume 1: Long Papers), Vancouver, BC, Canada, 30 July–4 August 2017; pp. 1913–1924. [CrossRef]
12. Yao, S.; Wan, X. Multimodal Transformer for Multimodal Machine Translation. In Proceedings of the 58th Annual Meeting of the Association for Computational Linguistics, Online, 5–10 July 2020; pp. 4346–4350. [CrossRef]
13. Elliott, D.; Frank, S.; Sima'an, K.; Specia, L. Multi30K: Multilingual English-German Image Descriptions. In Proceedings of the 5th Workshop on Vision and Language, Berlin, Germany, 27 June–1 July 2016; pp. 70–74.
14. Nakayama, H.; Tamura, A.; Ninomiya, T. A Visually-Grounded Parallel Corpus with Phrase-to-Region Linking. In Proceedings of the 12th Language Resources and Evaluation Conference, Marseille, France, 11–16 May 2020; pp. 4204–4210.
15. Elliott, D.; Kádár, Á. Imagination Improves Multimodal Translation. In Proceedings of the Eighth International Joint Conference on Natural Language Processing (Volume 1: Long Papers), Taipei, Taiwan, 27 November–1 December 2017; pp. 130–141.
16. Nishihara, T.; Tamura, A.; Ninomiya, T.; Omote, Y.; Nakayama, H. Supervised Visual Attention for Multimodal Neural Machine Translation. In Proceedings of the 28th International Conference on Computational Linguistics, 8–13 December 2020; pp. 4304–4314.
17. Imankulova, A.; Kaneko, M.; Hirasawa, T.; Komachi, M. Towards Multimodal Simultaneous Neural Machine Translation. In Proceedings of the WMT, Online, 19–20 November 2020.
18. Calixto, I.; Rios, M.; Aziz, W. Latent Variable Model for Multi-modal Translation. In Proceedings of the 57th Annual Meeting of the Association for Computational Linguistics, Florence, Italy, 28 July–2 August 2019; pp. 6392–6405.
19. Wang, X.; Wu, J.; Chen, J.; Li, L.; Wang, Y.F.; Wang, W.Y. VaTeX: A Large-Scale, High-Quality Multilingual Dataset for Video-and-Language Research. In Proceedings of the 2019 IEEE/CVF International Conference on Computer Vision (ICCV), Seoul, Republic of Korea, 27 October–2 November 2019; pp. 4580–4590.

20. Vaswani, A.; Shazeer, N.; Parmar, N.; Uszkoreit, J.; Jones, L.; Gomez, A.N.; Kaiser, Ł.; Polosukhin, I. Attention is all you need. In Proceedings of the Advances in Neural Information Processing Systems 30 (NIPS 2017), Long Beach, CA, USA, 4–9 December 2017.
21. Alinejad, A.; Siahbani, M.; Sarkar, A. Prediction Improves Simultaneous Neural Machine Translation. In Proceedings of the 2018 Conference on Empirical Methods in Natural Language Processing, Brussels, Belgium, 31 October–4 November 2018; pp. 3022–3027.
22. Arivazhagan, N.; Cherry, C.; Macherey, W.; Foster, G. Re-translation versus Streaming for Simultaneous Translation. In Proceedings of the 17th International Conference on Spoken Language Translation, Online, 9–10 July 2020; pp. 220–227.
23. Huang, P.Y.; Hu, J.; Chang, X.; Hauptmann, A. Unsupervised Multimodal Neural Machine Translation with Pseudo Visual Pivoting. In Proceedings of the 58th Annual Meeting of the Association for Computational Linguistics, Online, 5–10 July 2020; pp. 8226–8237.
24. Yang, P.; Chen, B.; Zhang, P.; Sun, X. Visual agreement regularized training for multi-modal machine translation. In Proceedings of the AAAI Conference on Artificial Intelligence, New York, NY, USA, 7–12 February 2020; Volume 34, pp. 9418–9425.
25. Zhang, X.; Su, J.; Qin, Y.; Liu, Y.; Ji, R.; Wang, H. Asynchronous Bidirectional Decoding for Neural Machine Translation. In Proceedings of the Thirty-Second AAAI Conference on Artificial Intelligence, (AAAI-18), the 30th innovative Applications of Artificial Intelligence (IAAI-18), and the 8th AAAI Symposium on Educational Advances in Artificial Intelligence (EAAI-18), New Orleans, LA, USA, 2–7 February 2018; pp. 5698–5705.
26. Zheng, Z.; Zhou, H.; Huang, S.; Mou, L.; Dai, X.; Chen, J.; Tu, Z. Modeling Past and Future for Neural Machine Translation. *Trans. Assoc. Comput. Linguist.* **2018**, *6*, 145–157. [CrossRef]
27. Zhou, L.; Zhang, J.; Zong, C. Synchronous Bidirectional Neural Machine Translation. *Trans. Assoc. Comput. Linguist.* **2019**, *7*, 91–105. [CrossRef]
28. Zheng, Z.; Huang, S.; Tu, Z.; Dai, X.Y.; Chen, J. Dynamic Past and Future for Neural Machine Translation. In Proceedings of the 2019 Conference on Empirical Methods in Natural Language Processing and the 9th International Joint Conference on Natural Language Processing (EMNLP-IJCNLP), Hong Kong, China, 3–7 November 2019; pp. 931–941.
29. Duan, C.; Chen, K.; Wang, R.; Utiyama, M.; Sumita, E.; Zhu, C.; Zhao, T. Modeling Future Cost for Neural Machine Translation. *IEEE/Acm Trans. Audio Speech Lang. Process.* **2021**, *29*, 770–781. [CrossRef]
30. Plummer, B.A.; Wang, L.; Cervantes, C.M.; Caicedo, J.C.; Hockenmaier, J.; Lazebnik, S. Flickr30k entities: Collecting region-to-phrase correspondences for richer image-to-sentence models. In Proceedings of the IEEE International Conference on Computer Vision, Santiago, Chile, 7–13 December 2015; pp. 2641–2649.
31. Koehn, P.; Hoang, H.; Birch, A.; Callison-Burch, C.; Federico, M.; Bertoldi, N.; Cowan, B.; Shen, W.; Moran, C.; Zens, R.; et al. Moses: Open Source Toolkit for Statistical Machine Translation. In Proceedings of the 45th Annual Meeting of the Association for Computational Linguistics Companion Volume Proceedings of the Demo and Poster Sessions, Prague, Czech Republic, 25–27 June 2007; pp. 177–180.
32. Papineni, K.; Roukos, S.; Ward, T.; Zhu, W.J. Bleu: A method for automatic evaluation of machine translation. In Proceedings of the 40th Annual Meeting of the Association for Computational Linguistics, Philadephia, PA, USA, 6–12 July 2002; pp. 311–318.
33. Denkowski, M.; Lavie, A. Meteor universal: Language specific translation evaluation for any target language. In Proceedings of the Ninth Workshop on Statistical Machine Translation, Baltimore, MD, USA, 26–27 June 2014; pp. 376–380.

Disclaimer/Publisher's Note: The statements, opinions and data contained in all publications are solely those of the individual author(s) and contributor(s) and not of MDPI and/or the editor(s). MDPI and/or the editor(s) disclaim responsibility for any injury to people or property resulting from any ideas, methods, instructions or products referred to in the content.

MDPI AG
Grosspeteranlage 5
4052 Basel
Switzerland
Tel.: +41 61 683 77 34

Mathematics Editorial Office
E-mail: mathematics@mdpi.com
www.mdpi.com/journal/mathematics

Disclaimer/Publisher's Note: The statements, opinions and data contained in all publications are solely those of the individual author(s) and contributor(s) and not of MDPI and/or the editor(s). MDPI and/or the editor(s) disclaim responsibility for any injury to people or property resulting from any ideas, methods, instructions or products referred to in the content.

www.ingramcontent.com/pod-product-compliance
Lightning Source LLC
LaVergne TN
LVHW070407100526
838202LV00014B/1407